Related titles

Nuclear Corrosion Science and Engineering
(ISBN 978-1-84569-765-5)

Handbook of Small Modular Nuclear Reactors
(ISBN 978-0-85709-851-1)

Understanding and Mitigating Ageing in Nuclear Power Plants
(ISBN 978-1-84569-511-8)

Electrochemistry in Light Water Reactors
(ISBN 978-1-84569-240-7)

Structural Materials for Generation IV Nuclear Reactors
(ISBN 978-0-08-100906-2)

Woodhead Publishing Series in Energy

Materials and Water Chemistry for Supercritical Water-cooled Reactors

David Guzonas

Radek Novotny

Sami Penttilä

Aki Toivonen

Wenyue Zheng

ELSEVIER

WP
WOODHEAD PUBLISHING
An imprint of Elsevier

Woodhead Publishing is an imprint of Elsevier
The Officers' Mess Business Centre, Royston Road, Duxford, CB22 4QH, United Kingdom
50 Hampshire Street, 5th Floor, Cambridge, MA 02139, United States
The Boulevard, Langford Lane, Kidlington, OX5 1GB, United Kingdom

Notices
Knowledge and best practice in this field are constantly changing. As new research and experience broaden our understanding, changes in research methods, professional practices, or medical treatment may become necessary.

Practitioners and researchers must always rely on their own experience and knowledge in evaluating and using any information, methods, compounds, or experiments described herein. In using such information or methods they should be mindful of their own safety and the safety of others, including parties for whom they have a professional responsibility.

To the fullest extent of the law, neither the Publisher nor the authors, contributors, or editors, assume any liability for any injury and/or damage to persons or property as a matter of products liability, negligence or otherwise, or from any use or operation of any methods, products, instructions, or ideas contained in the material herein.

Library of Congress Cataloging-in-Publication Data
A catalog record for this book is available from the Library of Congress

British Library Cataloguing-in-Publication Data
A catalogue record for this book is available from the British Library

ISBN: 978-0-08-102049-4 (print)
ISBN: 978-0-08-102050-0 (online)

Publisher: Joe Hayton
Acquisition Editor: Maria Convey
Editorial Project Manager: Natasha Welford
Production Project Manager: Swapna Srinivasan
Cover Designer: Victoria Pearson

Typeset by TNQ Books and Journals

Contents

Acknowledgements

While the words in the pages that follow are those of the authors, the ideas presented took form over many years, distilled from our own research and the work of our colleagues in various national and international collaborations. The authors are extremely grateful to our colleagues, past and present, in the Generation IV International Forum SCWR System for valuable and often enlightening discussions, information exchange, camaraderie and for their dedication to developing the SCWR concept: Sylvie Aniel, Patrick Arnoux, Yoon-Yeong Bae, Carlos-Kjell Guerard, Liisa Heikinheimo, Akos Horvath, Jinsung Jang, Junya Kaneda, Alberto Saez-Maderuelo, Hideki Matsui, Henri Paillère, Thomas Schulenberg, Gary Was, Katsumi Yamada, Marketa Zychova. They would especially like to acknowledge the many important contributions of the late Jan Kysela, who passed away in 2015 as this book was being conceived. Jan was a dedicated nuclear water chemist, a strong advocate for the SCWR, as well as a warm, generous man who was as passionate about art and music as he was about science. They would also like to acknowledge the valuable contributions of the participants in the SCWR Information Exchange Meetings sponsored by the International Atomic Energy Agency: A. Bakai, M. Fulger, B. Gong, Y. Huang, Q.J. Peng, V. Yurmanov, A. Zeeman, Q. Zhang. The many contributions of the late Roger Staehle are acknowledged here, in particular his passion for corrosion science, deep understanding of environmentally assisted corrosion and strong sense of the history of nuclear materials development.

Dave Guzonas would like to thank his many former colleagues at Chalk River Laboratories of Atomic Energy of Canada Limited (now Canadian Nuclear Laboratories): V. Anghell for helping in navigating the physics of the critical transition; K. Burrill for throwing him headfirst into SCWR research in 2001; S. Bushby and H. Khartabil for their early support for SCWR water chemistry research; R. Duffey for encouragement, probing questions, challenging discussions and always setting the bar higher; M. Edwards for engineering insights and debates on the nature of SCW; L. Leung for teaching a lowly chemist something about thermalhydraulics; L. Qiu for discussions on thermodynamics and solubilities in SCWR; J. Semmler for her willingness to make her laboratory, equipment and technologists available; C.R. Stuart for making water radiolysis a little less exotic; R. Tapping for many wide-ranging and insightful discussions and for encouragement; C. Turner for the afternoon brainstorming session on the possible structure of SCW at a surface; L. Deschenes, M. Godin and S. Rousseau for day-to-day lab operations, taking care of paperwork, supervising students, fixing leaks and pointing out errors; J. Wills for supervising the laboratories for many years and for being his close collaborator in the experiments, especially on the studies of coatings; as well as J. Ball, T. Chiasson, P. Chow, J. Festarini, W. Kupferschmidt, J. Licht, S. Livingstone, C. Macdonald, D. McCracken, D. Mancey, G. McRae, I. Muir, J. Pencer, D. Rodgers, R. Rogge, R. Sadhankar, S. Sell, J. Smith, R. Speranzini, Z. Tun, L. Walters, O.-T. Woo, D. Wren, M. Wright and M. Yetisir for valuable discussions and support. He would also like to acknowledge the contributions of the many summer and co-op students who struggled with polishing coupons, confusing data, power outages and leaking autoclaves: K. Chu, M. Chutumstid, H. Dole, S. Fessahaye, A. Fraser, M. Haycock, S. Jang, J. Michel, S. Sullivan, S. Wren and M. Yetisir.

Special thanks go out to Peter Tremaine (University of Guelph) and Jean-Paul Jay-Gerin (Universite de Sherbrooke) for those early collaborations on SCWR water chemistry, during which most of the ideas presented in Chapter 4 first took shape. F. Brosseau and K. Bissonette (University of Guelph) spent their co-op MSc work terms in Dave's lab, and their significant contributions are gratefully acknowledged. W. Cook (University of New Brunswick) elected to spend part of a sabbatical in Chalk River, fleshing out SCWR cycle chemistry and the relationships between supercritical water chemistry and corrosion processes. These insights led to a fruitful collaboration with I. Svishchev (Trent University) using molecular dynamics to study the structure of SCW at surfaces, turning blackboard speculation into hard science.

Dave is grateful to his colleagues in the International Association for the Properties of Water and Steam for many discussions on the nature of supercritical water and chemistry control in supercritical fossil-fired power plants, in particular the insights provided by Barry Dooley, Derek Lister, Digby Macdonald and Ian Woolsey. Dave also thanks T. Allen, D. Bartel and G. Was for discussions and laboratory tours; A. Bakai (Kharkov Institute of Physics and Technology) for collaborating on the irradiated loop and a very pleasant visit to Kharkov; J. Kysela (Nuclear Research Institute Řež plc) for many pleasant and enlightening discussions on water chemistry, Pilsner Urquell, apple strudel and jazz adventures worldwide; R. Staehle (Staehle Consulting) for educating him about corrosion, cracking and the early work on nuclear steam superheat reactors while exploring Petten and over dinners; Y. Katsumura and M. Lin for showing him the state-of-the-art pulse radiolysis facility in Japan and participating in workshops on SCW radiolysis in Canada; V. Yurmanov for spirited discussions on Russian water chemistry and sharing insights and data; L. Zhang for inviting him to tour his laboratories and give an impromptu seminar to his students, his many colleagues at CanmetMaterials for discussions and data and for teaching him a little about materials science; and P. Ampornrat, J. Bischoff, S. Teysserre and E. West for discussions on SCWR materials and chemistry at conferences when he was still a newcomer. Finally, Dave thanks Margaret and John for their encouragement, and most of all Jaleh for her encouragement, support, patience and love during the preparation of the book.

The contributions of Dave and Wenyue to this book would be much diminished without the work of their colleagues at Canadian universities. They sincerely thank the participants in the Canadian Materials and Chemistry projects funded by the NSERC/NRCan/AECL Gen-IV Energy Technologies Program, who provided data, participated in workshops and conferences, sent slides for update meetings, learned to take quality assurance seriously, sometimes responded to gentle nudges concerning research direction, and participated in spirited discussions: A. Alfantazi (University of British Columbia), A. Anderson (Saint Francis Xavier University), D. Artymowizc (University of Toronto), G. Botton (McMaster University), W. Chen (University of Alberta), W. Cook (University of New Brunswick) M. Daymond (Queen's University), K. Ghandi (Mount Allison University), R. Holt (Queen's University), X. Huang (Carleton University), R. Hui (NRC-IFCI), J.-P. Jay-Gerin (Universite de Sherbrooke), J. Kish (McMaster University), R. Klassen (Western University), J. Kozinski (York University), J. Luo (University of Alberta), D. Mitlin (University of Alberta), R. Newman (University of Toronto), J. Noel (Western University), P. Percival (Simon Fraser University), C. Pye (St. Mary's University), I. Svishchev (Trent University), P. Tremaine (University of Guelph), C. Wren (Western University). Many students and postdoctoral fellows participated in these projects and made valuable contributions; the following are thanked for sending data, discussing experiments, and putting up with sometimes drastic editing of reports and papers: C. Alcorn, H. Arcis, L. Applegarth, Y. Behnamian, R. Carvajal-Ortiz, K. Choudhry, Y. Jiao, D. Kallikragas, S. Mahboubi, J. Meesungnoen, J. Miles, R. Olive, R. Sanchez, S. Sanguanmith. This Canada-wide, 8-year collaboration was kept running smoothly through the efforts of T. Andersen (NSERC) and D. Brady, K. Huynh, C. MacKinlay, J. Poupore and S. Quinn (NRCan). The contributions of C. Rangacharyulu and J. Szpunar (University of Saskatchewan) and their students to SCWR research in Canada are also acknowledged.

Wenyue Zheng would like to acknowledge contributions from his colleagues at CanmetMaterials who showed unwavering support for the materials-related Generation IV work from 2007, when he initiated the nuclear material research programme at NRCan, up to 2017, when he moved to the University of Science and Technology Beijing. In particular, he acknowledges the excellent effort put in by Kevin Boyle, Jen Collier, Gordon Gu, Mike Greenwood, Selcuk Kuyucak, Jian Li, Jason Lo, Maggie Matchim, Xu Su, Yimin Zeng. Many students and postdoctoral fellows at various times contributed to the experimental and computational results produced at Canmet.

Aki Toivonen and Sami Penttilä want to thank, in addition to the abovementioned persons, all their colleagues at VTT who have enabled the experimental work and participation in international cooperation in corrosion and SCC experiments. Special thanks go to Liisa Heikinheimo who acted as the knowledge centre manager and delegated SCWR materials-related projects to us about 10 years ago; to laboratory engineers and technicians, P. Väisänen, T. Lehtikuusi, M. Mattila, J. Lukin, T. Ikäläinen, S. Peltonen, J. Ranta, who had to struggle with water chemistry issues and leaking autoclaves; to T. Saario, P. Kinnunen and P. Auerkari (VTT) for the insight they provided concerning the general corrosion processes related to the experimental work done within the last 10 years; as well as M. Bojinov (University of Chemical Technology and Metallurgy, Sofia, Bulgaria) and I. Betova (Technical University of Sofia, Bulgaria) for their valuable guidance and discussions on aspects of materials modelling under high-temperature water conditions.

Radek Novotny would like to thank to his colleagues at JRC L. Debarberis, P. Hähner, V. Ranguelova and O. Martin, who have enabled and supported the experimental work and participation in European and international projects, in particular regarding corrosion and SCC. Radek would like to acknowledge the laboratory engineers, technicians and students, T. Timke, T. Heftrich, A. Van de Sande, S. Ripplinger, A. Ruiz, Z. Szaraz, P. Janík, D. Prchal, K. Mathis, who contributed to the development of experimental facilities and to the production of high-quality data. Radek expresses special gratitude to P. Moilanen for many hours of discussions on the design of devices that would enable performance of mechanical tests in SCW, and J. Macak for his interest and enthusiasm in using electrochemistry to understand corrosion in subcritical water and SCW. Last but not least he would like to thank his wife Anna Kotoučová for her support and endless patience.

The technical contributions are only one part of the process of writing a book. Special thanks to Matthew and Danielle Edwards for their assistance with the preparation of Chapter 4. This book would not have been written without the strong support of the staff at Elsevier: Alex White, Sarah Hughes, and Cari Owen for getting the ball rolling, Natasha Welford, who guided the book to completion, Sheela Bernardine Josy for assistance with copyright clearance, and Swapna Srinivasan and the production team who did an excellent job of putting it all together.

List of abbreviations

AC	Alternating current
AEC	Atomic Energy Commission
AECL	Atomic Energy of Canada Limited
AOA	Axial offset anomaly
ASME	American Society of Mechanical Engineers
ASTM	American Society for Testing and Materials
ATEM	Analytical transmission electron microscopy
BONUS	BOiling NUclear Superheater
BOP	Balance-of-plant
BORAX–V	BOiling Reactor Experiment V
BWR	Boiling water reactor
CANDU	CANada Deuterium Uranium
CERT	Constant extension rate tensile
CF	Corrosion fatigue
CGR	Crack growth rate
CILC	Crud-induced localized corrosion
CNL	Canadian Nuclear Laboratories
CO	Chemical oxidation
CT	Compact tension
CU–EITF	Canada–Ukraine electron irradiation test facility
CVD	Chemical vapour deposition
CVR	Centrum Výzkumu Řež
CW	Cold work
DB	Double bellows
DCPD	Direct current potential drop
DHC	Delayed hydride cracking
DL-EPR	Double-loop electrochemical potentiokinetic reactivation
DO	Dissolved oxygen
dpa	Displacements per atom
EAC	Environmentally assisted corrosion
EBSD	Electron backscatter diffraction
EBR-II	Experimental breeder reactor
ECAP	Equal channel angular processing
ECP	Electrochemical corrosion potential
EDX or EDS	Energy dispersive X-ray spectroscopy
EGBE	Effective grain boundary energy
EO	Electrochemical oxidation
EPRI	Electric Power Research Institute

ESADA	Empire State Atomic Development Associates
F/M	Ferritic-martensitic
FAC	Flow accelerated corrosion
FEM	Finite element method
FIB	Focussed ion beam
FPP	Fossil power plant
FWHM	Full width at half maximum
GBE	Grain boundary engineering
GC	Gas chromatography
GDOES	Glow discharge optical emission spectroscopy
GE	General Electric
GETR	General Electric test reactor
GIF	Generation IV International Forum
HAADF	High-angle annular dark field
HDAC	Hydrothermal diamond anvil cell
HDR	Heissdampfreaktor
HKF	Helgeson–Kirkham–Flowers
HPLWR	High-performance light water reactor
HPOC	High-pressure optical cell
HWC	Hydrogen water chemistry
IAEA	International Atomic Energy Agency
IAPWS	International Association for the Properties of Water and Steam
IASCC	Irradiation-assisted stress corrosion cracking
IC	Ion chromatography
ICP	Inductively coupled plasma
KIPT	Kharkiv Institute of Physics and Technology
LET	Linear energy transfer
LINAC	Linear electron accelerator
LLRM	Liquidlike radiolysis model
M&C	Materials and chemistry
MA	Mill-annealed
MC	Monte Carlo
MCM	Mixed conduction model
MCNP	Monte Carlo N Particle
MCST	Maximum cladding surface temperature
MD	Molecular dynamics
MDL	Method detection limit
MS	Mass spectrometry
Mu	Muon
μSR	Muon spin spectroscopy
NEA	Nuclear Energy Agency
NPP	Nuclear power plant
NR	Norgett, Robinson and Torrens
NRX	National Research Experimental
NWC	Normal water chemistry
ODS	Oxide dispersion strengthening
OECD	Organization for Economic Cooperation and Development
OES	Optical emission spectrometry
OT	Oxygenated treatment

PCI	Pellet–cladding interaction
PCT	Peak cladding temperature
PDM	Point defect model
PHWR	Pressurized heavy water reactor
PKA	Primary knock-on atom
PMB	Project Management Board
PWR	Pressurized water reactor
R&D	Research and Development
RBMK	Reactor Bolshoy Moshchnosty Kanalny (high-power channel-type reactor)
RIS	Radiation-induced segregation
S/N	Signal-to-noise
SADE	Superheat advanced demonstration experiment
SAM	Scanning auger microscopy
SC	Supercritical
SCC	Stress corrosion cracking
SCW	Supercritical water
SCWO	Supercritical water oxidation
SCWR	Supercritical water-cooled reactors
SEM	Scanning electron microscopy
SFE	Stacking fault energy
SHS	Superheated steam
SICC	Strain-induced corrosion cracking
SIMFUEL	Simulated fuel
SIMS	Secondary ion mass spectrometry
SSRT	Slow strain rate testing
TEM	Transmission electron microscopy
TRIGA	Training, research, isotopes, general atomic
TT	Thermally treated
UKAEA	United Kingdom Atomic Energy Agency
US	United States
USA	United States of America
USC	Ultrasupercritical
VBWR	Vallecitos boiling water reactor
VdG	van de Graaff
VGB	Vereinijung der Groβkesselbesitzen
VHN	Vickers hardness number
VLRM	Vapourlike radiolysis model
VVER	Water moderated reactor
WCR	Water-cooled reactor
XAS	X-ray absorption spectroscopy
XPS	X-ray photoelectron spectroscopy
XRF	X-ray fluorescence
YSZ	Yttrium-stabilized zirconium oxide

Introduction

1

The use of supercritical water (SCW) as the coolant in a nuclear reactor is the logical evolution of the current generation of water-cooled reactors (WCRs)[1], which generate almost all of the electricity produced by nuclear power worldwide. Water is a familiar and relatively safe heat transfer medium, and many power utilities already operate both nuclear power plants (NPPs) and supercritical (SC) fossil power plants (FPPs). It was recognized in the 1950s and 1960s that operation of a WCR at a temperature above the critical temperature T_c would confer a much higher thermodynamic efficiency. The first SC FPPs were developed in the United States of America in the late 1940s and 1950s, Philo No. 6 starting operation in 1957 (621 °C, 31.0 MPa), followed by Eddystone No. 1 (649 °C, 34.5 MPa) and Breed No. 1 (566 °C, 24.1 MPa) in 1959. General Electric Company's Operation Sunrise (Cohen and Zebroski, 1959) to develop commercially viable boiling water reactor (BWR) concepts examined the merits of natural circulation boilers, forced circulation boilers and nuclear superheater boilers. Dollezhal et al. (1958, 1964) described a uranium-fueled graphite-moderated reactor designed to produce superheated steam (SHS)[2]. Although no WCR operating at both supercritical temperature and supercritical pressure was ever built, a significant effort was expended in the 1960s to explore the use of SHS as a reactor coolant.

The temperature–pressure (T–P) diagram of water in Fig. 1.1 illustrates the operating regions of current WCRs, SC FPPs, 1960s nuclear steam superheat and reheat concepts and modern supercritical water-cooled reactor (SCWR) concepts. The additional advantage of operating above the critical pressure, P_c, is the higher heat transfer coefficients achievable due to the increased coolant density, which allows higher mass velocity without the need for excessive pumping power. The thermalhydraulics aspects of SCWR cores are well described by Oka et al. (2010).

1.1 Early efforts

Superheated steam was first produced by direct nuclear means in 1959 in the Superheat Advanced Demonstration Experiment (SADE) at the Vallecitos Atomic Laboratory (Barnard, 1961). The Empire State Atomic Development Associates (ESADA) Inc. developed the ESADA Vallecitos Experimental Superheat Reactor (EVESR) (Barnard, 1961), a separate superheater concept. Saturated steam at about 6.9 MPa

[1] Water-cooled reactors include all reactor types operating with a light- or heavy-water coolant; boiling water reactors (BWRs), pressurized water reactors (PWRs), water cooled, water moderated reactors (VVERs), and pressurized heavy-water reactors (PHWRs).

[2] The relationship between supercritical water and superheated steam will be explored in Chapter 4.

Materials and Water Chemistry for Supercritical Water-cooled Reactors
https://doi.org/10.1016/B978-0-08-102049-4.00001-5

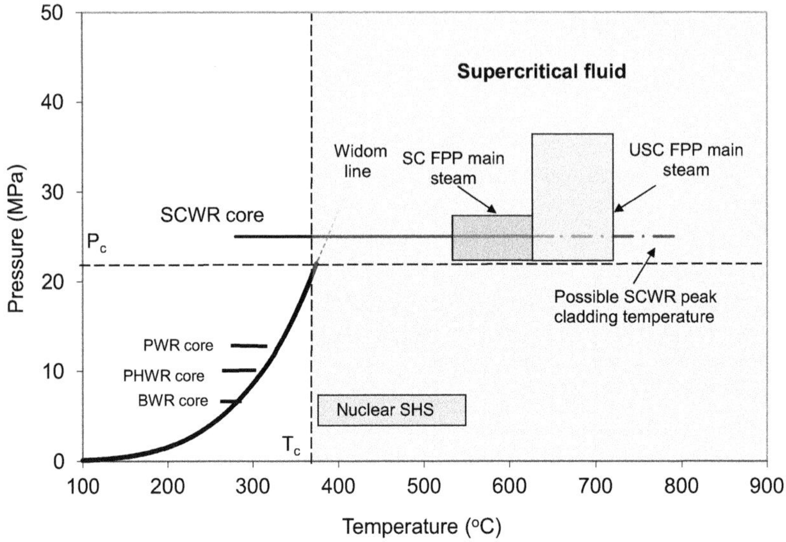

Figure 1.1 Temperature–pressure phase diagram for water illustrating the operating regions of current generation water-cooled reactors (WCRs), 1960s nuclear superheated steam (SHS) concepts, supercritical fossil power plants (SCFPPs) and the core of various proposed Generation IV supercritical water-cooled reactor (SCWR) concepts. For the SCWR core, the *solid horizontal line* covers the range of proposed core inlet and outlet temperatures; the *dot-dashed line* indicates the range of possible peak fuel cladding temperatures. The critical temperature T_c and critical pressure P_c are illustrated by the *dashed lines*. The liquid–vapour coexistence curve is shown by the *solid line*; a Widom Line (Chapter 4) is shown as the dashed line continuation of the coexistence curve. *BWR*, boiling water reactor; *PWR*, pressurized water reactor; *PHWR*, pressurized heavy-water reactor.
Adapted from Heikinheimo, L., Guzonas, D., Fazio, C., 2009. GenIV materials and chemistry research – common issues with the SCWR concept. In: 4th Int. Symp. on Supercritical Water-Cooled Reactors, Heidelberg, Germany, March 8–11, 2009, Paper 81.

from the Vallecitos BWR was supplied to the fuel-element section. Other designs considered used an integral superheater in which the boiling reactor and the superheater are inside a single vessel, for example, the Pathfinder integral superheat reactor designed and built by Allis Chalmers and which started operation in 1966. The United States Atomic Energy Commission also funded the construction of the BOiling Reactor Experiment V (BORAX–V) that started operation in 1962 and the BOiling NUclear Superheater (BONUS) that started operation in 1964 (Saltanov and Pioro, 2011). Ru and Staehle (2013a,b,c) provide a detailed summary of the data on materials performance obtained under these US programs.

In Germany, the Heissdampfreaktor (HDR) prototype 100 MW BWR with nuclear superheat was designed to operate at 457 °C and 9 MPa (Kornbichler, 1964). The plant reached criticality on 14 October 1969 and started commercial operation on 2 August, 1970, reaching a power of 23 MW_e. The core sustained damage soon after and the reactor was shut down in April 1971; the reasons were never published (Schulenberg et al., 2014). In Sweden, the Marviken boiling heavy-water reactor was designed with an option for nuclear superheat (Margen et al., 1964) using a separate superheater section that could be isolated for operation with saturated steam. The United Kingdom evaluated the utilization of supercritical steam in nuclear power reactors (Moore et al., 1964), examining a number of direct and indirect cycle pressure vessel and pressure tube concepts.

The Russians obtained significant experience with the operation of nuclear SHS channels in the pressure-tube BWRs at the I.V. Kuchatov Nuclear Power Station at Beloyarsk (referred to as the Beloyarsk NPP in this book as this name is more common), which operated for almost 3 decades at 500 °C and 8 MPa (Dollezhal et al., 1958, 1964; Emel'yanov et al., 1972). Two reactors (100 and 200 MW_e) were installed with identical steam parameters at the turbine inlet ($P_{in} = 8.8$ MPa, $T_{in} = 500-510$ °C). Unit 1 began operation in 1964 followed by Unit 2 in 1967. The dimensions and design of the two units were similar, employing boiling (evaporator) channels and superheater channels, but Unit 2 had a much simplified flow and core arrangement compared to Unit 1. In particular, the boiling channels in Unit 1 were part of a closed loop that exchanged heat through a steam generator to produce saturated steam for the superheat channels, while Unit 2 was a direct cycle. A detailed exposition of the operating experience at these plants can be found in Saltanov and Pioro (2011).

Table 1.1 lists some of the alloys used for in-core materials in these early reactors. The designations of alloys can vary with time and jurisdiction; as an example the alloy 'Inconel' has been variously named '600', 'Inconel 600' and 'Alloy 600'. In this book the convention 'Alloy 600' is adopted. Some alloys have multiple commonly used designations, e.g., HCM12A and T122 have the same composition. Where it will not lead to confusion, alloy designation will be the same as in the original reference.

Materials performance issues were recognized as being of paramount significance during these early developmental days. Novick et al. (1964) summarized many of the key issues that must be overcome during the development of a reactor using SHS, including 'probably most significant, the performance and corrosion resistance of fuel cladding at high temperatures in a reactor environment'. Marchaterre and Petrick (1960) noted that 'The major gap in supercritical water technology pertaining to a reactor system is the lack of information on the magnitude of the problems of deposition of radioactivity in the external system and of the build-up of internal crud under irradiation.' Chloride deposition from the feedwater during the boiling transition was a major issue during much of this early testing, eventually leading to failure by stress corrosion cracking (SCC), even with the best available efforts to remove chloride (Ru and Staehle, 2013a; Bevilacqua and Brown, 1963).

1.2 Recent developments

Interest in SCWR concepts underwent a renaissance in the 1990s. Researchers at the University of Tokyo published conceptual designs for thermal and fast spectrum SCWRs (Oka and Koshizuka, 1993; Oka, 2000; Oka et al., 2010); Dobashi et al. (1998)

Table 1.1 Summary of materials used in 1960s nuclear superheated steam (SHS) programs

Reactor	Fuel cladding alloy (SHS sections only)	References
Beloyarsk NPP	Kh18Ni10T: C max 0.1; Si max 0.8; Mn(1–2); Ni(10–11); S max 0.02; P max 0.035; Cr (17–19); 5(C – 0.02) < Ti < 0.6, equivalent to Type 321 SS EI-847: C(0.04–0.06); Mn(0.4–0.8); Si ≤ 0.4; S ≤ 0.010; P ≤ .015; Cr(15.0–16.0); Ni(15.0–16.0); Mo(2.7–3.2); Nb ≤ 0.9; N ≤ 0.025; B ≤ 0.001; Co ≤ 0.02; Cu ≤ 0.05; Bi ≤ 0.01; Pb ≤ 0.001; Ti ≤ 0.05	Emel'yanov et al. (1972)
BONUS	Type 348H SS	Saltanov and Pioro (2011)
BORAX 5	Type 304 SS	Saltanov and Pioro (2011)
VSER	Alloy 800, Type 347 SS, Type 348 SS, Type 310 SS, Type 304 SS, Alloy 600	Hazel et al. (1965)
Pathfinder	Type 316L SS	Saltanov and Pioro (2011)
HDR	Annular fuel: inner cladding Alloy 625, outer cladding 1.4981 S (X8CrNiMoNb16 16) or 1.4550 (X6CrNiNb18 10, equivalent to Type 347 SS)	Schulenberg and Starflinger (2012)

Only the materials used in the SHS sections are listed. Alloy compositions are given in wt%.

described a thermal spectrum SC light-water reactor (LWR) that had an average core outlet temperature of 455 °C. Russian researchers described a natural circulation integrated supercritical pressure WCR concept (B500SKDI) (Slin et al., 1992), an integral pressurized water reactor (PWR) with a core outlet pressure 23.5 MPa and coolant temperatures of 365 °C at the core inlet and 381.1 °C at the core outlet. Atomic Energy of Canada Limited[3] proposed a number of direct and indirect cycle concepts that used SCW at a nominal pressure of 25 MPa and core outlet temperatures ranging from 400 to 625 °C (Bushby et al., 2000).

The SCWR was selected by the Generation IV International Forum (GIF) as one of six innovative reactor concepts to be pursued by international collaborations (GIF, 2002; GIF, 2014). The GIF set four high-level goals for Gen IV concepts, requiring improvements in safety, economics, sustainability and nonproliferation. Locatelli et al. (2013) provide a useful overview of the recent status of the six Gen IV concepts. Pioro (2011) discusses the use of SCW as a coolant in nuclear reactors and includes a

[3] Now Canadian Nuclear Laboratories (CNL). CANDU – CANada Deuterium Uranium is a registered trademark of Atomic Energy of Canada Limited (AECL).

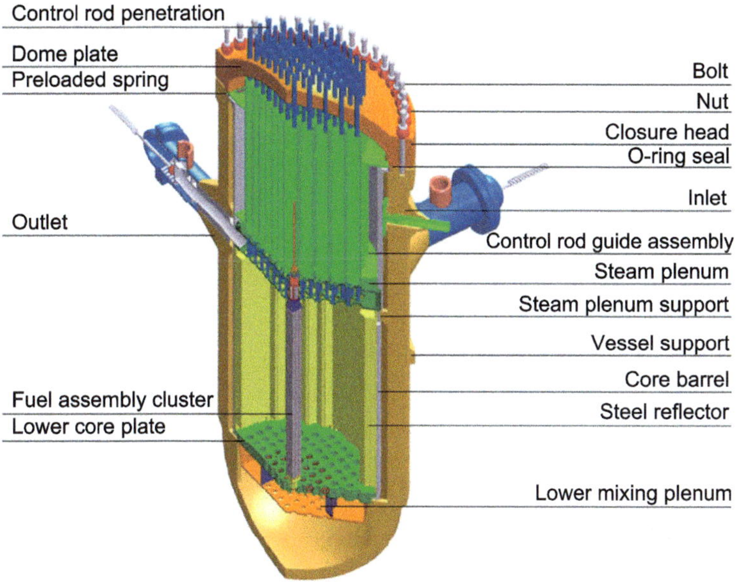

Figure 1.2a Schematic of the high-performance light water reactor core design concept (Schulenberg and Starflinger, 2012).

useful set of definitions of terminology. Schulenberg et al. (2012, 2014) are succinct summaries of the SCWR system and the research work undertaken by the various Project Management Boards (PMBs). The GIF SCWR Materials and Chemistry (M &C) PMB (Guzonas, 2009) coordinated the work being carried out by various national R&D programs (Buorngiorno et al., 2003; Anderson et al., 2009; Bae et al., 2007; Schulenberg and Starflinger, 2009). The GIF M&C Project Arrangement was signed by Canada, Euratom and Japan in 2010 and amended in 2017 to include the accession of China to the project.

Two types of core concept have been proposed: (1) a reactor pressure vessel containing the reactor core analogous to PWR and BWR designs (Fig. 1.2(a)) and (2) distributed pressure tubes or channels containing fuel bundles, analogous to conventional pressurized heavy-water reactors (PHWRs) and RBMK[4] nuclear reactors (Fig. 1.2(b)). Unlike BWRs, the various SCWR concepts do not provide recirculation in the core; in a BWR, a large fraction of the coolant volume is recirculated in the core, whereas in an SCWR all of the coolant flows to the turbines. This results in significant simplification in SCWR core design.

The majority of these Gen IV SCWR concepts are direct cycle in which the heated coolant passes directly to the turbines, although some indirect cycle concepts have been proposed (Cook and Fatoux, 2009; Alekseev et al., 2015). The balance-of-plant

[4] RBMK – Reactor Bolshoy Moshchnosty Kanalny (high-power channel type reactor).

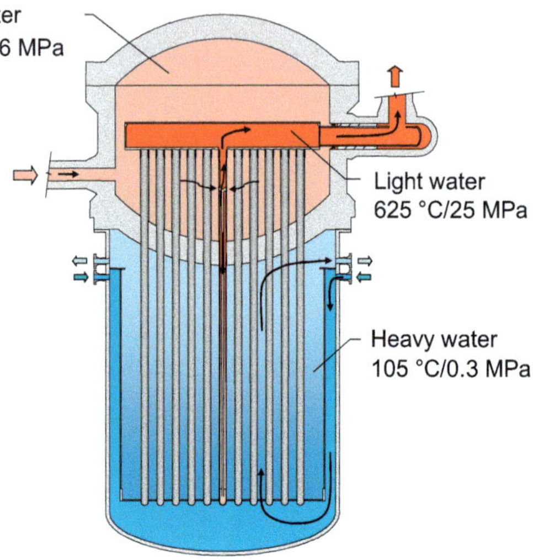

Figure 1.2b Schematic of the Canadian supercritical water-cooled reactor core concept (Yetisir et al., 2016).

(BOP) portions of the direct cycle concepts are similar to those found in existing SC FPPs.

The coolant system of most SCWR concepts consists of (1) feedtrain; (2) reactor core; and (3) main steam line, turbines and generator. Only a portion of the in-core piping, plus the main steam line and high-pressure turbines, is at $T > T_c$, and only the former components are irradiated. The rest of the system operates under conditions for which significant operating experience and a well-developed knowledge base exist. Organizations such as the International Association for the Properties of Water and Steam (IAPWS), the Electric Power Research Institute (EPRI), and Vereinijung der Groβkesselbesitzen (VGB) Powertech continue to coordinate, report and interpret the results of R&D in these areas. It is expected that SCWR feedtrain material selection will therefore be largely based on BWR and SCFPP best practices at the time of detailed design, and advances in material development for ultrasupercritical[5] (USC) FPPs.

The SCWR core presents two challenges for materials and water chemistry. The first is the presence of an intense radiation field from neutrons, γ-rays, etc. The effects of irradiation are twofold (Fig. 1.3). First, irradiation of the materials damages the atomic structure of the irradiated material, affecting the material's properties, as discussed in Chapter 3. Second, irradiation of the coolant leads to the breakdown of the water molecules (see Chapter 4), and the resulting radiolysis products can affect surface processes such as corrosion (general, Chapter 5 and localized, Chapter 6). Therefore, studies of unirradiated systems can only provide a partial picture of materials degradation phenomena in an SCWR core, as discussed in Chapter 2. The second challenge is the very large change in operating temperature across the core. The high-temperature gradient and the high operating pressure are well outside the operating

[5] The term ultrasupercritical is used by industry to denote SCFPPs operating at steam temperatures exceeding 566 °C (1050 °F). There is no physical significance to this temperature.

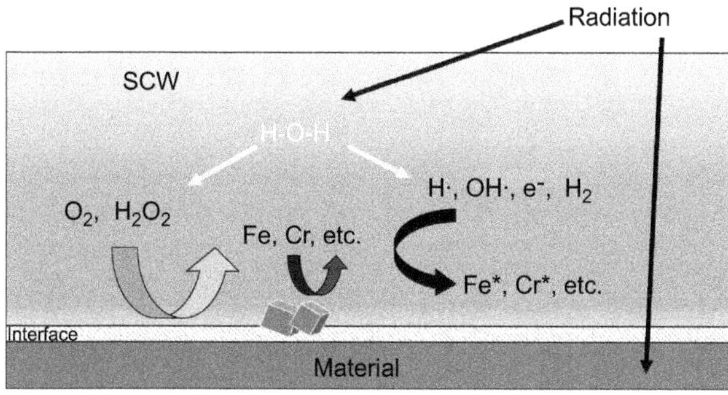

Figure 1.3 Interactions of radiation with in-core materials including the supercritical water (SCW) coolant. In this book, interactions of radiation with materials are covered in Chapter 3, interactions with the coolant in Chapter 4. The transient and stable species formed by interactions of radiation with the coolant can react with the material's surface (Chapters 5 and 6) and with other solution species.
Adapted from Katsumura, Y., Kiuchi, K., Domae, M., Wada, Y., Yotsuyanagi, T., 2003. Fundamental R&D program on water chemistry of supercritical pressure water under radiation field. In: Genes4/ANP3003, Kyoto, Japan, September 15–19, 2003, Paper 1178.

experience of current WCRs, so that WCR data are only of limited value, mainly identifying the important degradation phenomena.

1.2.1 Synergies

During the interval between the 1960s nuclear SHS concepts and the renaissance of SCWR research and development (R&D) in the 1990s, significant advances in materials (e.g., Lo et al. (2009)) in general, and specifically for SC and USC FPPs (Viswanathan et al., 2013) took place. FPPs with steam temperatures of 600–620 °C at pressures of 27–28 MPa have been in operation for many years with steam temperatures of 700–760 °C targeted for future plants. Materials development for FPPs continues to be an active area of research and is synergistic with SCWR materials R&D. This is especially relevant for in-core SCWR components not exposed to high radiation dose over their in-service life. For highly irradiated materials such as the fuel cladding, FPP data on parameters such as corrosion rate that are not strongly affected by irradiation are still valuable. However, the piping in an FPP boiler is significantly larger and thicker than the fuel cladding in an SCWR (Fig. 1.4). This has a significant effect on the allowable corrosion rate and on the significance of environmentally assisted cracking (EAC). In addition, the allowable concentrations of specific alloying elements (e.g., Co) in in-core materials may be restricted because they can be neutron-activated to hazardous isotopes (e.g., $^{59}Co(n,\gamma)^{60}Co$)[6]. In an

[6] This nomenclature for the nuclear reaction indicates (from left to right) the parent ^{59}Co reacting with an incident neutron (n) followed by the release of excess energy in the form of a gamma ray (γ) to form the product ^{60}Co.

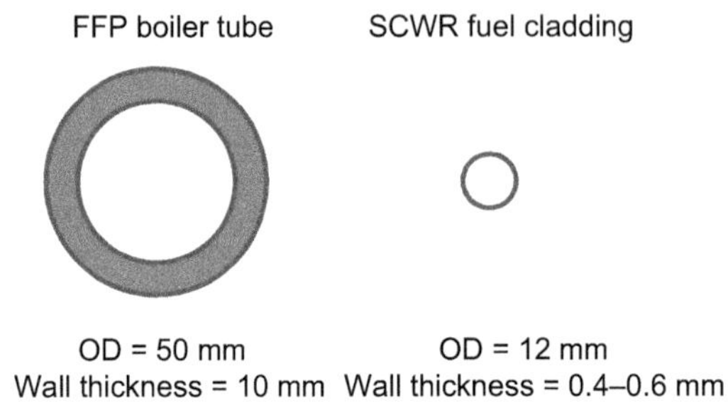

Figure 1.4 Comparison of typical fossil-fired plant (FFP) boiler tube dimensions to fuel rod dimensions in the reference supercritical water-cooled reactor (SCWR) design.
Adapted from Allen, T.R., Chen, Y., Ren, X., Sridhara, K., Tan, L., Was, G.S., West, E., Guzonas, D.A., 2012. Material performance in supercritical water. In: Konings, R.J.M. (Ed.), Comprehensive Nuclear Materials, vol. 5. Elsevier, Amsterdam, pp. 279–326.

SCFPP, the water passes through the inside of the piping, and the heat source (combusted gases) is on the outside. In an SCWR, the solid fuel is on the inside of the fuel cladding and the SCW coolant passes over the outer cladding surface.

While supercritical water oxidation chemistry conditions are much more aggressive than those expected in an SCWR core, corrosion and SCC data obtained under these conditions can provide valuable insights into corrosion phenomena in an SCWR core, especially during off normal chemistry conditions.

1.3 Supercritical water-cooled reactor materials requirements

Guzonas and Novotny (2014) have summarized the materials challenges for SCWR development. Performance criteria for in-core materials can be developed from the requirement that the material should not fail during the in-service life. For the fuel cladding, failure could occur by (Guzonas et al., 2016):

1. Through-wall penetration by general or localized corrosion;
2. Oxide build-up that impedes heat transfer;
3. EAC;
4. Stress exceeding the yield limits of the material caused by:
   a. Reduced yield limits as the material ages and is irradiated, including changes in strength caused by creep, swelling and microstructural changes that lead to fatigue and embrittlement and
   b. Increased stress as the material is irradiated, including internal pressurization due to fission gas release, and swelling within a constrained geometry.
5. Pellet-cladding interaction (PCI).

Buongiorno et al. (2003) provided a detailed summary of materials requirements for a pressure-vessel type SCWR and suggested candidate materials. They noted that for the US SCWR concept, all pressure vessel internal components (core internals) would be designed for periodic replacement so that the effects of high fluences (>20 displacements per atom, dpa) would not need to be considered. Zhang et al. (2012) outlined general technical requirements for an SCWR fuel cladding in terms of minimum yield strength, minimum creep strength, allowable in-service wall loss, phase stability and susceptibility to irradiation-induced effects. Schulenberg (2013) outlined key materials requirements for the fuel cladding and various core internals of the high-performance light-water reactor (HPLWR). Guzonas et al. (2016) and Zheng et al. (2016) described the evaluation of five candidate fuel cladding alloys against eight performance criteria developed for the Canadian SCWR concept.

To date, little attention has been given to PCI, relevant data under for similar fuel and fuel cladding materials being considered adequate for initial materials selection. Oka et al. (2010) noted that while the thermal expansion coefficient of the Zr alloy cladding used in LWRs is smaller than that of the UO_2 fuel pellets, the thermal expansion coefficients of the stainless steels or nickel-base alloys being considered for the SCWR are expected to be close to that of UO_2.

It is important to note that it is the peak cladding temperature (PCT), not the maximum coolant temperature, which is important for many of the materials properties. As noted by Oka et al. (2010) the critical heat flux is not the core design criterion because at supercritical pressure, where the coolant is a single-phase fluid, no dryout and burnout phenomena occur. Instead, the PCT (also denoted maximum cladding surface temperature) is taken as the design criterion. This temperature must therefore be limited such that fuel cladding integrity is maintained under all normal conditions and abnormal transients.

Two different SCWR fuel rod concepts have been proposed. In free-standing, internally pressurized concepts, new fuel rods would initially be pressurized to about 8 MPa using He, and the internal pressure would increase to about 25 MPa at the end of a cycle due to fission gas release. Internal pressurization imposes requirements on yield strength and creep strength and defines the minimum wall thickness; the maximum allowable corrosion penetration for the HPLWR fuel cladding is 140 μm after 20,000 h (Schulenberg, 2013). The Canadian SCWR fuel cladding is designed to collapse onto the fuel pellets, which supports the 25 MPa external pressure (Yetisir et al., 2016). The high-temperature mechanical strength and creep properties of the alloy become secondary factors in material selection, although fuel swelling due to irradiation will introduce a tensile hoop stress to the cladding. For the Canadian SCWR a maximum allowable corrosion penetration of 200 μm (including oxide penetration along grain boundaries) over the in-service life was specified, which leads to a value of ~130 μm after 20,000 h, similar to the value specified for the HPLWR.

During reactor operation, oxides will build-up on the cladding surface by corrosion and deposition of corrosion products originating in the feedtrain (Chapter 4). Small amounts of oxide formation on the cladding surface will increase the roughness and may enhance convective heat transfer to the coolant. However, significant oxide build-up reduces heat transfer efficiency by conduction through the cladding leading to increases in cladding and fuel temperatures. Oxide exfoliation (spalling) is also a concern, as the fragments of oxide released could block flow through the fuel assembly if they became trapped by wire wrap or spacers or could damage the high-pressure turbine. Therefore a limit on total oxide thickness (corrosion film plus deposited oxide) is required.

For a fuel cladding with a 0.4–0.6 mm wall thickness to meet the in-core service life the crack depth during service should be less than 30% of the wall thickness, as the reduced wall thickness under irradiation would mean much reduced toughness for rupture after crack initiation. The 'tolerable' crack growth rate is therefore roughly 0.05 mm per year (approximately 1.4×10^{-9} mm/s). The extensive operating experience of current WCRs demonstrates that avoiding high residual stress and cold-working plus good water chemistry control will reduce the likelihood of SCC.

1.3.1 Candidate materials

Various aspects of materials performance for SCWRs have been reviewed (Allen et al., 2012). The worldwide programs to study corrosion in SCW have examined many material classes, including ferritic-martensitic (F/M) steels, austenitic steels, Ni-base alloys, Zr-base alloys and Ti-base alloys, and corrosion data exist for more than 90 alloys (Gu et al., 2010). Zirconium-base alloys have been considered for in-core SCWR applications because of their good neutron economy; however, at SCWR temperatures current commercial alloys lack the high-temperature strength for load-bearing applications and exhibit unacceptably high corrosion rates. The Canadian SCWR concepts retain the use of zirconium alloy pressure tubes by using an insulated fuel channel design (Yetisir et al., 2013); in this design, the pressure tube, which forms the in-core pressure boundary, operates at about the moderator temperature (~80 °C), where the mechanical strength is higher and the corrosion rate is lower. Table 1.2

Table 1.2 Summary of the performance of supercritical water-cooled reactor candidate material classes with respect to the major in-core degradation modes

Alloy type	Corrosion resistance		Radiation damage resistance		High-temperature mechanical integrity
	General corrosion	Stress corrosion cracking	Swelling	Embrittlement	
Austenitic steel	High	Medium	Low	High	Medium
Ferritic-Martensitic steel	Medium	High	High	High	Low
Ni-base	High	Medium	Low	Medium	High
Ti-base	Medium - High	Limited data			Medium - High

Adapted from Heikinheimo, L., Guzonas, D., Fazio, C., 2009. GenIV materials and chemistry research – common issues with the SCWR concept. In: 4th Int. Symp. on Supercritical Water-Cooled Reactors, Heidelberg, Germany, March 8–11, 2009, Paper 81.

summarizes the performance of the various classes of alloys evaluated against the key SCWR degradation mechanisms; these degradation mechanisms will be described in subsequent chapters.

Of the commercially available alloys evaluated, at this time only austenitic stainless steels and some nickel-based alloys can meet all of the proposed performance requirements. Table 1.3 lists candidate fuel cladding materials that have been considered for recent SCWR concepts. The list is not exhaustive and does not include Ti- and Zr-base alloys that have been tested, nor does it include F/M steels, which, although well studied are not currently considered viable for in-core use. Allen et al. (2012) lists other alloys examined for SCWR use, and Table 1 in Ru and Staehle (2013a) lists alloys that were examined in early fossil and nuclear superheat studies, including alloys not in Table 1.3.

1.4 Summary

The GIF SCWR Materials and Chemistry PMB identified two major knowledge gaps to be overcome before the safe and reliable performance of an SCWR could be ensured (Guzonas, 2009):

1. Insufficient data are available for any single alloy to unequivocally ensure its performance in an SCWR, especially for alloys to be used for in-core components.
2. Current understanding of SCW chemistry is inadequate to specify a chemistry control strategy, as the result of the large changes in physical and chemical properties of water through the critical point, coupled with the as yet poorly understood effects of water radiolysis.

In recent years significant advances have been made in closing these knowledge gaps, the result of various international research and development programs. In 2010, 2012, and 2015, Japan, the EU, and Canada respectively, presented their SCWR design concepts. At that time, these countries had each completed several phases of SCWR research and development (the Japanese efforts have subsequently been on hold after the Fukashima accident).

The remainder of this book presents the current state of knowledge of SCWR chemistry and materials. Chapter 2 describes experimental facilities used for SCWR materials and chemistry studies, and some of the experimental challenges. Chapter 3 discusses mechanical properties (e.g., tensile and creep), introduces the effects of irradiation on these properties and summarizes relevant data. Chapter 4 starts by addressing the question 'What is SCW?' and then discusses how the nature of the in-core environment affects corrosion and EAC performance. The final two chapters focus on general corrosion and EAC.

SCWR research and development is an on-going endeavour, and this book is by its nature a snapshot of the current state of knowledge during the roughly 18 months over which it was written. New research results continue to be presented at various conferences, such as the International Symposia on SCWR (held every 2 years), at IAEA-sponsored technical symposia and in the peer-reviewed literature. As a result, it is inevitable that some important data have been omitted because the authors were

Table 1.3 Compositions of candidate fuel cladding alloys tested for use in supercritical water-cooled reactor concepts

Alloy	Fe	Cr	Ni	Mo	Mn	Si	C	Ti	Others
ODS PM2000	bal	19						0.5	Y_2O_3: 0.5, Al: 5.5
ODS MA956	bal	20.0	–	–	≤0.30		<0.1	0.2–0.6	Y_2O_3: 0.34
Austenitic stainless steels									
1.4970	bal	15	15.3	1.18	1.68	0.53	0.095	0.45	
316L	bal	16.6	10	2.0–3.0	1.9	0.65	0.022	–	
316L(N)	bal	18	12	2.0–3.0	2.0	1.0	0.030	–	N: 0.16
316Ti	bal	16.6	12.1	2.03	1.15	0.45	0.032	0.38	
316L + Zr (H1)	bal	16.54	10.71	2.22		0.46	0.006		P: 0.016, Zr: 0.56
347	bal	18.0	11.0	–	2.0	1.0	0.08	–	Nb:10xC[a]
321	bal	18	8				0.08		Ti: 0.6
AL6XN	bal	20.4	23.8	6.23	0.42	0.34	0.02	–	Co: 0.24, P: 0.024, N: 0.21
SAVE25	bal	22.5	18.5	0.1	0.5	0.2	0.07		W: 1.5, Cu: 3.25, Nb: 0.45, N: 0.2, V: 0.04
Sanicro 25	bal	22.5	25		0.5	0.2	0.1		Co:1.5, Cu: 3.0, W: 3.6, Nb: 0.5, N: 0.23
800H	bal	22.5	34.8	–	1.59	0.95	0.08	–	Al: 0.45

310	bal	24.5	20.2	0.29	1.17	0.33	0.048	–	
310S FG (T3F)	bal	24.74	21.92	–	–	0.25	0.099	0.81	N: 0.0006 P: <0.005
310S FG (T6F)	bal	25.03	22.81	2.38	–	–	–	0.41	Nb: 0.26 N: 0.002
310S + Zr (H2)	bal	25.04	20.82	0.51	–	0.51	0.034	–	P:0.016, Zr: 0.59
HR3C	bal	25	21	0.1	1.1	0.4	0.07	–	N: 0.25, Nb: 0.45, V: 0.07
Ni-based									
C276	5.35	15.88	bal	15.64	0.52	0.03		0.01	
214	3	16	75	–	0.5	0.2	0.05	–	Al: 4.5, Y: 0.01 B: 0.01, Zr: 0.1
625	4.9	22.6	bal	9.8	0.43	0.47	0.09	0.45	Al: 0.47, Nb: 3.7 Cu: 0.68
Hastelloy C									
690	9.2	29	72		1 max	0.35			Al: 0.02

This list is not exhaustive; specifically, F/M steels have been omitted. Allen et al. (2012) provides details on this alloy class.
[a]Nb + Ta.

unaware of its existence. The content of the chapters reflects the authors' own research interests, and the authors acknowledge the limits of their knowledge.

This book is intended as a reference for researchers active in the field and those who are planning to enter the field, but should also find a home with researchers in the SCFPP industry. It assumes a good undergraduate level knowledge of materials science, thermodynamics, corrosion science, chemical kinetics, as well as the fundamentals of nuclear power generation. While intended for readers at the postgraduate level or higher, much of the content should be accessible to an enthusiastic 4th-year undergraduate student.

References

Alekseev, P., Semchenkov, Y., Sedov, A., Sidorenko, V., Silin, V., Mokhov, V., Nikitenko, M., Mahkin, V., Churkin, A., 2015. Conceptual proposals on reactor VVER-SCW developed on the basis of technologies of VVER and steam-turbine installations at supercritical parameters. In: 7th International Symposium on Supercritical Water-Cooled Reactors (ISSCWR-7), March 15−18, 2015, Helsinki, Finland, Paper ISSCWR7-2055.

Allen, T.R., Chen, Y., Ren, X., Sridharan, K., Tan, L., Was, G.S., West, E., Guzonas, D.A., 2012. Material performance in supercritical water. In: Konings, R.J.M. (Ed.), Comprehensive Nuclear Materials, vol. 5. Elsevier, Amsterdam, pp. 279−326.

Anderson, T., Brady, D., Guzonas, D., Khartabil, H., Leung, L., Quinn, S., Zheng, W., 2009. Canada's NSERC/NRCan/AECL Generation IV energy technologies program. In: 4th Int. Symp. on SCWRs, March 8−11. Heidelberg, Germany.

Bae, Y.Y., Jang, J., Kim, H.Y., Yoon, H.Y., Kang, H.O., Bae, K.M., 2007. Research activities on a supercritical pressure water reactor in Korea. Nucl. Eng. Technol 39 (4), 273−286.

Barnard, J., 1961. The Reactor and Plant Design of the ESADA and EVESR. ASME paper 61-WA-223.

Bevilacqua, F., Brown, G.M., 1963. Chloride Deposition from Steam onto Superheater Fuel Clad Materials. General Nuclear Engineering Corporation Report GNEC 295.

Buongiorno, J., Corwin, W., MacDonald, P., Mansur, L., Nanstad, R., Swindemann, R., Rowcliffe, A., Was, G., Wilson, D., Wright, I., 2003. Supercritical Water Reactor (SCWR): Survey of Materials Experience and R&D Needs to Assess Viability. INEEL/EXT-03−00693Rev1, 2003.

Bushby, S.J., Dimmick, G.R., Duffey, R.B., Burrill, K.A., Chan, P.S.W., 2000. Conceptual designs for advanced, high temperature CANDU reactors. In: Proc. 8th Int. Conference on Nuclear Engineering, Baltimore, MD, April 2−6, 2000. ICONE- 8470.

Cohen, K., Zebroski, E., 1959. Operation Sunrise. Nucleonics 17 (3), 63−70.

Cook, W.G., Fatoux, W., 2009. A CANDU-SCWR with a steam generator: thermodynamic assessment and estimation of fouling rates. In: 4th Int. Symp. on SCWRs, March 8−11, Heidelberg, Germany.

Dobashi, K., Kimura, A., Oka, Y., Koshizuka, S., 1998. Conceptual design of a high temperature power reactor cooled and moderated by supercritical light water. Ann. Nucl. Energy 25, 487−505.

Dollezhal, N.A., Krasin, A.K., Aleshchenkov, P.I., Galanin, A.N., Grigoryants, A.N., Emelyanov, I.Ya., Kugushev, N.M., Minashin, M.E., Mityaev, U.I., Florinsky, B.V., Sharpov, B.N., 1958. Uranium-graphite reactor with superheated high pressure steam. In:

Proc. 2nd United Nations Int. Conf. on the Peaceful Uses of Atomic Energy, Geneva, September 1–13, 1958, vol. 8, pp. 398–414.

Dollezhal, N.A., Emel'yanov, I.Ya., Aleshchenkov, P.I., 1964. Development of superheating power reactors of Beloyarsk nuclear power station type. In: Proceedings of the Third International Conference on the Peaceful Uses of Atomic Energy, Geneva, August 31–September 9, 1964, vol. 6, paper 309, pp. 256–265.

Emel'yanov, I., Shatskaya, O.A., Rivkin, E.Yu., Nikolenko, N.Y., 1972. Strength of construction elements in the fuel channels of the Beloyarsk power station reactors. Atomnaya Energiya 33 (3), 729–733 (in Russian) Translated in Soviet. Atomic Energy 33 (3), 842–847.

GIF, 2002. A Technology Roadmap for Generation IV Nuclear Energy Systems. GIF-002–00.

GIF, 2014. Technology Roadmap Update for Generation IV Nuclear Energy Systems. https://www.gen-4.org/gif/jcms/c-9352/technology-roadmap.

Gu, G.P., Zheng, W., Guzonas, D., 2010. Corrosion database for SCWR development. In: 2nd Canada-China Joint Workshop on Supercritical Water-Cooled Reactors (CCSC-2010) Toronto, Ontario, Canada, April 25–28, 2010.

Guzonas, D.A., 2009. SCWR materials and chemistry – status of ongoing research. In: Proceedings of the GIF Symposium, Paris, France, September 9–10, 2009.

Guzonas, D., Novotny, R., 2014. Supercritical water-cooled reactor materials – Summary of research and open issues. Prog. Nucl. Energy 77, 361–372.

Guzonas, D., Edwards, M., Zheng, W., 2016. Assessment of candidate fuel cladding alloys for the Canadian supercritical water-cooled reactor concept. J. Nucl. Eng. Radiat. Sci. 2, 011016.

Hazel, V.E., Boyle, R.F., Busboom, H.J., Murdock, T.B., Skarpelos, J.M., Spalaris, C.N., 1965. Fuel Irradiations in the ESADE-VBWR Nuclear Superheat Loop. General Electric Atomic Power report GEAP-4775.

Heikinheimo, L., Guzonas, D., Fazio, C., 2009. GenIV materials and chemistry research – common issues with the SCWR concept. In: 4th Int. Symp. on Supercritical Water-Cooled Reactors, Heidelberg, Germany, March 8–11, 2009, Paper 81.

Katsumura, Y., Kiuchi, K., Domae, M., Wada, Y., Yotsuyanagi, T., 2003. Fundamental R&D program on water chemistry of supercritical pressure water under radiation field. In: GENES4/ANP3003, Kyoto, Japan, September 15–19, 2003, Paper 1178.

Kornbichler, H., 1964. Superheat reactor development in the Federal Republic of Germany. In: Proc. 3rd Int. Conf. on the Peaceful Uses of Atomic Energy, Geneva, August 31–September 9, 1964, vol. 6, pp. 266–276.

Lo, K.H., Shek, C.H., Lai, J.K.L., 2009. Recent developments in stainless steels. Mater. Sci. Eng. R 65, 39–104.

Locatelli, G., Mancini, M., Todeschini, N., 2013. Generation IV nuclear reactors: current status and future prospects. Energy Policy 61, 1503–1520.

Marchaterre, J.F., Petrick, M., 1960. Review of the Status of Supercritical Water Reactor Technology. Argonne National Laboratory. Atomic Energy Commission Research and Development Report, ANL 6202.

Margen, P.H., Leine, L., Nilson, R., 1964. The design of the Marviken boiling heavy-water reactor with nuclear superheat. In: Proc. 3rd Int. Conf. on the Peaceful Uses of Atomic Energy, Geneva, August 31–September 9, 1964, vol. 6, pp. 277–288.

Moore, R.V., Barker, A., Bishop, J.F.W., Bradley, N., Iliffe, C.E., Nichols, R.W., Thorn, J.D., Tyzack, C., Walker, V., 1964. The Utilisation of Supercritical Steam in Nuclear Power Reactors, Final Report of the Supercritical Steam Panel. United Kingdom Atomic Energy Authority, TRG Report 776(R).

Novick, M., Rice, R.E., Graham, C.B., Imhoff, D.H., West, J.M., 1964. Development of nuclear superheat. In: Proc. 3rd Int. Conf. on the Peaceful Uses of Atomic Energy, vol. 6, p. 225. Geneva, August 31–September 09, 1964.

Oka, Y., 2000. Review of high temperature water and steam cooled reactor concepts. In: Proc. 1st Int. Symp. on SCWR. Tokyo, Japan, November 6–8, 2000, Paper 104 (2000).

Oka, Y., Koshizuka, S., 1993. Concept and design of a supercritical-pressure, direct-cycle light water reactor. Nucl. Technol. 103 (3), 295–302.

Oka, Y., Koshizuka, S., Ishiwatari, Y., Yamaji, A., 2010. Super Light Water Reactors and Super Fast Reactors: Supercritical-Pressure Light Water Cooled Reactors. Springer, New York.

Pioro, I., 2011. The potential use of supercritical water-cooling in nuclear reactors. In: Krivit, S.B., Lehr, J.H., Kingery, T.B. (Eds.), Nuclear Energy Encyclopedia: Science, Technology, and Applications. John Wiley and Sons, Inc.

Ru, X., Staehle, R.W., 2013a. Historical experience providing bases for predicting corrosion and stress corrosion in emerging supercritical water nuclear technology: Part 1–Review. Corrosion 69 (3), 211–229.

Ru, X., Staehle, R.W., 2013b. Historical experience providing bases for predicting corrosion and stress corrosion in emerging supercritical water nuclear technology: Part 2 – Review. Corrosion 69 (4), 319–334.

Ru, X., Staehle, R.W., 2013c. Historical experience providing bases for predicting corrosion and stress corrosion in emerging supercritical water nuclear technology: Part 3 – Review. Corrosion 69 (5), 423–447.

Saltanov, E., Pioro, I., 2011. World experience in nuclear steam reheat. In: Tsvetkov, P. (Ed.), Nuclear Power -Operation, Safety and Environment. InTech, ISBN 978-953-307-507-5. Available from. http://www.intechopen.com/books/nuclear-power-operation-safetyand-environment/world-experience-in-nuclear-steam-reheat.

Schulenberg, T., 2013. Material requirements of the high performance light water reactor. J. Supercrit. Fluids 77, 127–133.

Schulenberg, T., Starflinger, J., 2009. European research project on the high performance light water reactor. In: 4th International Symposium on Supercritical Water-cooled Reactors, Heidelberg, Germany, March 8–11, 2009, Paper 54.

Schulenberg, T., Starflinger, J (Eds.), 2012. High Performance Light Water Reactor: Design and Analyses. KIT Scientific Publishing, Karlsruhe.

Schulenberg, T., Matsui, H., Leung, L., Sedov, A., 2012. Supercritical water cooled reactors. In: GIF Symposium Proceedings/2012 Annual Report, San Diego California, November 14–15, 2012, NEA No. 7141.

Schulenberg, T., Leung, L.K.H., Oka, Y., 2014. Review of R&D for supercritical water cooled reactors. Prog. Nucl. Energy 77, 282–299.

Slin, V.A., Voznessensky, V.A., Afrov, A.M., 1992. The light water integral reactor with natural circulation of the coolant at supercritical pressure B 500 SKDI. In: Proc. ANP'92, Tokyo, Japan, October 25–29, 1992, vol. 1. Session 4.6, 1 7.

Viswanathan, R., Henry, J.F., Tanzosh, J., Stanko, G., Shingledecker, J., Vitalis, B., Purgent, R., 2013. US program on materials technology for ultra-supercritical coal power plants. J. Mater. Eng. Perform. 22 (10), 2904–2915.

Yetisir, M., Gaudet, M., Rhodes, D., 2013. Development and integration of Canadian SCWR concept with counter-flow fuel assembly. In: 6th International Symposium on Supercritical Water-cooled Reactors (ISSCWR-6), March 3–7, 2013, Shenzhen, Guangdong, China, Paper ISSCWR6-13059.

Yetisir, M., Xu, R., Gaudet, M., Movassat, M., Hamilton, H., Nimrouzi, M., Goldak, J.A., 2016. Various design aspects of the Canadian supercritical water-cooled reactor core. J. Nucl. Eng. Radiat. Sci. 2, 011007.

Zhang, L., Bao, Y., Tang, R., 2012. Selection and corrosion evaluation tests of candidate SCWR fuel cladding materials. Nucl. Eng. Design 249, 180–187.

Zheng, W., Guzonas, D., Boyle, K.P., Li, J., Xu, S., 2016. Materials assessment for the Canadian SCWR core concept. JOM 68, 456–462.

Experimental methodologies **2**

Performing experiments in water at the elevated temperatures and pressures relevant to light-water reactors (LWRs) is challenging and becomes even more so under the conditions expected in a supercritical water-cooled reactor (SCWR) core. The Generation IV International Forum Materials and Chemistry Project Management Board (GIF M&C PMB) proposed a four-tier program for materials R&D in support of SCWR concepts (Fig. 2.1), the degree of complexity and cost increasing as one moves toward the apex and the extent of control and the available knowledge base decreasing. Tests using irradiated materials or simulated radiolysis conditions (Level 3) significantly increase the experimental challenges, and in-reactor tests (Level 4) are technically difficult and expensive.

This chapter discusses experimental facilities and methodologies for materials and water chemistry testing for SCWR development, focussing mainly on aspects unique to working in SCW.

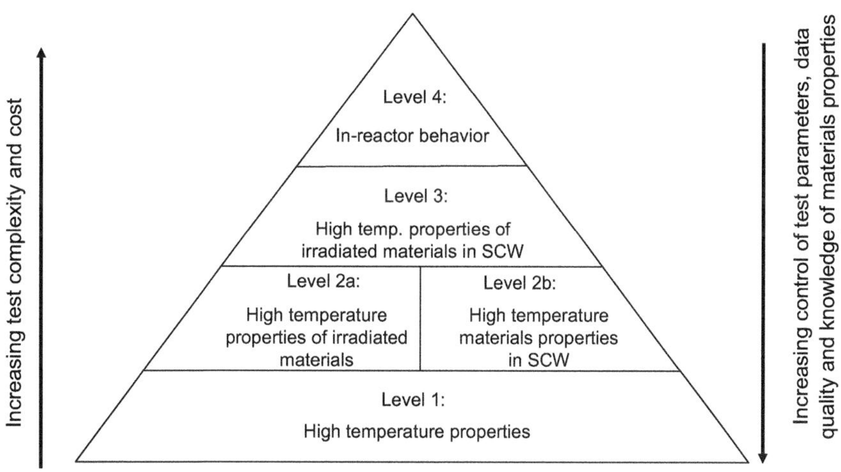

Figure 2.1 Schematic of the four-tier test program for the Generation IV International Forum supercritical water-cooled reactor (SCWR) materials work package.
Adapted from Guzonas, D.A., 2009. SCWR materials and chemistry – status of on-going research. In: Proceedings of the GIF Symposium, Paris, France, September 9–10, 2009.

Materials and Water Chemistry for Supercritical Water-cooled Reactors
https://doi.org/10.1016/B978-0-08-102049-4.00002-7

2.1 Test facilities for corrosion and environmentally assisted cracking studies

For a given alloy, the key experimental variables affecting general corrosion and environmentally assisted cracking (EAC) in SCW are temperature, water density (pressure), dissolved oxygen (DO) concentration, water purity, flow rate and surface preparation. Test facilities must be capable of controlling and monitoring the first five of these parameters for periods of up to several thousand hours. The water chemistry employed in tests carried out in support of SCWR development has typically been low conductivity, additive-free water with DO concentrations ranging from nominally deaerated ($<10\ \mu g \cdot kg^{-1}$) to fully air saturated at 25 °C ($\sim 8\ mg \cdot kg^{-1}$); in some tests hydrogen peroxide (H_2O_2) has been added to obtain higher DO concentrations (e.g., $20\ mg \cdot kg^{-1}$).

Corrosion and EAC experiments in SCW can be performed in static autoclaves and capsules, refreshed autoclaves and flow systems. The latter can range in size from small-volume bench-scale systems to large, engineering-scale systems. In capsule tests, the capsule body itself is the test section. At the high temperatures required for SCWR testing (up to 800 °C for some concepts), the materials used for loop and autoclave construction typically have corrosion rates similar to those of the materials being tested, complicating data interpretation because the surface areas of the test coupons are typically much smaller that the surface area of the loop or autoclave piping exposed to the same conditions. Therefore, the test fluid will contain dissolved corrosion products introduced from corrosion of loop and autoclave piping surfaces in addition to those released by corrosion of the test coupons. Measurements of metal concentrations in the test solution removed from Hastelloy C-276 and Alloy 625 autoclaves (the autoclave body acting as the test specimen) before and after exposure to SCW conditions showed that significant amounts of Ni, Mo and W were released into solution (Fig. 2.2; Guzonas and Cook, 2012). The ratio of the concentration of element 'X' in solution to the concentration of 'X' in the alloy was used to identify those elements preferentially released into the solution; for Hastelloy C-276, Mo and W were preferentially released, and for Alloy 625, Al, Mn and Mo were preferentially released. These dissolved species can deposit onto test specimen surfaces by precipitation or direct incorporation into the growing corrosion film[1]. Daigo et al. (2007a,b) found that Cr released from an autoclave body by corrosion can be transported by the test solution and deposit on test specimen surfaces of test in the autoclave; in their experiments this improved the corrosion resistance of the test alloy.[2] The observed reduction in corrosion rate was attributed to (1) changes in the solubility of Cr_2O_3 due to temperature gradients in the loop and (2) the possible formation of soluble Cr^{6+} species under the highly oxidizing conditions of these tests. SIMS depth

[1] These impurity or corrosion product transport and deposition phenomena are the basis for activity transport in current generation WCRs and will be discussed further in Chapter 4.

[2] Tests were carried out at 400 °C and 30 MPa in 0.01 M H_2SO_4 and 0.025 M oxygen (approximately $800\ mg \cdot kg^{-1}$) for the development of SCWO processes.

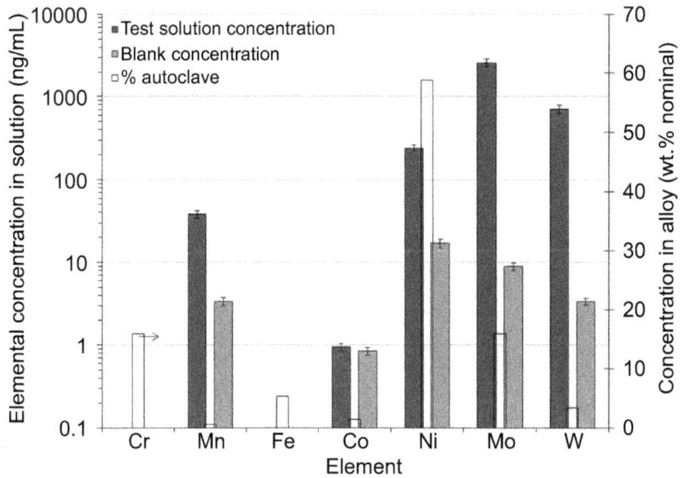

Figure 2.2 Concentrations of various metals in solution after 280 h exposure of a Hastelloy C-276 autoclave containing deionized water at 450 °C. The concentrations of the various elements in the alloy and the concentrations in the water (*blank*) before the test are also shown. Adapted from Guzonas, D.A., Cook, W.G., 2012. Cycle chemistry and its effect on materials in a supercritical water-cooled reactor: a synthesis of current understanding. Corrosion Sci. 65, 48–66.

profiles of 403 SS coupons exposed to SCW (Guzonas and Cook, 2012) showed evidence for corrosion product incorporation into surface films from sources other than the base metal. The $^{52}Cr/^{56}Fe$ ratio indicated that a Cr-rich inner layer formed below an Fe-rich outer layer, the outer oxide becoming progressively more Cr-rich towards the oxide-solution interface, attributed to Cr incorporation into the outermost oxide from sources such as the autoclave body. A weak signal for $^{90}Zr^{16}O$ (released from the zirconia tree used to hold the test coupons) was also observed in the outer part of the oxide, clear evidence for migration of material from one surface to another in SCW. Therefore, while scoping studies can be carried out exposing a number of different alloys in the same test vessel, it seems prudent that qualification testing be carried out using one alloy per test or a small number of alloys of similar composition.

Tests in static autoclaves or capsules are relatively simple, but water chemistry control (e.g., DO concentration) is difficult because the vessel is sealed. In addition, significant and possibly unrepresentative concentrations of dissolved corrosion products can build up in an autoclave or capsule over time due to corrosion of test specimens and autoclave body.

The build-up of corrosion products and other impurities can be minimized using a refreshed autoclave or loop, which uses ion exchange media and filtration to remove dissolved and particulate species (respectively). In 'once-through' loops, water is either added from a make-up tank (Fig. 2.3(a)), or cooled and purified to remove impurities prior to reintroduction to the test section (Fig. 2.3(b)). Flow systems ensure that test samples are exposed to water with a known and controlled chemistry, but the

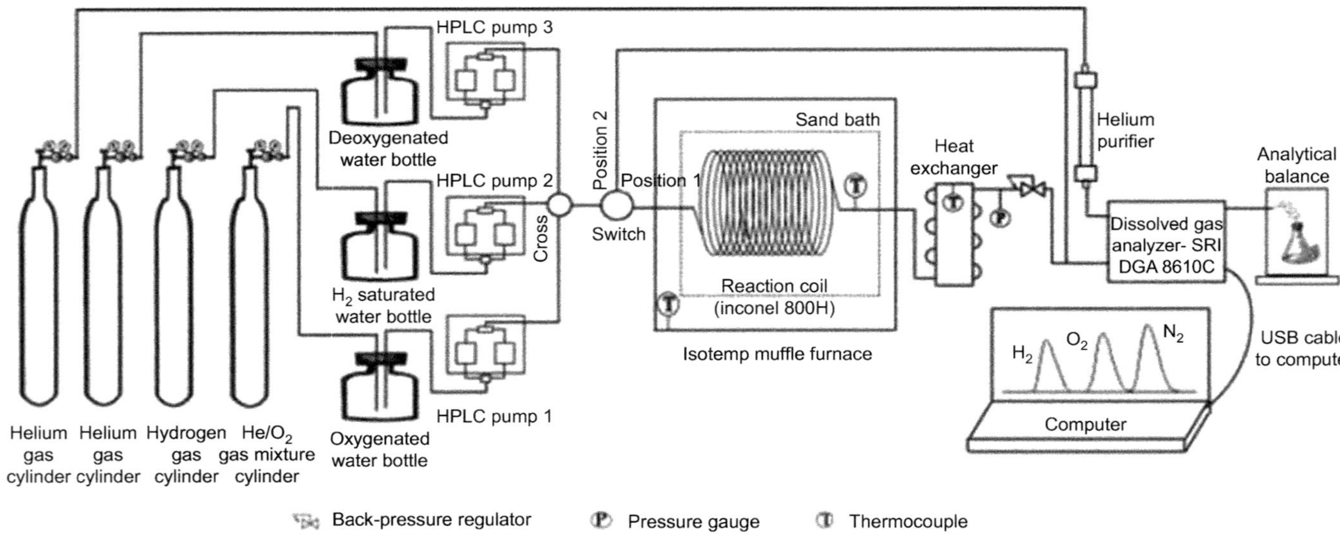

Figure 2.3a Schematic diagram of a bench-scale once-through flow system (Choudhry et al., 2014). All of the water that passes through the test section is used for sampling or discarded.

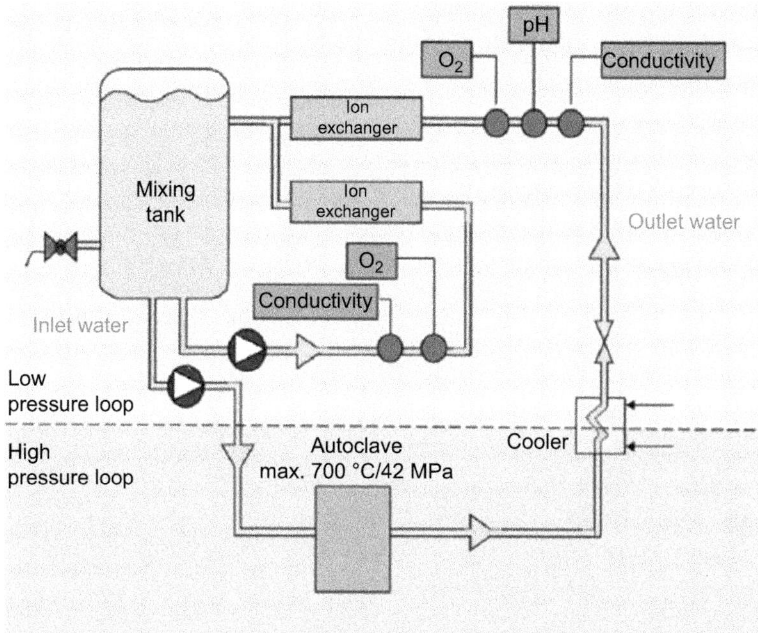

Figure 2.3b Schematic diagram of the 'once-through' recirculating flow loop at VTT (Penttilä et al., 2013). All of the water is purified after passing through the test section (autoclave).

experimenter must carefully consider the appropriate water chemistry conditions used; failures in operating reactors often occur during off-normal or transient chemistry conditions rather than during normal operation. The heat transport coolants of operating water-cooled reactors (WCRs) contain impurities (dissolved corrosion products, anionic impurities (Cl^-, SO_4^{2-})) that can affect the corrosion response; similar impurities may not be present in a once-through loop at representative concentrations unless they are deliberately introduced upstream of the test section. In recirculating loops the water is circulated through the loop circuit with or without purification[3]. The primary disadvantage of loops is that they are typically complex systems requiring significant engineering and operational oversight.

Operating temperatures and pressures of corrosion test facilities are limited by the strength of the autoclave and piping materials. Measurements at temperatures greater than 700 °C are more easily performed at low pressures, and many facilities for testing in high-temperature steam have been described in the literature (for example, Ruther et al., 1966, Bsat and Huang, 2015, Holcomb et al., 2016). Such facilities, including in-reactor systems, were used extensively during the development of nuclear superheated steam reactors (Fitzsimmons et al., 1961; Spalaris, 1963; Hazel et al., 1965). Test facilities now exist that are capable of corrosion studies at temperatures up to

[3] A side-stream purification system is often used to reduce impurity concentrations.

800 °C at $P > P_c$ (Behnamian et al., 2016) albeit for only short duration tests. Choudhry et al. (2014) have reported the results of tests of ~500 h duration at 750 °C using the bench-scale once-through facility depicted in Fig. 2.3(a).

The purity of the water can be assessed by measurements of conductivity before and after the heated test section, bearing in mind that (1) conductivity does not measure particle concentrations and is unaffected by uncharged dissolved metal hydrolysis species such as $Fe(OH)_2$ and (2) the conductivity of **cooled** samples taken after a test section may not accurately reflect water purity due to the sharp changes in corrosion-product solubility through the critical point, and the possible precipitation of colloids that may be thermodynamically unstable but which may be slow to redissolve. It has been found (as expected) that the water conductivity at the outlet of the test section is almost always higher than the inlet conductivity. Ampornrat and Was (2007) reported the test section outlet conductivities from nine corrosion experiments of the F/M alloys T91, HCM12A and HT-9 (inlet conductivity 0.06 $\mu S \cdot cm^{-1}$); at 400 °C the outlet conductivity was 0.075 $\mu S \cdot cm^{-1}$, increasing to 0.12 $\mu S \cdot cm^{-1}$ at 600 °C, indicative of corrosion product release. Unfortunately, since the much larger surface area loop surfaces also release corrosion products (cf. Fig. 2.2), attempts by various authors to use conductivity data to quantify corrosion product release and for mass balance have been largely unsuccessful.

Under SCW conditions, the solvent density can be varied by changing either the temperature at constant pressure or changing the pressure at constant temperature. While the first case is more relevant to an SCWR, this type of experiment complicates data interpretation as both the temperature (which affects the kinetics of the process being studied, e.g., through an Arrhenius-type dependence) and the solvent density (which affects the concentrations of rate-determining species in the solvent) change during a test series. At the relatively low pressures (low in the sense that $P/P_c \approx 1$) relevant to the SCWR, the largest density variation occurs at temperatures just above T_c (Fig. 2.4). Experiments performed at constant T in the vicinity of T_c in which the pressure (density) is varied provide a means of exploring the effects of SCW density on corrosion. At 25 MPa the variation in density above 500 °C is relatively small, and experiments carried out above this temperature can be considered to be at essentially constant density (about 80 $kg \cdot m^{-3}$). The temperature regimes in the core can be divided into two sections, named in analogy to supercritical fossil power plant (SCFPP) boilers; the 'evaporator', a near-critical temperature region where the temperature is increased through T_c, and the 'superheater', where the temperature is increased to the core outlet temperature. While there is no phase change in the 'evaporator' region, these terms do provide a useful shorthand for discussion of temperature/density regimes.

The water flow rate or autoclave refresh time is a key parameter of flow systems, determining the rates of transport of reactants to the test specimens and of corrosion products away from the test specimens. Novotny et al. (2013b) reported that at 500 and 550 °C, 25 MPa, flow rate 15–17 $L \cdot h^{-1}$ (autoclave refresh time ~6 min), inlet DO concentration of 1985–2020 $\mu g \cdot kg^{-1}$ and inlet conductivity <0.1 $\mu S \cdot cm^{-1}$, the outlet water conductivity slowly decreased from 0.4 $\mu S \cdot cm^{-1}$ at the beginning of the test to 0.12 $\mu S \cdot cm^{-1}$, and the autoclave outlet DO concentration steadily

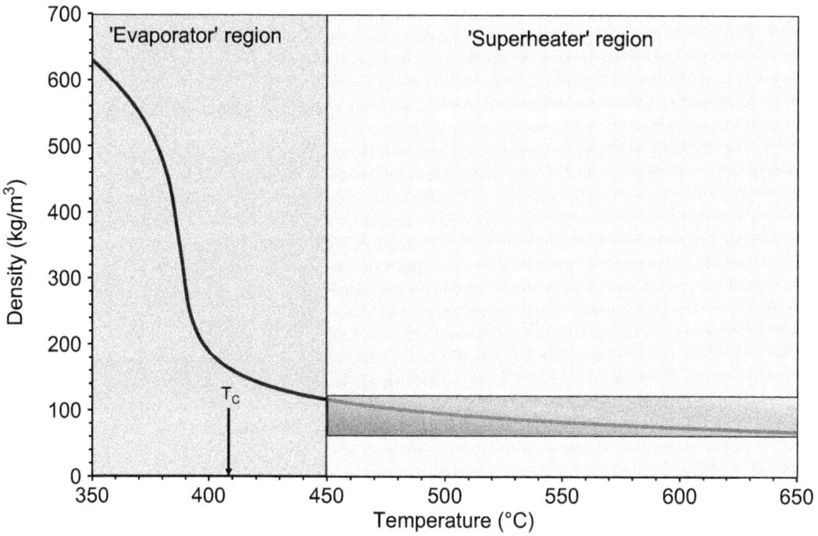

Figure 2.4 Variation of density with temperature at 25 MPa, the nominal operating pressure of current supercritical water-cooled reactor concepts. At temperatures near T_c (the near-critical or 'evaporator' region) the density changes rapidly with temperature, while at temperatures above ~450 °C (the 'superheater' region), the variation is small; in the shaded region the density decreases by only ~30% over a 200 °C temperature range.

increased during the initial stage of the test from 0 to ~1800 $\mu g \cdot kg^{-1}$, after which it remained almost constant at 1800 ± 10 $\mu g \cdot kg^{-1}$ for the rest of the heating sequence.

As part of an international interlaboratory comparison exercise, Guzonas et al. (2016) examined the dependence of weight gain on autoclave refresh time (Fig. 2.5). Several of the test loops participating in the exercise monitored the DO concentration before and after the test section; one laboratory reported a systematic loss of about 600 $\mu g \cdot kg^{-1}$ between inlet and outlet, assumed to be due to corrosion of the corrosion coupons and also the high-temperature sections of the loop, including the autoclave itself. Comparison of data from descaled corrosion coupons exposed in a static autoclave and in a flow loop suggested that transport of corrosion products in the static autoclave was not a limiting factor.

2.2 Test specimens

2.2.1 Galvanic effects

As in any corrosion testing in an aqueous environment, care must be taken to avoid galvanic coupling between adjacent corrosion coupons and also with the test section or autoclave; the coupon holder should be constructed of an insulating material. This can be a challenge in SCW, as many ceramics including ZrO_2 can dissolve at very high temperatures. Concerns about galvanic effects are mitigated to some extent at higher temperatures ($T > 500$ °C) because of the very low ionic conductivity of SCW under these test conditions.

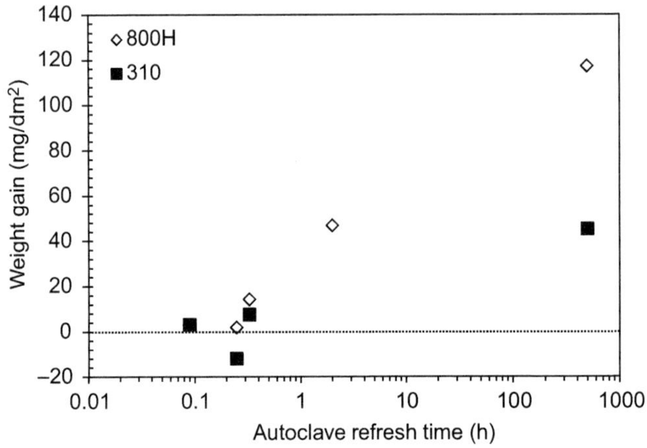

Figure 2.5 Weight gain as a function of autoclave refresh time for Alloy 800H and 310 SS (1200 grit surface finish) exposed to supercritical water (550 °C, 25 MPa, dissolved oxygen 8 $mg \cdot kg^{-1}$, water pH ~7 at room temperature, water resistivity at room temperature 18 $M\Omega \cdot cm$, 500 h duration).
Adapted from Guzonas, D., Penttilä, S., Cook, W., Zheng, W., Novotny, R., Sáez-Maderuelo, A., Kaneda, J., 2016. The reproducibility of corrosion testing in supercritical water – results of an international interlaboratory comparison exercise. Corros. Sci. 106 147–156 https://doi.org/10.1016/j.corsci.2016.01.034.

2.3 Corrosion rate measurements

2.3.1 Surface finish

It has long been recognized that the surface finish of an alloy affects the corrosion rate in high-temperature aqueous environments (discussed in detail in Chapter 5). A complete description of a surface should consider four factors (Berge, 1997): geometrical factors such as roughness; chemical factors such as compositional differences between bulk and surface; structural factors such as grain size and cold working and mechanical factors such as residual surface/bulk stresses. Ruther et al. (1966) noted that, in SCW, surface preparations that gave a strain-free surface resulted in maximum general corrosion while surface finishes that resulted in severe cold work reduced general corrosion (see Chapter 5). Cold work will also have an effect on stress corrosion cracking (SCC) (see Chapter 6).

The choice of an appropriate surface finish should be borne in mind, particularly when preparing coupons for surface analysis as a highly smooth polished surface will result in a higher weight gain than a coarsely ground surface. Great care should be taken when comparing weight gain data for samples with different surface finishes, especially at $T \gg T_c$, as differences in weight change of 1–2 orders of magnitude have been observed for different surface finishes. Details on coupon surface finish should be included with all reported corrosion test results.

2.3.2 Weight gain versus weight loss

The metal loss of a material over the design life of a component (e.g., the total decrease in pipe wall thickness) is a key performance requirement for development of SCWR concepts. Typically, the results of corrosion tests carried out under SCWR conditions are reported as the change in weight of the samples (weight after − weight before), for example, as milligrammes mass change per square decimetre ($mg \cdot dm^{-2}$) of surface area exposed to the test solution. Coupon weights should be determined using a balance with an accuracy of 0.1 mg (10^{-4} g) or better. It is recommended that weighing of a specimen be repeated five times (at a minimum, three times), and the datum of the weight taken as the average of these measured values.

While weight change measurements facilitate relative comparison between different alloys, it is not an unequivocal measure of metal loss since some of the oxidized metal can dissolve or exfoliate and be carried away by the test solution, or material released from other system surfaces can be transported to, and deposit on, the coupon surface. Longton (1966) discussed in detail the differences between weight gain and weight loss measurements, emphasizing that weight gain measurements can only yield reliable information when it is certain that no oxide is lost to the system, noting that oxide losses of up to 50% had been reported. Mass balance data (Guzonas and Cook, 2012) for 304 SS exposed to SCW at 500 °C with a low concentration of DO showed that the postexposure weight change underestimated the actual corrosion (metal loss) by a factor of 5–6, the metal contained in the oxide film being insufficient to account for the weight loss. This was indicative of corrosion product release to the coolant; although the postexposure weight change was essentially constant, the corrosion rate (as determined by descaled weight loss) decreased with time. Novotny et al. (2015) found that oxide spallation significantly affected weight change in loop tests, noting that for the alloys studied (316L, 347H, 08Cr18Ni10Ti) weight change measurements consistently underestimated the real oxide layer growth.

Fig. 2.6 shows a simplified cross-sectional schematic of oxide growth during exposure to SCW illustrating three possible scenarios (oxide solubility and growth are discussed in detail in Chapters 4 and 5, respectively). If the solubility of the surface film is high, then the film will completely dissolve into the solution, the coupon will lose weight and the weight change measured will directly correspond to the metal loss. For intermediate cases (low but nonnegligible oxide solubility) the measured weight change cannot easily be related to the metal loss unless the loss to solution can also be quantified. If the solubility of the surface film is negligible, then all of the oxide formed will remain on the surface, the coupon will gain weight and the weight gain will correspond to the metal loss by a factor that can be derived from the film composition. In addition to these solubility-based phenomena, oxide can exfoliate leading to weight loss, or metal species released from test vessel surfaces can be transported to the surface and deposit leading to weight gain. If, and only if, all of the metal oxidized

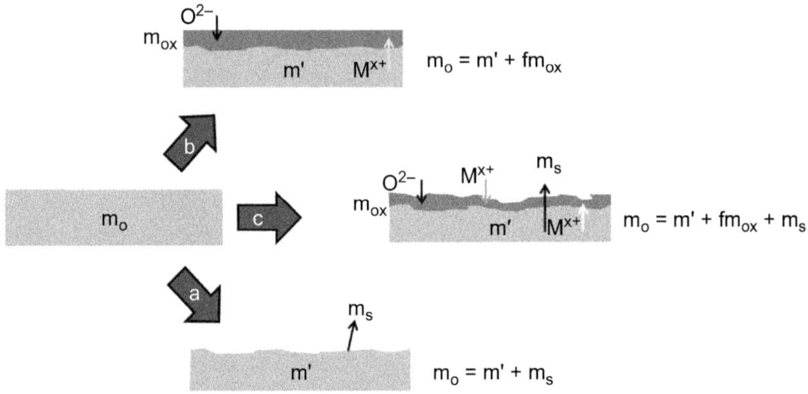

Figure 2.6 Effect of oxide solubility on weight change. m_O, original coupon mass; m', mass of base material remaining after corrosion; m_{ox}, mass of oxide; m_s, mass of metal in solution; f, a factor to convert oxide mass to mass of metal.
Adapted from Guzonas, D.A., 2015. The physical chemistry of corrosion in a supercritical water-cooled reactor. In: Proceedings of the 16th Int. Conf. on the Properties of Water and Steam: Water, Steam and Aqueous Solutions Working for the Environment and Industry, September 1–5, 2013, London, UK. Paper PWS-116.

during the corrosion process remains in the surface oxide layers **and** no additional species deposit from solution can the weight change and the true metal loss be directly related. The only reliable measurement of metal loss is to measure the weight loss obtained by descaling the coupons (e.g., ASTM 61-03) after exposure to the test environment to remove oxidized corrosion products or to directly measure the thickness of the remaining metal in the coupon from a postmortem cross section. In the latter case the pretest thickness of the coupons must be known.

There are a number of well-established procedures for descaling, although finding a suitable descaling method for some alloys can be challenging. Unfortunately, descaling destroys the corrosion film so any required surface characterisation must be done prior to descaling or multiple coupons must be used in a test with one or more coupons retained for surface analysis. Combining knowledge of the surface oxide structure obtained from surface analysis and the oxide weight, the mass of iron, nickel, chromium and other metals in the corrosion film can be estimated and compared with the mass of material lost from the surface as measured by descaling. Differences between the two imply corrosion products have either been transported away from the sample during the test or deposited on the coupon from the test solution (Guzonas and Cook, 2012).

Metal loss can also be inferred from metallographic examination of cross sections, as long as the original metal surface prior to exposure can be located. Such examinations can determine if oxide exfoliation occurred, and if there is no dissolution, the weight change can be inferred from the oxide thickness. Such a calculation requires that the oxide composition and porosity be known. If the oxide thickness is uneven it may be better to use the maximum penetration as a measure of the corrosion, as in the case of intergranular corrosion.

2.3.2.1 Reproducibility

The reproducibility of published corrosion data in support of SCWR development can be difficult to ascertain as error bars are not always included in data plots, and experimental uncertainties are often not discussed. In addition, testing (especially scoping tests) may often include only a few coupons of a particular alloy, and these coupons may have been prepared with different surface finishes for various types of postexposure characterization. Differences in test facilities and test conditions (temperatures, exposure time, water chemistry, surface treatment) can make comparisons difficult. Guzonas and Cook (2012) found relatively good agreement between published data sets for the ferritic steels T91 and T92 exposed to SCW at 500 °C as a function of exposure time from tests carried out under generally similar conditions but different DO concentrations (see Chapter 5, Fig. 5.20). Similar good agreement is found for other alloys for which a large number of studies exist, such as the common 300-series stainless steels, Alloy 800H and Alloy 625.

The GIF SCWR Materials and Chemistry PMB organized an international interlaboratory comparison test (round-robin test) to study the reproducibility of general corrosion data obtained in different test facilities (Guzonas et al., 2016). Such a comparison was considered necessary given the large number of research groups world-wide performing corrosion testing in support of SCWR development. Each participating signatory provided coupons machined to the appropriate dimensions (Alloy 800H, 08Cr18Ni10T SS, 310 SS) to the other participants (up to seven laboratories), who each followed a standard procedure developed by the group to prepare, characterize and test the coupons. Data reporting was limited to weight change only. Considerably more variation in weight gain was observed than expected. The scatter in the weight change data for a given alloy was small within the same laboratory (typically less than ±20% of the average value) but large between different laboratories (typically greater than ±100% of the average value). The variation appeared to be largely attributable to differences between the test facilities not considered during the development of the test procedures, particularly the autoclave refresh time (Fig. 2.5). A key conclusion was that care must be taken when comparing data obtained in different test facilities. Details of the facilities and the experimental pre/post-treatments must be reported and taken into consideration when assessing the data. It was recommended that standard test methodologies for corrosion tests in support of SCWR development be developed.

The GIF SCWR M&C PMB is planning a second phase of interlaboratory comparison corrosion tests to address some of the deficiencies identified in the first phase.

2.3.3 Electrochemical methods

Corrosion studies in support of current generation WCRs often use in situ electrochemical methods to characterize corrosion (Lillard, 2012). As will be discussed in

Chapter 4, the low dielectric constant of SCW at temperatures well above T_c favours ion association leading to a decrease in ionic conductivity. At temperatures in the vicinity of T_c, various electrochemical methods (electrochemical polarization (Kritsunov and Macdonald, 1995), electrochemical noise analysis (Liu et al., 1994), film impedance and film resistance (Betova et al., 2007) have been successfully used. Hettiarachchi et al. (1993a,b) described the development of sensors for pH and potential in SCW for monitoring the water chemistry of SCFPPs. Liu and Macdonald (1995) demonstrated the feasibility of monitoring dissolved hydrogen concentrations in SCW using a *Pd/Pt* relative resistance sensor at 410 °C and 27.6 MPa. A major difficulty in the use of noncontact electrochemical methods at very high temperatures is the lack of a stable reference electrode whose potential is environment-independent and can be related to a rational thermodynamic scale. Van Nieuwenhove (2012) described a reference electrode consisting of an yttrium-stabilized zirconium oxide (YSZ) ceramic tube filled with a mixture of iron and iron oxide powder surrounding a Fe conductor. At high temperatures, zirconium oxide is an oxygen ion conductor, allowing electrochemical reactions of the oxygen ions at the ceramic/water and ceramic/iron oxide interfaces, thereby determining the potential of the inner Fe conductor. Such an electrode has been successfully used in an SCW loop to measure electrochemical potential as well as electrochemical impedance from 230 to 604 °C (Novotny et al., 2017). The limitations of electrochemical measurements in SCW are discussed in Chapters 4 and 5.

2.3.4 Other in situ analyses

After exposure to SCW, test specimens have been characterized using a range of ex situ techniques ranging from weight change measurements to various surface analytical methods such as scanning electron microscopy (SEM), energy dispersive X-ray spectroscopy (EDX or EDS), transmission electron microscopy (TEM), glow discharge optical emission spectroscopy (GDOES), X-ray photoelectron spectroscopy (XPS), Raman spectroscopy and scanning auger microscopy (SAM). Weight change measurements and characterization of oxide films are typically ex situ measurements performed after the sample has been cooled, rinsed and dried, and in the case of techniques such as SEM or TEM, subjected to vacuum conditions. The possibility of oxide film changes during (1) cooling due to the large changes in oxide solubility through the critical point and (2) dehydration of the oxide layers during evacuation must be borne in mind.

In situ characterization methods, while experimentally challenging, avoid these issues. Maslar et al. (2001, 2002) reported the use of in situ Raman spectroscopy to characterize oxide layers on chromium and stainless steel, respectively, at 25.2 MPa and 496 °C. A major experimental challenge is the choice of optical window material (discussed further in Section 2.4). Measurements of the concentrations of metals in the test solution downstream of the test section provide data on metal release rates, which can be used to infer the relative corrosion rates. Han and Muroya (2009) reported data on the release of ^{60}Co radiotracer from irradiated 304 SS samples as a function of time and temperature in SCW. They collected the material released on ion exchange resins and used gamma spectrometry to quantify the amounts released. Choudhry et al. (2014) used a bench-scale flow system (Fig. 2.3(a)) to monitor oxygen consumption, hydrogen generation and metal release during the corrosion of a number of alloys at temperatures up to 750 °C. Concentrations of dissolved metals in the test solution can be measured by standard analytical methods such as inductively coupled plasma (ICP) optical emission spectrometry (OES), ICP mass spectrometry (MS) or adsorptive stripping voltammetry, and anion concentrations can be measured by ion chromatography (IC). Hydrogen and oxygen concentration can be measured using gas chromatography (GC).

2.4 Measurements of thermodynamic properties

Models for the behaviour of dissolved species in SCW for predicting the effects of corrosion product transport require data on the temperature and concentration dependencies of ion speciation, equilibrium constants and other thermodynamic parameters. Various spectroscopic methods can be used to quantitatively measure the concentrations of aqueous species in high-temperature solutions. The aforementioned choice of window material also can be an issue for these types of measurements. Recent advances in high pressure optical cells (HPOCs) for use in geochemical studies (Chou et al., 2005) enable Raman studies to be carried out at temperatures up to 500 °C and pressures up to 100 MPa. Raman spectroscopy can provide detailed structural information, especially when coupled with modern computational methods that allow the energies, high-temperature structures and thermodynamic stabilities of compounds to be evaluated (Pye et al., 2014). Applegarth et al. (2015) described the use of fused silica HPOCs[4] to study the thermal decomposition of several noncomplexing anions (often used as internal standards for Raman) at 30 MPa from 250 to 400 °C. Small-volume hydrothermal diamond anvil cells (HDACs) (Bassett et al., 1993; Solferino and Anderson, 2012) have enabled in situ X-ray fluorescence (XRF), X-ray absorption spectroscopy (XAS) and Raman spectroscopy measurements of metal oxide behaviour in SCW at temperatures up to 600 °C. Solferino and Anderson (2012) discussed the control of oxygen fugacity in HDACs, including the use of hydrogen peroxide to vary the fugacity in a controlled manner. Yan et al. (2013) used in situ Co K-edge XAS measurements to study Co^{2+} adsorption on Fe_3O_4 nanoparticles in aqueous fluids to 500 °C and approximately 220 MPa. Yan et al. (2011) used X-ray spectroscopy to study the speciation of Mo^{6+} in SCW contained in an HDAC.

Frequency-dependent electrical conductivities of solutions of aqueous strontium hydroxide and strontium chloride have been measured from 295 to 625 K at 20 MPa, over a very wide range of ionic strength using a high-precision, high-temperature, high-pressure conductance flow AC conductivity instrument (Arcis et al., 2014). The flow cell used was built at the University of Delaware by Hnedkovsky et al. (2005) with improvements on the original designs of Zimmerman et al. (1995)

[4] An additional consideration is that acidic or alkaline conditions can result in reactions of the solvent with the window material.

and Sharygin et al. (2002) to allow for the measurement of more corrosive solutions. The cell has been used at temperatures as high as 673 K at 28 MPa with ionic strengths as low as 10^{-5} mol·kg^{-1} (Zimmerman et al., 2012).

2.5 Stress corrosion cracking

Many methods to study SCC and crack growth resistance exist, each with advantages and disadvantages. Typically, long-term autoclave tests are used to quantify crack initiation and growth rates. After sample failure, the fracture surface is examined by microscopy. In recent years the use of advanced sample preparation (focussed ion beam, FIB Li, 2006; Li et al., 2008) and microscopy techniques such as analytical TEM (ATEM) and electron backscatter diffraction (EBSD) to probe processes at the crack tip has become more common (Lozano-Perez et al., 2014). SCC testing under SCW conditions is more challenging than general corrosion testing because the apparatus used to apply strain must either fit entirely in the autoclave or a means must be provided for vessel penetration to allow movement through the sealed vessel wall, making the apparatus small, expensive or both. Behnamian et al. (2013) studied SCC using capsules, calculating the stresses on the capsule wall from the volume of water added and the temperature. Constant-load tests use a C-ring or O-ring and a set of springs fixed within a hard surface and a tightening nut; the C-ring or O-ring expands into the springs reducing the stress increase from expansion. Swift et al. (2015) described a constant load C-ring assembly for use in SCW using Inconel 718 Belleville washers acting as a spring to deliver a near-constant load to the sample. A mechanistic model was used to calculate the force applied by the spring due to the thermal expansion to enable prediction of the stress at the apex of the C-ring.

Novotny et al. (2013a) described a pneumatic servo-controlled system for fracture toughness, SCC, corrosion fatigue, tensile and electrochemical measurements with high sensitivity and accuracy. As no moving parts penetrate the pressure boundary, friction force at the sealing element location is avoided enabling more accurate load control than is possible with conventional servo-hydraulic and step-motor driven devices. This allows use of smaller specimens, facilitating testing irradiated materials in hot cells, in-reactor measurements and crack growth rates studies in SCW. A prototype miniature bellows-based loading device (Fig. 2.7) was developed as an integral

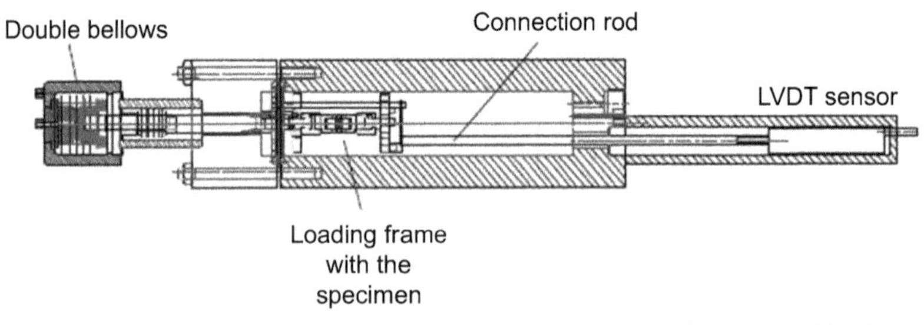

Figure 2.7 Schematic drawing of the miniature autoclave double bellows–based loading device (Novotny et al., 2013a).

part of a miniature autoclave for IASCC testing in hot cell laboratories and future in-reactor facilities. The double bellows (DB) device eliminates the main drawback of single bellows-based loading devices, namely, the very high initial pressure inside the bellows (~25 MPa for SCWR testing). Servo-valve and bellows failure may be unacceptable, particularly for hot-cell or in-reactor operation. Pentillä et al. (2017) described a DB-based pneumatic loading unit with a miniature autoclave capable of performing loading for tensile-type testing in SCW to measure crack growth rates. Calibrations for the pneumatic loading unit were performed and a verification test using a miniature autoclave material testing system with 5 mm disk compact tension (DC(T)) precracked 316 SS specimens was performed at temperatures of 288 and 550 °C at 25 MPa.

An important consideration is the way in which SCC susceptibility is quantified. Two different measures have been used, one based on the amount of intergranular cracking (%IG) of the fracture surface following failure and the other based on quantification of cracks on the sample surface (density, length, length/unit area, depth). As discussed by Allen et al. (2012) these measurements do not always agree.

2.6 Testing under irradiated conditions

Radiation can affect the degradation of in-core materials by (1) direct interaction with the material leading to displacement of atoms and nuclear reactions such as neutron activation or (2) by indirect interaction by water radiolysis (Fig. 1.3) and the subsequent interaction of radiolysis products with the alloy. Therefore, in-reactor testing in SCW under the expected water chemistry conditions must be the final step in SCWR materials testing (Level 4 in Fig. 2.1). However, many of the irradiation-induced changes in bulk mechanical properties can be studied using irradiations under non-SCWR conditions. For these studies, it is more important to ensure that irradiations are performed in a facility that reproduces the expected SCWR neutron spectrum (correct balance of fast and thermal neutrons) and that has a neutron flux sufficient to achieve the target dose in a reasonable time (Walters et al., 2017).

2.6.1 Materials testing

As part of the US nuclear steam reheat development program, a number of alloys were irradiated in test reactors under superheated steam conditions. For example, fuel assemblies with 0.406 mm thick fuel cladding were exposed to superheated steam at temperatures up to 738 ± 83 °C for 10,292 h (Comprelli et al., 1969) in the SADE test loop in the EVESR. There is currently no in-reactor test facility for SCWR studies capable of operating using SCW. At Centrum Výzkumu Řež (CVR), an SCW loop (Fig. 2.8) has been constructed for in-reactor materials and water chemistry experiments (Ruzickova et al., 2011). The loop is divided to two functional parts: irradiation channel and auxiliary circuits. The irradiation channel was designed for operation at temperatures up to 600 °C at 25 MPa, and it is planned to place it in one of the test sections of the LVR-15 test reactor. The auxiliary circuits condition the water to the

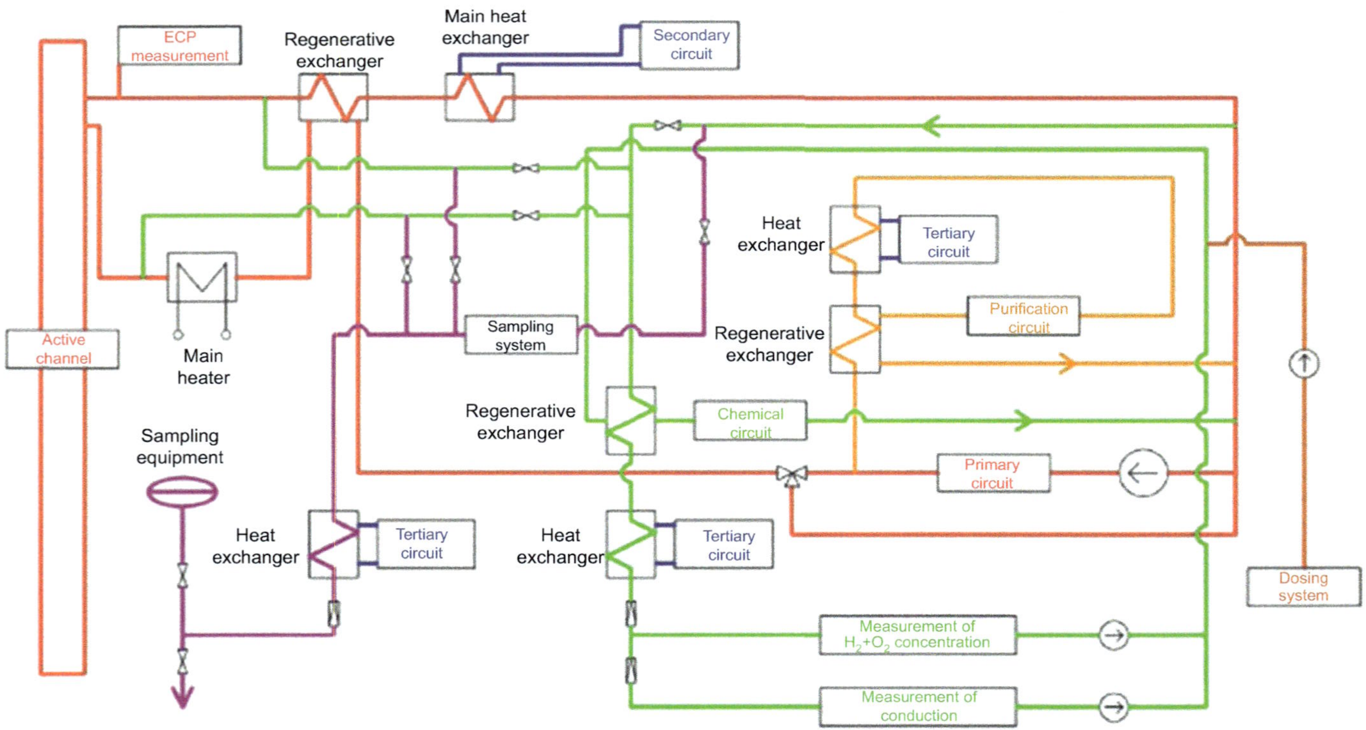

Figure 2.8 Flowchart of the SCW loop at CVR (Ruzickova et al., 2011). The active channel will be inserted in the reactor core and contains the test section(s). The auxiliary circuits are the primary circuit (red), dosing system (dark orange), measuring system (green), purification system (orange) and cooling circuits. The dosing system enables the addition of chemicals (gasses and solutions). Chemistry monitoring (blue) in the primary circuit is provided by a side-stream circuit. The medium circulating in the loop is partially purified by a system of mechanical filters and ion exchange media to maintain the required chemical conditions in the primary circuit.

desired channel inlet parameters, as well as enabling monitoring, control and chemical addition. A second, conceptually similar loop has been designed for SCWR fuel qualification tests (Vojacek et al., 2015). The design of the active channel is limited by space in the reactor core, availability of materials certified for nuclear use under the proposed test conditions and nuclear safety considerations. The stainless steel 08Cr18Ni10 T was proposed as the alloy for the active channel limiting the maximum temperature to ~450 °C. As of mid-2017, neither the materials test loop nor the fuel qualification loop had been installed in-reactor.

While in-reactor irradiations can require several years accumulate target damage levels and require special precautions, facilities and instrumentation for radioactive sample handling, ion irradiations conducted at a well-defined energy, dose rate and temperature (Was and Averback, 2012) can quickly reach damage levels in the 1−100 dpa range and produce little or no residual radioactivity, simplifying sample handling and significantly reducing cycle length and cost. By quantifying the effects of both neutron and ion irradiation experiments, conditions can be designed such that ion irradiation yields the same equivalent measure of radiation effect as neutron irradiation. High-energy ($E > 1$ MeV) self-similar ion irradiation combined with nanomechanical testing techniques was recently used to study radiation-induced hardening of some candidate SCWR alloys (Klassen and Rajakumar, 2015).

An SCW convection loop with an irradiation cell coupled to a 10 MeV, 10 kW linear electron accelerator (linac) (Bakai et al., 2011) was constructed at the Kharkiv Institute of Physics and Technology (Fig. 2.9). The loop is equipped with chemistry monitors (oxygen concentration, conductivity), although these had to be located some distance from the irradiation cell. Bakai et al. (2016) presented the results of the postirradiation microstructure and hardness characterization of surface oxide layers on selected specimens including a highly irradiated specimen (up to $\sim 10^{21}$ electrons$\cdot$cm^{-2} fluence and ≈ 16 keV$\cdot$atom^{-1} specific deposited energy E_{dep}) of 12Cr18Ni10T SS cut from the irradiation cell piping. Due to the rather short range (~1 cm) of ~10 MeV electrons in the materials and lateral shaping of the scanning flux and consequent large energy deposition in the coupons, there was significant variation in heating at different areas of the test coupon surface. The combined effects of temperature and irradiation the spatial variation of these quantities were separated by modelling using Monte Carlo simulated 3D energy absorption maps in the irradiated zone based on e-beam spatial and energy distribution data from linac operation, coupled to finite element method calculations of thermal fields.

2.6.2 Water radiolysis studies

As will be discussed in Chapter 4, modelling SCW radiolysis requires knowledge of the dependence of reaction rates on parameters such as fluid temperature and density. This kinetic information is typically obtained using pulse radiolysis (Spinks and Woods, 1976), in which a short-duration pulse of electrons or other ions irradiates a sample and the behaviour of the resulting transient radiolysis species monitored, often using optical spectroscopy. A number optical cell designs for use at high pressure and temperature have been implemented (Fujisawa et al., 2004; Takahashi et al., 2000).

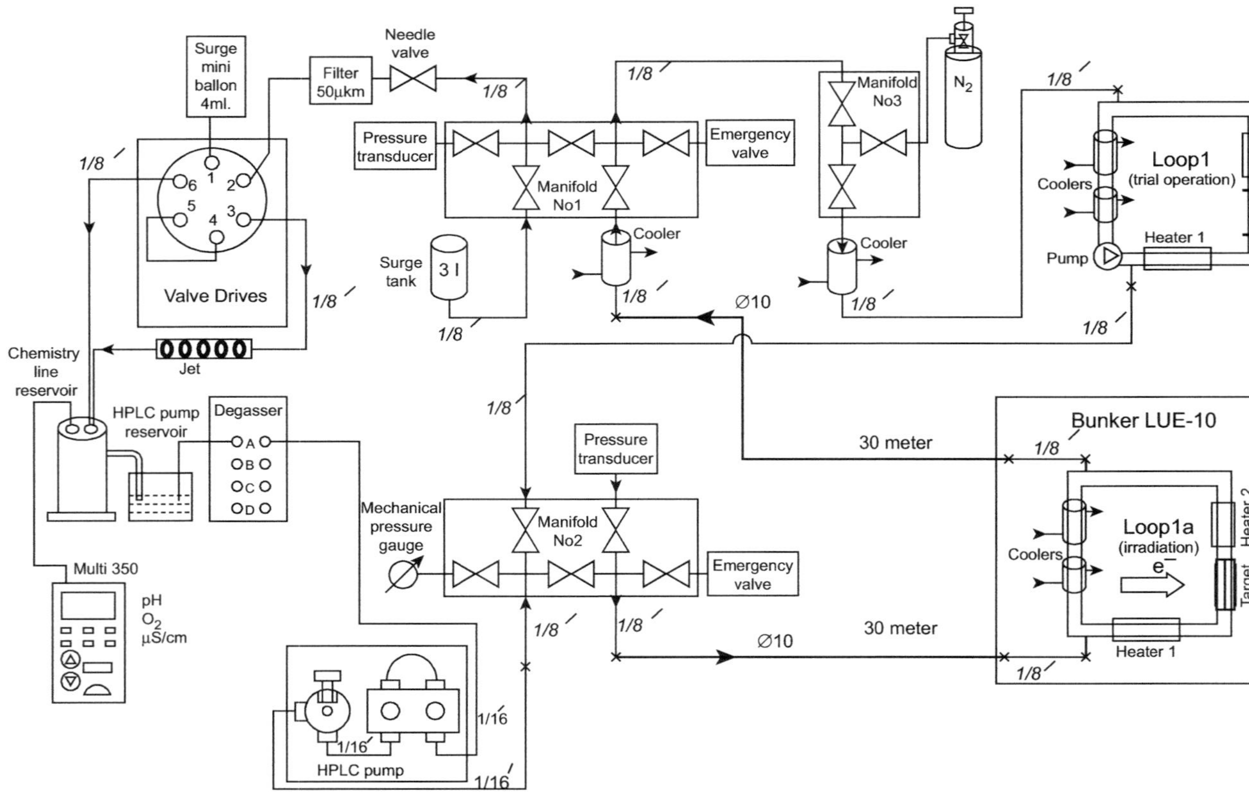

Figure 2.9 Schematic of the Canada-Ukraine Electron Irradiation Test Facility (Bakai et al., 2011). The loops are equipped with tubular continuous-flow water coolers, a main horizontal lip-type heater (Heater 1, $\leq$6 kW$_e$) and a vertical startup heater (Heater 2, $\leq$3 kW$_e$) to control the natural circulation direction. The 12Cr18Ni10T SS piping had an internal diameter of 32 mm and a wall thickness of 4 mm; the piping wall thickness increased to 6 mm inside Heater 1.
Figure courtesy of A. Bakai.

Janik et al. (2007) reported the results of β-radiolysis experiments performed at the Notre Dame Radiation Laboratory using a custom-made SCW irradiation block and 2.5 MeV electrons from a 3.0 MeV van de Graaff accelerator.

An SCW loop inserted into the 1 MW TRIGA[5] reactor at the University of Wisconsin was used to measure neutron/gamma water radiolysis yields (Bartels et al., 2006). The loop design, complex due to the stringent safety requirements, consisted of four sections: irradiated volume/void and neutron shielding; lead gamma shield; voided heating section; and water shielding. A number of different designs and materials for the irradiation volume were used; materials included Hastelloy C-276, Alloy 800 and titanium. The Monte Carlo N Particle transport code was used to model the reactor and understand the radiation environment in the irradiated volume at full reactor power to optimize the experiment. When the SCW loop at CVR is installed, it will be a valuable test facility for water radiolysis studies.

Pommeret (2007) described a small (1.27 cm^3) Inconel 718 cell for pulse radiolysis studies using an accelerator, capable of operation up to 100 MPa, 500 °C. Katsumura et al. (2010) and Muroya et al. (2012) (Fig. 2.10) described the development and use of an S-band linac to determine the time-dependent radiolysis yields of hydrated electrons at elevated temperatures, indirectly using methyl viologen scavenging of e_{aq} and

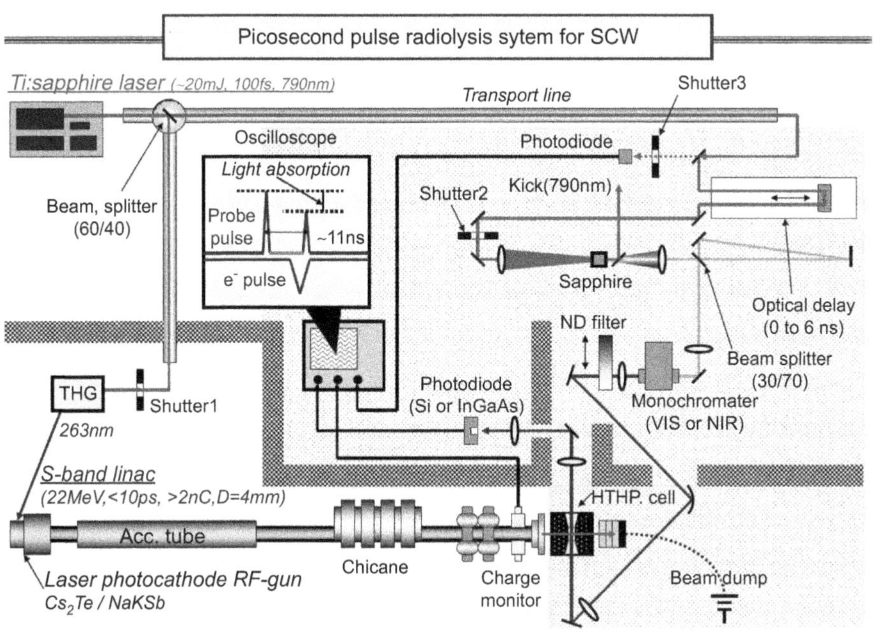

Figure 2.10 Schematic of the picosecond time-resolved pulse radiolysis system at the Nuclear Professional School, The University of Tokyo (Katsumura et al., 2010).

[5] Training, Research, Isotopes, General Atomic.

conventional nanosecond pulse radiolysis measurements, and directly using a picosecond time-resolved pulse radiolysis system. To overcome signal-to-noise (S/N) ratio limitations at SCW temperatures, a laser-driven photocathode RF-gun combined with a femtosecond Ti:sapphire laser was developed. The system produced an 11 ps (FWHM) pulse of 22 MeV electrons, which was incident on the water, and the absorption of the resulting transient species was measured using a femtosecond laser pulse. The overall time resolution was about 60 ps, the analysing light covering wavelengths from the visible to near infrared continuously with a relatively good S/N ratio in SCW.

Muons have been used to study radiolysis in SCW at temperatures up to 450 °C (Alcorn et al., 2014), using muonium (Mu) as an analogue of the hydrogen atom (Ghandi and Miyake, 2010).[6] In muon spin spectroscopy (μSR) experiments a beam of positive muons is directed to a sample cell, the beam momentum adjusted such that the muons pass through the cell window and stop in the sample. The muon beam is spin-polarized and a weak magnetic field is applied transverse to the muon spin direction, causing the stopped muons to precess at frequencies characteristic of their environment. Chemical reactions that change the magnetic environment experienced by the muons results in spin dephasing, which can be measured. Alcorn et al. (2014) discuss the possible interferences from paramagnetic species such as metal ions released by corrosion of the test cell, noting that in their work such interferences typically occurred during the first ~24 h, until a stable oxide layer was formed.

Mayanovic et al. (2012) reported the observation of X-ray induced radiolysis of Fe and W ions in an HDAC exposed to synchrotron radiation; this was an unintended consequence of the sample irradiation but does provide a novel means to study radiolysis effects. For example, Men et al. (2014) used XAS to characterize the radiolysis-induced redox behaviour of a Cr standard solution in an HDAC at temperatures from 300 to 500 °C. The concentrations of steady-state radiolysis products, hydrogen, oxygen and hydrogen peroxide, can also be monitored using a cell placed inside a gamma irradiator.

References

Alcorn, C.D., Brodovitch, J.-C., Percival, P.W., Smith, M., Ghandi, K., 2014. Kinetics of the reaction between H· and superheated water probed with muonium. Chem. Phys. 435, 29–39.

Allen, T.R., Chen, Y., Ren, X., Sridharan, K., Tan, L., Was, G.S., West, E., Guzonas, D.A., 2012. Material performance in supercritical water. In: Konings, R.J.M. (Ed.), Comprehensive Nuclear Materials, vol. 5. Elsevier, Amsterdam, pp. 279–326.

Ampornrat, P., Was, G.S., 2007. Oxidation of ferritic-martensitic alloys T91, HCM12A and HT-9 in supercritical water. J. Nucl. Mater. 371, 1–17.

[6] Muonium consists of a positive muon (μ+, ~2.2 μs lifetime, mass ~0.11 μ) that has captured an electron to form a bound atomic state analogous to a hydrogen atom (Ghandi and Miyake, 2010). As the reduced masses of Mu and H are nearly the same, Mu can be considered a light isotope of hydrogen•, reacting similarly to hydrogen• although the rate constants may differ due to kinetic isotope effects (Walker, 1981).

Applegarth, L., Alcorn, C., Bissonette, K., Noël, J., Tremaine, P.R., 2015. Non-complexing anions for quantitative speciation studies using Raman spectroscopy in fused silica high pressure optical cells under hydrothermal conditions. Appl. Spectrosc. 69, 972–983.

Arcis, H., Zimmerman, G.H., Tremaine, P.R., 2014. Ion-pair formation in aqueous strontium chloride and strontium hydroxide solutions under hydrothermal conditions by AC conductivity measurements. Phys. Chem. Chem. Phys. 16, 17688–17704.

Bakai, A.S., Boriskin, V.N., Dovbnya, A.N., Dyuldya, S.V., Guzonas, D.A., 2011. Supercritical water convection loop (NSC KIPT) for materials assessment for the next generation reactors. In: Proc. of the 5th Int. Symposium on Supercritical Water-Cooled Reactors. Vancouver, Canada, on CD-ROM, Paper 051.

Bakai, O.S., Guzonas, D.A., Boriskin, V.M., Dovbnya, A.M., Dyuldya, S.V., 2016. Combined effect of irradiation, temperature, and water coolant flow on corrosion of Zr–, Ni–Cr-,and Fe–Cr-based alloys. J. Nucl. Eng. Radiat. Sci. 2, 021007.

Bartels, D.M., Anderson, M., Wilson, P., Allen, T., Sridharan, K., 2006. Supercritical Water Radiolysis Chemistry Supercritical Water Corrosion. Progress Report Available at: http://nuclear.inl.gov/deliverables/docs/uwnd_scw_ii_sep_2006_v3.pdf.

Bassett, W.A., Shen, A.H., Bucknam, M.J., Chou, I.-M., 1993. A new diamond anvil cell for hydrothermal studies to 2.5 GPa and −190 °C to 1200 °C. Rev. Sci. Instrum. 64, 2340–2345.

Behnamian, Y., Li, M., Zahiri, R., Kohandehghan, A., Mitlin, D., Chen, W., Luo, J.-L., Zheng, W., Guzonas, D., 2013. Stress corrosion cracking of austenitic alloys in supercritical water. In: The 6th International Symposium on Supercritical Water-Cooled Reactors, March 03–07, 2013, Shenzhen, Guangdong, China, Paper ISSCWR6-081.

Behnamian, Y., Mostafaei, A., Kohandehghan, A., Shalchi Amirkhiz, B., Serate, D., Sun, Y., Liu, S., Aghaie, E., Zeng, Y., Chmielus, M., Zheng, W., Guzonas, D., Chen, W., Luo, J.-L., 2016. A comparative study of oxide scales grown on stainless steel and nickel-based superalloys in ultra-high temperature supercritical water at 800 °C. Corros. Sci. 106, 188–207. https://doi.org/10.1016/j.corsci.2016.02.0040010-938X.

Berge, P., 1997. Importance of surface preparation for corrosion control in nuclear power stations. Mater. Perform 36, 56–62.

Betova, I., Bojinov, M., Kinnunen, P., Penttilä, S., Saario, T., 2007. Surface film electrochemistry of austenitic stainless steel and its main constituents in supercritical water. J. Supercrit. Fluids 43, 333–340.

Bsat, S., Huang, X., 2015. Corrosion behavior of 310 stainless steel in superheated steam. Oxid. Metals 84, 621–631.

Chou, I.M., Burruss, R.C., Lu, W., 2005. A new optical capillary cell for spectroscopic studies of geologic fluids at pressures up to 100 MPa. In: Chen, J., Wang, Y., Duffy, T.S., Shen, G., Dobrzhinetskaya, L.P. (Eds.), Advances in High Pressure Technology for Geophysical Applications. Elsevier, Amsterdam.

Choudhry, K.I., Carvajal-Ortiz, R.A., Kallikragas, D.T., Svishchev, I.M., 2014. Hydrogen evolution rate during the corrosion of stainless steel in supercritical water. Corros. Sci. 83, 226–233.

Comprelli, F.A., Busboom, H.J., Spalaris, C.N., 1969. Comparison of radiation damage studies and fuel cladding performance for Incoloy-800. In: Irradiation Effects in Structural Alloys for Thermal and Fast Reactors. Am. Soc. Test. Mater. 1969, 400–413. ASTM STP 457.

Daigo, Y., Watanabe, Y., Sue, K., 2007a. Effect of chromium ion from autoclave material on corrosion behavior of nickel-based alloys in supercritical water. Corrosion 63, 277–284.

Daigo, Y., Watanabe, Y., Sue, K., 2007b. Corrosion mitigation in supercritical water with chromium ion. Corrosion 63, 1085–1093.

Fitzsimmons, M.D., Pearl, W.L., Siegler, M., 1961. A Simulated Superheat Reactor Corrosion Facility. General Electric Atomic Power Report GEAP-3778.

Fujisawa, T., Maru, E., Amita, F., Harada, M., Uruga, T., Kimura, Y., 2004. Development and application of a multipurpose optical flow cell under supercritical condition of water. In: 14th International Conference on the Properties of Water and Steam in Kyoto, 445, p. 2004.

Ghandi, K., Miyake, Y., 2010. Muon interactions with matter. In: Hatano, Y., Katsumura, Y., Mozumder, A. (Eds.), Charged Particle and Photon Interactions with Matter: Recent Advances, Applications and Interfaces. Taylor & Francis, pp. 169–208.

Guzonas, D., 2009. SCWR materials and chemistry – status of ongoing research. In: Proceedings of the GIF Symposium, Paris, France, September 9–10, 2009.

Guzonas, D.A., Cook, W.G., 2012. Cycle chemistry and its effect on materials in a supercritical water-cooled reactor: a synthesis of current understanding. Corros. Sci. 65, 48–66.

Guzonas, D.A., 2015. The physical chemistry of corrosion in a supercritical water-cooled reactor. In: Proceedings of the 16th Int. Conf. on the Properties of Water and Steam: Water, Steam and Aqueous Solutions Working for the Environment and Industry, September 1–5, 2013. London, UK. Paper PWS-116.

Guzonas, D., Penttilä, S., Cook, W., Zheng, W., Novotny, R., Sáez-Maderuelo, A., Kaneda, J., 2016. The reproducibility of corrosion testing in supercritical water – results of an international interlaboratory comparison exercise. Corros. Sci. 106, 147–156. https://doi.org/10.1016/j.corsci.2016.01.034.

Han, Z., Muroya, Y., 2009. Development of a new method to study elution properties of stainless materials in subcritical and supercritical water. In: 4th International Symposium on Supercritical Water-cooled Reactors, Heidelberg, Germany, March 8–11, 2009. Paper No. 75.

Hazel, V.E., Boyle, R.F., Busboom, H.J., Murdock, T.B., Skarpelos, J.M., Spalaris, C.N., 1965. Fuel Irradiations in the ESADE-VBWR Nuclear Superheat Loop. General Electric Atomic Power report GEAP-4775.

Hettiarachchi, S., Song, H., Emerson, R., Macdonald, D.D., 1993a. Measurement of pH and Potential in Supercritical Water. Volume 1 Development of Sensors. Electric Power Research Institute. Report TR-102277 Volume 1.

Hettiarachchi, S., Makela, K., Song, H., Macdonald, D.D., 1993b. The viability of pH measurements in supercritical aqueous systems. J. Electrochem. Soc. 139, L3–L4.

Hnedkovsky, L., Wood, R.H., Balashov, V.N., 2005. Electrical conductances of aqueous Na_2SO_4, H_2SO_4, and their mixtures: limiting equivalent ion conductances, dissociation constants, and speciation to 673 K and 28 MPa. J. Phys. Chem. B 109, 9034–9046.

Holcomb, G.R., Carney, C., Doğan, Ö.N., 2016. Oxidation of alloys for energy applications in supercritical CO_2 and H_2O. Corros. Sci. 109, 22–35.

Janik, D., Janik, I., Bartels, D.M., 2007. Neutron and β/γ radiolysis of water up to supercritical conditions. 1. β/γ yields for H_2, H· atom, and hydrated electron. J. Phys. Chem. A 111, 7777–7786.

Katsumura, Y., Lin, M., Muroya, Y., Meesungnoen, J., Uchida, S., Mostafavi, M., 2010. Radiolysis of high temperature and supercritical water studied by pulse radiolysis up to 400 °C. In: 8th Intl Radiolysis, Electrochemistry & Materials Performance Workshop, Quebec City, Canada, October 8, 2010.

Klassen, R.J., Rajakumar, H., 2015. Combined effect of irradiation and temperature on the mechanical strength of Inconel 800H and AISI 310 alloys for in-core components of a GEN-IV SCWR. In: The 7th International Symposium on Supercritical Water-Cooled Reactors (ISSCWR-7) March 15–18, 2015, Helsinki, Finland Paper ISSCWR7-2006.

Kriksunov, L.B., Macdonald, D.D., 1995. Corrosion testing and prediction in SCWO environments. In: Proceedings of the ASME Heat Transfer Division. 317-2, p. 281.
Li, J., 2006. Focussed ion beam microscope. J. Metal 58 (3), 27–31.
Li, J., Elboujdaini, M., Gao, M., Revie, R.W., 2008. Investigation of plastic zones near SCC tips in a pipeline after hydrostatic testing. Mater. Sci. Eng. A 486, 496–502.
Lillard, S., 2012. Corrosion and compatibility. In: Konings, R.J.M. (Ed.), Comprehensive Nuclear Materials, vol. 5. Elsevier, Amsterdam, pp. 1–16.
Liu, C., Macdonald, D.D., Medina, E., Villa, J.J., Bueno, J.M., 1994. Probing corrosion activity in high subcritical and supercritical water through electrochemical noise analysis. Corrosion 50, 687–694.
Liu, C., Macdonald, D.D., 1995. An advanced Pd/Pt relative resistance sensor for the continuous monitoring of dissolved hydrogen in aqueous systems at high subcritical and supercritical temperatures. J. Supercrit. Fluids 8, 263–270.
Longton, P.B., 1966. The Oxidation of Iron- and Nickel-Based Alloys in Supercritical Steam: A Review of the Available Data. United Kingdom Atomic Energy Authority. TRG Report 1144(C).
Lozano-Perez, S., Dohr, J., Meisnar, M., Kruska, K., 2014. SCC in PWRs: learning from a bottom-up approach. Metallurg. Mater. Trans. E 1A, 194–210.
Maslar, J.E., Hurst, W.S., Bowers Jr., W.J., Hendricks, J.H., Aquino, M.I., Levin, I., 2001. In situ Raman spectroscopic investigation of chromium surfaces under hydrothermal conditions. Appl. Surf. Sci. 180, 102–118.
Maslar, J.E., Hurst, W.S., Bowers Jr., W.J., Hendricks, J.H., 2002. In situ Raman spectroscopic investigation of stainless steel hydrothermal corrosion. Corrosion 58, 739–747.
Mayanovic, R.A., Anderson, A.J., Dharmagunawardhane, H.A.N., Pascarelli, S., Aquilanti, G., 2012. Monitoring synchrotron X-ray-induced radiolysis effects on metal (Fe, W) ions in high-temperature aqueous fluids. J. Synchrotron Radiat. 19, 797–805.
Men, S., Anderson, A.J., Mayanovic, R.A., 2014. In situ X-ray absorption spectroscopy study of radiolysis-induced redox in chromium nitrate aqueous solution under supercritical condition. In: 2014 Canada-China Conference on Advanced Reactor Development (CCCARD-2014), April 27–30, 2014. Niagara Falls, Ontario, Canada, Paper CCCARD2014-014.
Muroya, Y., Sanguanmith, S., Meesungnoen, J., Lin, M., Yan, Y., Katsumura, Y., Jay-Gerin, J.-P., 2012. Time-dependent yield of the hydrated electron in subcritical and supercritical water studied by ultrafast pulse radiolysis and Monte-Carlo simulation. Phys. Chem. Chem. Phys. 14, 14325–14333.
Van Nieuwenhove, R., 2012. Proceedings of a IAEA Technical Meeting Held in Halden, Norway, 21–24 August 2012, In-pile Testing and Instrumentation for Development of Generation-IV Fuels and Materials, Session 1 (Instrumentation Development), Development and Testing of Instruments for Generation-IV Materials Research at the Halden Reactor Project. IAEA TECDOC-CD-1726. http://www-pub.iaea.org/MTCD/Publications/PDF/TE-CD-1726/PDF/IAEA-TECDOC-CD-1726.pdf.
Novotny, R., Moilanen, P., Hähner, P., Siegl, J., Hausild, P., 2013a. Development of crack growth rate SCC devices for testing in super critical water (SCW) environment. In: 6th International Symposium on Supercritical Water-Cooled Reactors, March 3–7, 2013, Shenzhen, Guangdong, China. Paper ISSCWR6-13070.
Novotny, R., Janik, P., Nilsson, K.F., Siegl, J., Hausild, P., 2013b. Pre-qualification of cladding materials for SCWR fuel qualification testing facility. In: 6th International Symposium on Supercritical Water-Cooled Reactors, March 3–7, 2013, Shenzhen, Guangdong, China. Paper ISSCWR6-13069.

Novotny, R., Janik, P., Toivonen, A., Ruiz, A., Szaraz, Z., Zhang, L., Siegl, J., Haušild, P., Penttilä, S., 2015. European Project 'Supercritical Water Reactor – Fuel Qualification Test' (SCWR-FQT): Summary of General Corrosion Tests. In: The 7th International Symposium on Supercritical Water-Cooled Reactors ISSCWR-7 March 15–18, 2015. Finland, Helsinki.

Novotny, R., Macak, J., Ruiz, A., Timke, T., 2017. Electrochemical impedance measurements in sub-critical and super-critical water. In: 8th International Symposium on Supercritical Water-Cooled Reactors (ISSCWR-8), March 13–15, 2017, Chengdu, China.

Penttilä, S., Toivonen, A., Li, J., Zheng, W., Novotny, R., 2013. Effect of surface modification on the corrosion resistance of austenitic stainless steel 316L in supercritical water conditions. J. Supercrit. Fluids 81, 157–163.

Penttilä, S., Moilanen, P., Karlsen, W., Toivonen, A., 2017. Miniature autoclave and double bellows loading device for material testing in future reactor concept conditions – case supercritical water. ASME J. Nucl. Eng. Radiat. Sci. https://doi.org/10.1115/1.4037897.

Pommeret, S., 2007. Radiolysis of supercritical water: an experimental and theoretical challenge. In: International Workshop on Supercritical Water Coolant Radiolysis, June 14–15, 2007. Řež, Czech Republic.

Pye, C.C., Cheng, L., Tremaine, P.R., 2014. Overview on the investigation of metal speciation under supercritical water-cooled reactor coolant conditions by ab-initio calculations, spectroscopy and conductivity measurements. In: 19th Pacific Basin Nuclear Conference (PBNC 2014), Vancouver, Canada, August 24–28, 2014.

Ruther, W.E., Schlueter, R.R., Lee, R.H., Hart, R.K., 1966. Corrosion behavior of steels and nickel alloys in superheated steam. Corrosion 22, 147–156.

Ruzickova, M., Vsolak, R., Hajek, P., Zychova, M., Fukac, R., 2011. Supercritical water loop for in-pile materials testing. In: Proc. of the 5th Int. Symposium on Supercritical Water-Cooled Reactors, Vancouver, Canada, on CD-ROM, vol. 11.

Sharygin, A.V., Wood, R.H., Zimmerman, G.H., Balashov, V.N., 2002. Multiple ion association versus redissociation in aqueous NaCl and KCl at high temperatures. J. Phys. Chem. B 106, 7121–7134.

Solferino, G., Anderson, A.J., 2012. Thermal reduction of molybdite and hematite in water and hydrogen peroxide-bearing solutions: insights on redox conditions in hydrothermal diamond anvil cell (HDAC) experiments. Chem. Geol. 322–323, 215–222.

Spalaris, C.N., 1963. Finding a corrosion resistant cladding for superheater fuels. Nucleonics 21, 41–49.

Spinks, J.W.T., Woods, R.J., 1976. An Introduction to Radiation Chemistry. John Wiley and Sons, Toronto.

Swift, R., Cook, W., Bradley, C., Newman, R.C., Zheng, W., 2015. Validation of constant load C-ring apex stresses for SCC testing in supercritical water. In: 7th International Symposium on Supercritical Water-Cooled Reactors ISSCWR-7, March 15–18, 2015, Helsinki, Finland Paper ISSCWR7-2088.

Takahashi, K., Cline, J.A., Bartels, D.M., Jonah, C.D., 2000. Design of an optical cell for pulse radiolysis of supercritical water. Rev. Sci. Instr. 71, 3345.

Vojacek, A., Ruzickova, M., Schulenberg, T., 2015. Design of an in-pile SCWR fuel qualification test loop. In: The 7th International Symposium on Supercritical Water-cooled Reactors, ISSCWR-7, March 15–18, 2015, Helsinki, Finland. Paper ISSCWR7-2065.

Walker, D.C., 1981. Muonium. A light isotope of hydrogen. J. Phys. Chem. 85, 3960.

Walters, L., Wright, M., Guzonas, D.A., 2017. Irradiation issues and material selection for Canadian SCWR components. In: 8th International Symposium on Supercritical Water-Cooled Reactors, ISSCWR-8, March 13–15, 2017, Chengdu, China.

Was, G.S., Averback, R.S., 2012. Radiation damage using ion beams. In: Konings, R.J.M. (Ed.), Comprehensive Nuclear Materials, vol. 5. Elsevier, Amsterdam, pp. 195–221.

Yan, H., Mayanovic, R.A., Anderson, A.J., Meredith, P.R., 2011. An in situ X-ray spectroscopic study of Mo^{6+} speciation in supercritical aqueous solutions. Nucl. Instr. Methods Phys. Res. A 649, 207–209.

Yan, H., Mayanovic, R.A., Demster, J.W., Anderson, A.J., 2013. In-situ monitoring of the adsorption of Co^{2+} on the surface of Fe_3O_4 nanoparticles in high-temperature aqueous fluids. J. Supercrit. Fluids 81, 175–182.

Zimmerman, G.H., Gruskiewicz, M.S., Wood, R.H., 1995. New apparatus for conductance measurements at high temperatures: conductance of aqueous solutions of LiCl, NaCl, NaBr, and CsBr at 28 MPa and water densities from 700 to 260 kg m^{-3}. J. Phys. Chem. 99, 11612–11625.

Zimmerman, G.H., Arcis, H., Tremaine, P.R., 2012. Limiting conductivities and ion association constants of aqueous NaCl under hydrothermal conditions: experimental data and correlations. J. Chem. Eng. Data 57, 2415–2429.

Radiation effects and mechanical properties

3

All in-core components will experience irradiation by α and β particles, neutrons and high-energy photons (γ-rays) (Fig. 1.3) resulting in damage at the atomic level in the form of ionization and microstructural degradation due to the development of vacancies, interstitials and voids. These microscopic defects induce changes in physical and mechanical properties and are a major factor in determining long-term component reliability.

There are many excellent references on radiation damage and related effects on materials (for example, alloys, graphite, oxides) (Boothby, 2012; Garner, 2012; Dai et al., 2012; Stoller, 2012; Zinkle, 2012); readers are encouraged to start with earlier work in this field so as to appreciate the evolution, over the past decades, of knowledge of these phenomena. Because of a lack of in-reactor test facilities for supercritical water-cooled reactor (SCWR) development, there has been little work to date on irradiation of candidate SCWR in-core materials. As a consequence, only a brief summary of the essential aspects of radiation damage and its effects on materials relevant to SCWR development can be presented here. For highly penetrating types of radiation such as neutrons, the effects of irradiation on a material occur throughout the material resulting in changes in bulk material properties; the fluid in which the material is situated (e.g., SCW) is of little consequence. Therefore, information obtained from irradiations of these alloys in other fluids under similar conditions of temperature, fluence, etc. can provide important insights into the expected damage modes.

3.1 Primary radiation damage

When an atom is struck by an incident particle, the primary knock-on atom (PKA) will be displaced from its normal position in the lattice if the energy transferred during the collision is greater than a certain threshold level, E_d, which is typically ~25 eV ($1\ eV = 1.6 \times 10^{-19}$ J) but depends on the type of metal and on the direction (the [hkl] value[1]) of the resulting Frenkel pair[2]. The initial damage defects are formed in pairs because a vacancy is created when the displaced atom and the recoiling atom, after striking a number of other atoms via elastic collisions, finds finally itself in an

[1] For a basic primer on hkl notation, crystal structure and lattice defects consult any solid-state physics textbook, e.g., Kittel (1976).
[2] A stable vacancy-interstitial pair.

Materials and Water Chemistry for Supercritical Water-cooled Reactors
https://doi.org/10.1016/B978-0-08-102049-4.00003-9

interstitial position. If the resulting vacancy and the recoil atom are not sufficiently separated, the two point defects will mutually annihilate. However, if they are separated by more than the vacancy capture radius, a Frenkel pair is formed. As the recoiling atoms tend to produce a series of collisions themselves, a PKA usually creates a cascade of displacements. The primary collisions and the resulting cascade occur in about 10^{-11} s, while the diffusion of the resulting defects to sink sites takes anywhere from milliseconds to months (as in the case of void formation and growth).

Kinchin and Pease (1955) proposed a linear displacement theory that models the binary collision of a particle and an atom as a knock-on event of two hard spheres. The number of displaced atoms (N_d) per PKA with an energy E is (Gittus, 1978):

$$N_d = E/(2E_d) \tag{3.1}$$

where E_d is the threshold displacement energy. Because of inelastic energy loss (due to interatomic potentials and electronic stopping mechanisms) and the need to account for the athermal recombination of unstable near-neighbour point-defect pairs, Norgett et al. (1975) proposed an improved method of estimating the displacement dose rate, widely known as the NRT approach (after the authors' initials). The NRT model is the basis for the standard method (ASTM, 2009) for estimating displacement cascade damage:

$$N_D = K_d \cdot E_D/2E_d \tag{3.2}$$

where E_D is the damage energy, which represents the effective energy transferred by the incident particle to the recoiling atom and E_d is the threshold displacement energy. The efficiency factor, K_d, is usually assumed to be chemistry and temperature independent with a value of 0.8 in the standard model. However, later work has shown that it is also energy dependent and can be as low as 0.3 depending on the damage energy level. For stainless steels, the threshold displacement energy is taken to be 40 eV (ASTM, 2009; Broeders and Konobeyev, 2004). A 40 eV threshold displacement energy value has also been reported for pure nickel (Averback and Diaz de la Rubia, 1997).

Material property changes are commonly correlated with displacement damage using the concept of displacements per atom (dpa). While in reactor cores there is sufficient thermally activated diffusion to enable many of the vacancies and interstitials to recombine (so that the retained displacement damage is only a fraction of the dpa value) the dpa value has been found to be an effective tool for quantifying radiation damage. Fig. 3.1 shows the results from a molecular dynamics (MD) simulation of the primary damage process in Au (OECD, 2015) depicting many of the key features of the primary damage process.

With the formation of interstitials at the outskirts of the damage zone (the cascade zone), vacancy clusters are pushed into the centre when the atoms in the cascade recrystallize from the outside towards the centre to return to the normal lattice arrangement. This is similar to the zone-refining process well known to metallurgists. As these defects rearrange themselves into energetically favourable configurations, dislocation

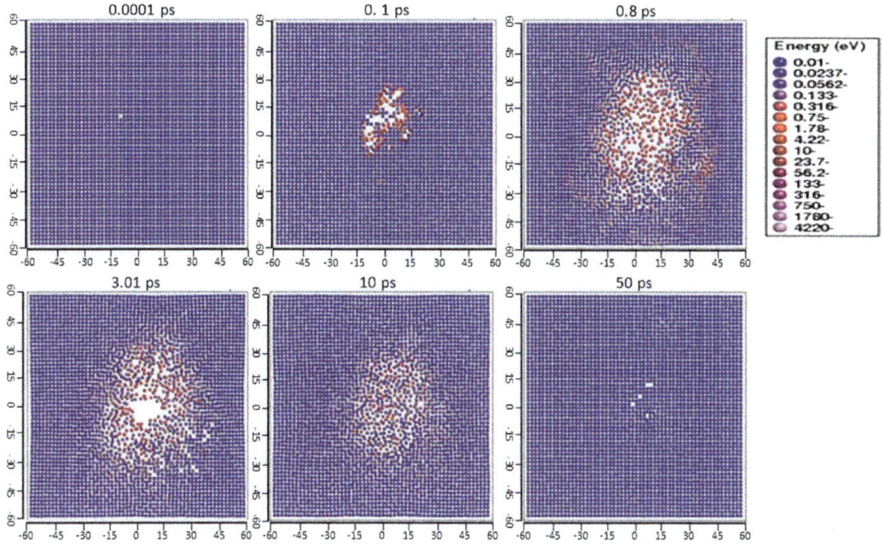

Figure 3.1 Collision cascade induced by 10 KeV recoil in Au at 0K. The circles indicate atom positions in a 1 unit cell thick cross-section through the centre of the simulation cell, and the colour scale the kinetic energy of the atoms. A very large number of atoms are initially displaced, as predicted by the NRT method, but when the cascade cools (after ~10 picoseconds), almost all return to normal crystal positions (although many atoms do not return to their original position), and hence the number of atom displacements is much larger than the number of stable defects produced.
Adapted from OECD, 2015. Primary Radiation Damage in Materials, NEA/NSC/DOC(2015)9. https://www.oecd-nea.org/science/docs/2015/nsc-doc2015-9.pdf.

loops form. In face-centred cubic (fcc) metals, aggregation of defects initially occurs on {111} planes to form a Frank loop[3]. This early stage nucleation and growth of defect clusters is well described by Gittus (1978). Clusters of vacancies can also form when there is increased mobility, forming sufficiently large defects visible as 'black dots' on TEM micrographs. These microscopic features impede the motion of dislocations and therefore the irradiated metal will harden.

At temperatures from 0.2 to 0.6 of the melting temperature (T_m, in K), longer-range aggregation of microscopic defects is facilitated by faster diffusional processes so that voids can, with accumulation of radiation dose, readily form and grow. Transmutation-induced small gas bubbles (such as He) are also sites for void nucleation. The growth of these voids creates void swelling, which not only changes the dimensions of a metal component but also reduces its ductility.

All in-core components are subjected to irradiation, the total dose sustained depending on the location of the component with respect to the fuel, the in-core residence

[3] A curved dislocation line pinned at both ends that is located between two normal lattice planes. Frank loops are faulted loops because their Burgers vectors are less than the normal interplane spacing.

time, the nature of the intervening materials (which provide shielding) and the neutron spectrum. To date there has been little published information on the irradiation conditions of SCWR in-core components. Walters et al. (2017) estimated the flux, fluence and temperature to which the various in-core components of the Canadian SCWR concept would be exposed. The pressure tube, which is close to the fuel and has a long (75 year) in-core residence time, is the component that will experience the highest dose. However, the SCWR fuel cladding will experience the most challenging combination of conditions (high temperature and fluence). The conditions of interest for the fuel cladding applications in the Canadian SCWR design, as an example of an SCWR concept, are temperature[4], 350–850 °C (Nava-Domínguez et al., 2016); design stress, negligible under normal operating conditions[5] (Xu et al., 2016b); design life, up to 30,000 h (Xu et al., 2014; Yetisir et al., 2017); total flux, $\sim 2.1 \times 10^{18}$ $n \cdot m^{-2} \cdot s^{-1}$; total flux with $E > 1$ MeV, $4.6–4.9 \times 10^{18}$ $n \cdot m^{-2} \cdot s^{-1}$; total fluence[6], $5.1–5.4 \times 10^{25}$ $n \cdot m^{-2}$. Walters et al. (2017) reported that the fuel cladding would experience total damage of 8.1–8.5 dpa and He generation of between 19.3 and 63.9 appm depending on the alloy chosen and the location in the fuel assembly. Fewer details are available for other SCWR concepts. Ehrlich et al. (2004) reported an average damage of 27.6 dpa and a maximum damage of 61.8 dpa after 45,000 h for High-Performance Light-Water Reactor (HPLWR) core materials. They also reported He generation from the boron (n,α) reaction of 60–85 appm in SS 1.4970 and 23.7 appm in Alloy 718 after 3 years. Matsui et al. (2007) listed irradiation test requirements for materials considered for the Japanese SCWR concept, indicating neutron fluences of 2×10^{21} $n \cdot m^{-2}$ fast and 8×10^{25} $n \cdot m^{-2}$ thermal.

The range of temperatures experienced by materials in an SCWR core will be much larger than in current generation water-cooled reactors. Because of thermal diffusion, a range of radiation damage types can develop over the service life time of fuel cladding depending on the cladding temperature as illustrated in Fig. 3.2 for the Canadian SCWR concept (Fig. 2.4 allows a rough correlation with locations in pressure vessel concepts).

The Canadian SCWR concept uses an insulated Zr alloy pressure tube, and the irradiation properties of both the insulator material (yttria-stabilized zirconia) and the pressure tube alloy (Excel) are of interest for this concept. The Excel pressure tube will operate at a temperature of 120–230 °C in a heavy-water moderator. Annealed Excel has been proposed as this material condition can minimize irradiation creep and growth rates (Chow and Khartabil, 2008). Walters and Donohue (2016) and Walters et al. (2017) have summarized the available data on this alloy relevant to the Canadian SCWR concept. Kr-ion irradiations of Excel (Idrees et al., 2013) over the temperature range 100–400 °C and up to 10 dpa showed formation of c-loops and the precipitation of the omega phase, which signified the potential for irradiation growth of this alloy.

[4] 850 °C represents the estimated maximum peak cladding temperature.

[5] The fuel cladding of the Canadian SCWR concept is designed to collapse onto the fuel so that the coolant pressure is supported by the fuel. During an accident the internal pressure of the fuel will produce a hoop stress of up to 40 MPa in the absence of coolant pressure, depending on the fuel burn-up.

[6] Flux and fluence vary with fuel pin location in the core.

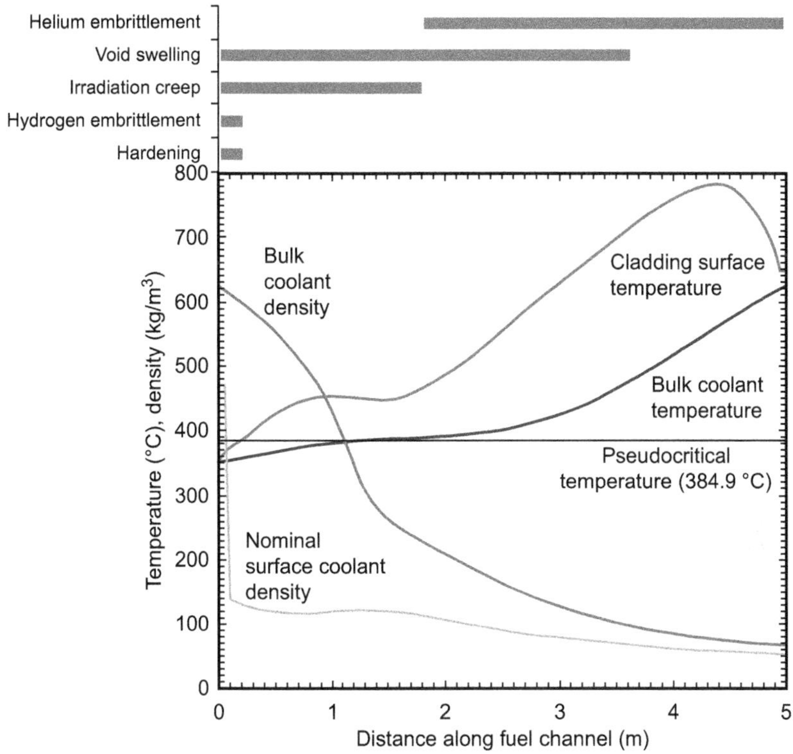

Figure 3.2 Modes of irradiation damage expected to be of significance over specific temperature regimes inside a Canadian SCWR core; bulk and surface temperature and coolant density profiles along a fuel channel in the concept are also shown (Zheng et al., 2016).

While annealed Excel has been shown to have considerable resistance to delayed hydride cracking (DHC), it is known to have high DHC velocities that would result in rapid crack growth once initiated.

As stainless steels and nickel-base alloys are the main alloys of choice for SCWR in-core use in pressure vessel concepts, and for the majority of the fuel assembly in the Canadian SCWR concept, the remainder of this chapter will focus on representative alloys from these alloy families.

3.2 Effects on mechanical properties

3.2.1 Hardening

There are hundreds of publications on irradiation hardening of materials including fcc (e.g., austenitic stainless steels, nickel alloys, copper and copper alloys), hexagonal close-packed (hcp) (e.g., Zr and Zr alloys) and body-centred cubic (bcc) (e.g., carbon steels, ferritic steels, etc.) alloys. Starting from very low radiation dose, hardening is

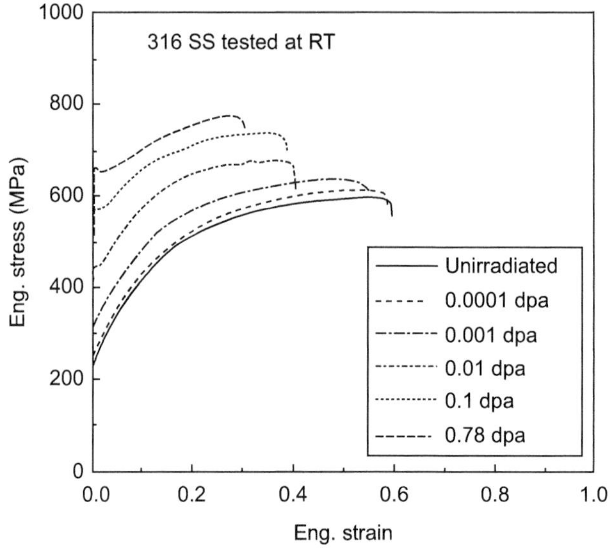

Figure 3.3 Engineering stress–strain curves for annealed 316 SS irradiated in the High Flux Isotope Reactor (HFIR) mixed spectrum reactor at 60–100 °C and tested at 25 °C (Garner, 2012).

one of the earliest indication of irradiation-induced microstructural change in alloys. With the increase in yield strength, elongation decreases and the work-hardening rate decreases. Fig. 3.3 is an example of this type of effect in a stainless steel. This is a typical result of constrained plastic flow by radiation-induced microscopic defects such as dislocation loops, microvoids and bubbles (Garner, 2012).

Dose rate seems to play a role in hardening. Studies of different types of annealed stainless steels (304, 304L, 316L, 316LN, Prime Candidate Alloy) showed that such hardening tends to saturate at about 5 dpa (Pawel et al., 1999). However, recent results of Xu et al. (2016a) demonstrated that the hardness of Xe-irradiated specimens saturated at about 2 dpa, whereas in the case of proton irradiation, the saturation dose was much higher, likely more than 7 dpa. This discrepancy was thought to be a result of differences in damage distribution between proton and Xe irradiation.

In a study of cold-worked 316 stainless steel retrieved from the reflector region of the Experimental Breeder Reactor (EBR-II) (dose range 1–56 dpa), Allen et al. (2002) found that hardening did not appear to saturate when the dose was less than 20 dpa. The temperature of the irradiation was 371–390 °C and the dose rate was $0.8\text{–}3.3 \times 10^{-7}$ dpa·s^{-1}. Voids, precipitates, and loops all contributed to the hardening. Quantitatively, the measured increase in yield strength attributed to the formation of loops, voids, and loops plus voids could be described as the following sum (Allen et al., 2002):

$$\Delta\sigma_y = \Delta\sigma_y \text{ dislocations} + \sqrt{\left(\Delta\sigma_y^{\text{voids}}\right)^2 + \left(\Delta\sigma_y^{\text{loops}}\right)^2 + \left(\Delta\sigma_y^{\text{precipitates}}\right)^2} \tag{3.3}$$

For dislocations, the yield strength increment is given by:

$$\Delta\sigma_y = M \cdot \alpha \cdot \mu \cdot b \cdot \sqrt{\rho_d} \tag{3.4}$$

where ρ_d is the dislocation density (line length per unit volume), M is the shear stress required to activate slip on a slip plane, α is the barrier strength, μ is the shear modulus, and b is the Burgers vector of a moving dislocation.

Recent work using Fe ion irradiation by Hardie et al. (2013) revealed that a high dose rate produced larger dislocation loop densities and sizes in the Fe–Cr alloy, and precipitation (Cr type) was found in the alloy irradiated at a low dose rate. Pure Fe and Fe–Cr alloys containing 5, 10 and 14 wt% Cr were studied in their work. They used Fe^+ ions at a maximum energy of 2 MeV at temperatures of 300, 400 and 500 °C to a dose of 0.6 dpa at dose rates of up to 6×10^{-4} dpa·s^{-1} and 3×10^{-5} dpa·s^{-1}. All materials exhibited an increase in hardness after irradiation at 300 °C. After irradiation at 400 °C, hardening was observed only in Fe–Cr alloys and not in pure Fe. After irradiation at 500 °C, no hardening was observed in any of the materials. Klassen and Rajakumar (2015) irradiated Alloy 800H and 310 SS using 8.0 MeV Fe^{4+} ions at doses up to 15 dpa, then subjected the samples to isothermal annealing at 400 and 500 °C. Both alloys experienced significant radiation hardening, but the authors found rapid thermal recovery of the radiation damage. Hazel et al. (1965) reported microhardness values preirradiation and postirradiation for fuel cladding samples exposed in the expanded superheat advanced demonstration experiment (ESADE) loop of the Vallecitos boiling water reactor (VBWR) (Table 3.1), also finding that irradiation hardening is annealed out at these elevated temperatures. These data suggest that irradiation hardening will only be an issue in the 'evaporator' region of the SCWR core where the coolant temperatures are less than 400 °C, as shown in Fig. 3.2.

Table 3.1 Microhardness data for irradiated fuel cladding specimens

ESH element	Material	Exposure time (d)	Microhardness (Vickers hardness number)	
			Preirradiation	**Postirradiation**
1.1-A-2	Alloy 800	106	164–165	162–168
1.1-BONUS	348 SS	106	150–153	147–150
1.1-F	310 VM	106	128–131	115–116
1.1-H	Alloy 800	135	141–145	165–179
1.1-I	Alloy 600	135	154–156	184–196
1-B	Alloy 800	29	141–145	153–159

Irradiations were performed in the expanded superheat advanced demonstration experiment loop of the Vallecitos boiling water reactor. Estimated maximum temperatures during irradiation were between 538 and 698 °C. ESH-1 and ESH-1.1 are experimental irradiation identifiers.

From Hazel, V.E., Boyle, R.F., Busboom, H.J., Murdock, T.B., Skarpelos, J.M., Spalaris, C.N., 1965. Fuel Irradiation in the ESADA-VBWR Nuclear Superheat Loop. General Electric Atomic Power Report GEAP-4775

3.2.2 Ductility

Concurrent with hardening, irradiated alloys tend to exhibit flow localization with increasing dose. On the fracture surface, flat facets of 'channel fracture' can be seen of irradiated stainless steels, the result of dislocations clearing out the Frank loops on the deformation planes (Fig. 3.4). Concentration of dislocation motion to these narrow slip bands suppresses uniform deformation, and on stress–strain curves necking occurs prematurely leading to much reduced uniform elongation. Intersection of these dislocation channels with grain boundaries can create microvoids for initiation of intergranular cracks (Stephenson and Was, 2016).

On a more microscopic level, dislocation gliding along a limited number of slip planes removes small barriers on their paths and creates an easy tunnel for further plastic flow. The slip bands in this channel are often a fraction of a micrometer in width, and the local strains within the bands can be a few hundred percent even when the bulk elongation of the irradiated alloy is only a few percent (Byun and Hashimoto, 2006). Examples of TEM micrographs of such channelling in 316 stainless steel, iron, vanadium and Zr-4 alloy are provided in Fig. 3.5. In 316 steel, the blocks between the defect-cleared channels had little dislocation activity in the unchannelled microstructure. The active {111}<110> slip systems for channelling are the same as those with uniform deformation.

Fcc metals can deform by twinning at high strain levels, and radiation also significantly increases the tendency for twin formation (Lee et al., 2001a, b). Byun and Hashimoto (2006) grouped the deformation microstructures in stainless steel in terms of the equivalent stress range:

1. Dislocation tangles as the dominant morphology at low equivalent stresses <400 MPa.

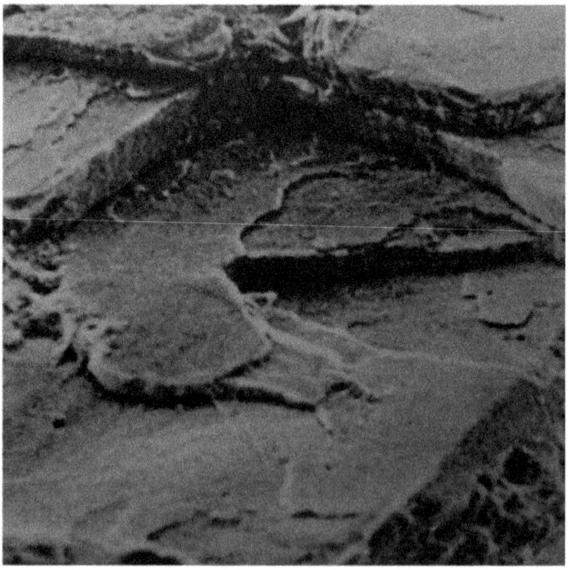

Figure 3.4 An example of channel fracture as seen in copper (Gittus, 1978).

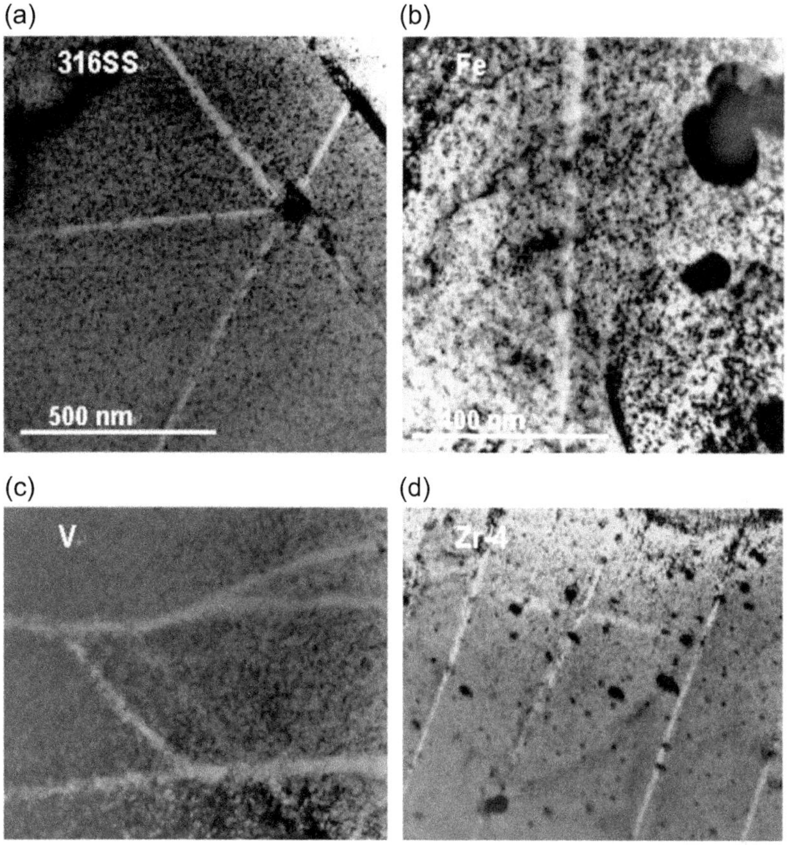

Figure 3.5 Channelled microstructures after neutron irradiation at ~80 °C and deformed at room temperature: (a) straight channels and their interception in 316 SS at 0.78 dpa, 5% strain; (b) single channel in iron at 0.89 dpa, 0.2% nominal strain; (c) river-pattern channel network in vanadium at 0.69 dpa, 10% strain; (d) parallel channels in Zr-4 at 0.8 dpa, 6% strain. The vanadium specimen was deformed by disk-bend technique, while the others were deformed by uniaxial tensile loading.
Reproduced from Byun, T., Hashimoto N., 2006. Strain localization in irradiated materials. Nucl. Eng. Technol. 38 (7), 619–638.

2. Small, isolated stacking faults smaller than about 1 μm in the stress range from about 400 to 600 MPa.
3. Bands of large stacking faults (>1 μm) and twin bands as the dominant feature at stresses > 600 MPa.

There are many other mechanisms whereby radiation-induced microstructural or chemical changes can reduce the ductility of an alloy, for example, formation of He bubbles and formation and growth of voids.

A potentially serious challenge for the application of austenitic stainless steels or nickel-based alloys in the SCWR core is production of He, which may lead to helium-bubble formation and high-temperature helium embrittlement (Schroeder, 1988; Dai et al., 2012). He can influence void swelling (Johnston et al., 1974; Wolfer et al., 1982; Garner, 2012). When nickel-bearing alloys are exposed to a high thermal neutron flux, He is generated via the sequential $^{58}Ni(n,\gamma)^{59}Ni(n,\alpha)^{56}Fe$ reactions (Greenwood, 1983; Greenwood and Oliver, 2006). Similarly, hydrogen is formed by $^{58}Ni(n,\gamma)^{59}Ni(n,p)^{59}Co$. These transmutation reactions also produce additional recoils, which may contribute significantly to the accumulated primary irradiation damage. In the course of these reactions, the nonradioactive naturally occurring ^{58}Ni isotope and the intermediate isotope ^{59}Ni are eventually consumed. As a result, He generation saturates at a thermal fluence of approximately 10^{24} $n \cdot cm^{-2}$. The amount of helium generated at this saturation fluence is enormous (~80,000 appm He).

The amount of He produced depends on the amount of nickel in the alloy and also on the neutron flux spectrum. The thermal fluence for the Canadian SCWR fuel cladding is less than 10^{21} $n \cdot cm^{-2}$ after three fuel cycles (3.5 years) largely as a result of the use of a thorium–plutonium fuel in the reference concept. Walters et al. (2017) reported that He generation in the fuel cladding was calculated to range from 19.3 to 23.7 appm for 310 SS to 50.8–63.9 appm for Alloy 625. This should be contrasted with the much larger values reported for calculated He yield for the Japanese concept (Table 3.2). Bysice et al. (2016) compared He generation for candidate fuel cladding alloys for the Canadian SCWR concept using a Pu-Th fuel and an LEU-Th fuel and found that the high thermal neutron capture cross-section for Pu isotopes relative to ^{238}U results in a roughly 2-fold reduction in He generation for the Pu-Th fuel. He production was found to increase roughly linearly with the Ni content of the alloy.

Small amounts of boron are added to some alloys to improve creep performance, and it is also present in many alloys as a residual element. ^{10}B has a large cross-section for the (n,α) reaction and boron is known to segregate to grain boundaries. The transmutation of ^{10}B to He could therefore result in localized high concentrations of He, which could promote bubble formation at grain boundaries. The transmutation of ^{10}B to He is complete in 1–3 years depending on the neutron spectrum and fluence. The effect of the loss of ^{10}B (about 20% natural abundance) could also affect long-term

Table 3.2 Calculated He yield for the candidate alloys for the Japanese supercritical water-cooled reactor concept (Matsui et al., 2007)

	He yield (appm) for key transmutation reactions in alloys containing nickel		
Material	$^{58}Ni(n,\gamma)^{59}Ni(n,\alpha)^{56}Fe$	$^{58}Ni(n,\alpha)^{55}Fe$	$^{60}Ni(n,\alpha)^{57}Fe$
310S SS	105	17	2
316L SS	68	11	1
Alloy 690	338	56	5
Pure Ni	583	96	9

creep performance. Therefore boron concentration should be minimized in candidate alloys where possible. Erhlich et al. (2004) suggest the use of B enriched in ^{11}B, which is available and has a much lower (n,α) cross-section.

Welding for repairs or maintenance can become problematic for alloys that undergo He formation. Helium concentrations as low as 1 appm have been found to make welding of austenitic stainless steels and Ni-based alloys difficult.

Ductility data for alloys irradiated at the expected peak cladding temperatures are limited, particularly for the very high temperatures predicted for the Canadian SCWR concept. Thiele et al. (1984) reported the ductility of Alloy 800 and other materials at various temperatures after irradiation in HTR I at 375 °C to a thermal fluence of 1.1×10^{21} n·cm^{-2}, a dose of about 1.4 dpa. The ductility dropped rapidly for all alloys as the temperature increased from 600 to 850 °C. At 700 °C most of the alloys studied showed elongation in the high single digits or to just above 10%. At 850 °C most materials showed elongation below 5%.

Comprelli et al. (1969) compared radiation damage and fuel performance for Alloy 800 in tests carried out in the VBWR, the EVESR, the General Electric Test Reactor (GETR) and the EBR-II. Experimental conditions in these facilities varied with temperatures between 538 and 704 °C, exposure times from 200 to 5000 h, fast neutron fluences from 0.17 to 67×10^{20} n·cm^{-2} and thermal neutron fluences (not always given) of $0.3\text{–}2.5 \times 10^{21}$ n·cm^{-2}; the environments included Ar, Ar + He, sodium and superheated steam (SHS). Of particular relevance are tests of precut tension specimens inserted into a fuel bundle in the EVESR and exposed to SHS. Specimens placed in the velocity booster[7] were irradiated at the axial peak flux location at ~454 °C. Two specimens were cut from the outer cladding of the peak burn-up rod and were irradiated at temperatures of 538–621 °C. The results (Table 3.3) show that the major effect of irradiation in a thermal reactor is to reduce high-temperature ductility. Differences in ductility between the specimens exposed in the velocity booster and cut from the fuel cladding were attributed to differences in irradiation temperature or stress–strain history.

Rowcliffe et al. (2009) note that the alloy PE16 has been used extensively as a fuel cladding with good results. They note that approximately 75% of the core of the Prototype Fast Reactor (PFR) at Dounreay was loaded with PE16 fuel pins that operated between 400 and 650 °C to approximately 80 dpa without a single failure. They concluded that with correct fuel design, it is possible to avoid subjecting the fuel cladding to a temperature–stress regime where intergranular failures occur. The maximum fission gas pressure was estimated at <70 MPa, a factor of two to three times less than the stress required to induce unstable growth of grain boundary He bubbles.

3.2.3 Irradiation-assisted stress corrosion cracking

Detailed discussions on stress corrosion cracking (SCC) are found in Chapter 6 of this book; this section briefly touches on aspects related to irradiation effects.

[7] A sealed steel tube placed in the central cavity of each fuel element designed to increase the velocity of the steam passing over the inner heat transfer surface.

Table 3.3 Tensile data at 704 °C from specimens irradiated in superheated steam in EVESR Mark II

Condition	Specimen type	Number of specimens	0.2% Offset yield stress ($N\cdot m^{-2}$)	Ultimate stress ($N\cdot m^{-2}$)	Uniform elongation (%)	Total elongation (%)
Annealed	Control	1	1.75×10^8	3.47×10^8	17.7	34.4
Cold-Worked	Control	1	4.38	4.87	4.0	14.95
Annealed	Velocity Booster	3	1.655–1.765	2.96–3.15	11.5–16.8	12.5–18.5
Cold-Worked	Velocity Booster	2	2.85–3.90	3.18–4.25	3.4–3.6	4.3–6.5
Annealed	Cladding	2	1.205–2.14	2.59–2.63	2.6–5.95	7.4–8.1

From Comprelli, F.A., Busboom, H.J., Spalrais, C.N., 1969. Comparison of radiation damage studies and fuel cladding performance for Incoloy-800. In: Irradiation Effects in Structural Alloys for Thermal and Fast Reactors, ASTM STP 457, American Society for Testing and Materials, pp. 400–413.

In addition to the microstructural defects (point, and linear, i.e., dislocations, and voids) discussed in the preceding sections, irradiation also causes changes in materials chemistry at the microscopic level such as radiation-induced segregation (RIS) and precipitation, both of which affect SCC susceptibility. The extent of this damage depends on the irradiation dose, flux, irradiation spectrum and the temperature at which the metal is irradiated due to the thermal nature of the diffusional processes involved. For example, RIS in stainless steels can lead to depletion of Cr in grain boundaries, one of the known causes for intergranular cracking in stainless steels (IAEA, 2011), discussed in detail in the next section. The formation of deformation channels in irradiated stainless steels and nickel-based alloys can lead to strain localization at grain boundaries, a factor with a high correlation for irradiation-assisted stress corrosion cracking (IASCC) in light-water reactor (LWR) environments (Jiao and Was, 2011).

For the subcritical temperature range, Hojná (2013) summarized the IASCC experiences in LWRs in the context of lifetime extension of current generation boiling water reactors (BWRs) and pressurized water reactors (PWRs). In general, the influence of irradiation on SCC becomes more pronounced with increasing damage dose. Constant load tests of irradiated 315 stainless steel at 340 °C (Freyer et al., 2007), which has Cr up to 19 wt% and Ni up to 9.5 wt%, showed that the threshold stress for IASCC decreased as the damage increased from 10 dpa to about 30 dpa. These effects are less important for the fuel cladding; stainless steel fuel cladding has been widely used in LWRs. The La Crosse BWR operated for 20 years using 348 SS fuel cladding (Strasser et al., 1982). While about 1.3% of the rods failed during the first five cycles of operation (almost all at higher than core-average burn-up) the failure rates during subsequent cycles were reduced through changes such as a limit on maximum burn-up and tighter operating restrictions (e.g., on control rod movements). Failure was attributed to a combination of internal stresses from pellet-cladding interaction and corrosion of the outer surface of the cladding (exposed to the coolant) resulting from the high concentration of oxidants in the coolant due to water radiolysis. This operating experience demonstrates that optimization of design (e.g., design and residual stresses, in-service lifetime (irradiation time)) can minimize the risk of IASCC (Chapter 6).

In contrast to the abundance of studies of IASCC in subcritical water (using LWR chemistries), there are few data for IASCC under expected SCW in-core water chemistry (Chapter 4). The higher temperature will have a considerable effect on the microstructural response of alloys to mechanical loading, with or without irradiation. Zhou et al. (2009) compared the SCC susceptibility of Alloys 316L, D9, Alloy 800H and Alloy 690 using constant extension rate tensile testing in pure SCW at 400 and 500 °C after the samples were irradiated by protons to 7 dpa. In all cases the irradiated alloys showed greater crack depths than the unirradiated specimens. For Alloy 690 and 316L SS irradiated to 7 dpa, greater crack depth was seen at 500 °C than at 400 °C. It should be pointed out that in these tests the SCC samples were strained to failure and it was not clear at what point of the stress–strain curve the cracking initiated. Under neutron irradiation, the ductility of stainless steels is significantly reduced.

3.2.4 Void swelling

Void swelling is a degradation mechanism that can lead to significant dimensional instability and premature failure (Johnston et al., 1974; Harries, 1979; Wolfer et al., 1982; Garner, 2012). Understanding the material response and metallurgical factors that influence the occurrence of void swelling is therefore critical. In many materials, including austenitic stainless steels, swelling does not saturate. Volumetric expansion is isotropic in the absence of applied stress or constraint. Conversely, in the presence of an applied stress or constraint, anisotropic void swelling may be observed (Lauritzen et al., 1987). In general, sufficient displacement damage in the form of mobile vacancies and vacancy clusters must be present to develop void nucleation and to drive growth. At high temperatures (>700 °C), primary irradiation damage, sink density and strength, and vacancy cluster stability are relatively low, which leads to a reduced driving force for swelling. At low temperatures (<300 °C), defect diffusion is limited and void swelling is therefore less likely. Void swelling is most pronounced between 350 and 700 °C. Fig. 3.2 shows that most of the materials in the SCWR core will be at temperatures within this range. A recent reevaluation of past studies on austenitic stainless steels has shown that void swelling follows two stages, a transient or incubation stage with negligible volume change and a stage of steady swelling. At dose rates relevant to PWR components, the low void swelling observed in steels was attributed to the irradiation-induced precipitation (at $T < 370$ °C) of fine precipitates (e.g., G phase, carbides, Y' phase) that acted as sinks for irradiation-induced vacancies (Chung, 2006).

Nickel has a strong influence on void swelling, although the extent of influence depends on the irradiation temperature. At lower irradiation temperatures, Ni concentrations between 0 and 24 wt% appear to have little influence on the transient damage dose. At higher irradiation temperatures, increasing Ni concentration, up to approximately 40 wt%, generally prolongs the incubation stage that preceeds steady-state swelling (Johnston et al., 1974; Johnston et al., 1976; Garner, 1984; Garner and Brager, 1985). Chromium has a less prominent influence on void swelling; increasing chromium content has been found to decrease transient damage dose for all nickel concentrations (Garner and Kumar, 1987). While higher nickel alloys appear to have some benefit for void swelling, He generation in these alloys must also be considered.

For austenitic stainless steels, the steady-state swelling rate is approximately 1% per dpa and is independent of alloy chemistry, temperature and atomic displacement rate. Fig. 3.6 shows that the void swelling rate in a 20% cold-worked 316 stainless steel as a function of irradiation temperature and dose is constant and approximately equal to 1% per dpa at steady state. The length of the transient stage has been found to be influenced by a variety of factors including major and minor alloying additions (Garner and Brager, 1985; Garner et al., 1987; Allen et al., 2000; Gessel and Rowcliffe, 1977; Herschbach et al., 1993), helium levels (Ayrault et al., 1981; Singh et al., 1981; Murphy, 1987; Ohnuki et al., 1988), microstructure and phase stability, applied stress (Porter et al., 1983), prior cold work, irradiation temperature (Kozlov et al., 2004; Okita et al., 2000) and damage dose rate (Kiritani, 1989; Okita et al., 2000; Okita et al., 2002; Allen et al., 2006).

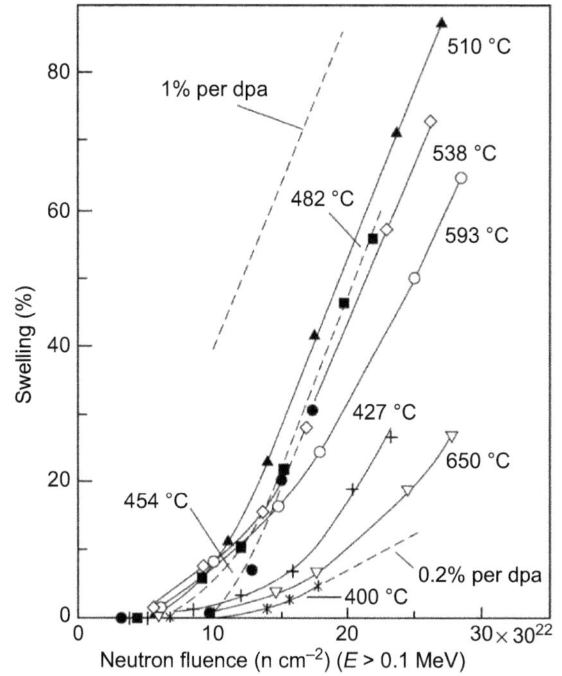

Figure 3.6 Swelling determined by density change as a function of temperature and dose, as observed in 20% cold-worked AISI 316 irradiated in the Experimental Breeder Reactor (EBR) fast reactor. All measurements at a given temperature were made on the same specimen after multiple exposures with subsequent reinsertion into the reactor. This procedure minimized specimen-to-specimen data scatter and assisted in a clear visualization of the posttransient swelling rate.
Reproduced from Garner, F.A., 2012. Radiation damage in austenitic steels. In: Konings, R.J.M. (Ed.), Comprehensive Nuclear Materials. Elsevier, Oxford, 33–95, ISBN:9780080560335, https://doi.org/10.1016/B978-0-08-056033-5.00065-3; originally by Garner, F.A., Gelles, D.S., 1990. In: Proceedings of Symposium on Effects of Radiation on Materials: 14th International Symposium, ASTM STP 1046, vol. II, pp. 673–683.

Zheng et al. (2017) studied the effect of retained hydrogen on the defect microstructure of AL-6XN SS (an alloy being considered for SCWR use) specimens irradiated with low energy (100 keV) H^{2+}. A high concentration of hydrogen was injected into the specimens, and displacement damage dose up to 7 dpa was achieved. Irradiation-induced dislocation loops and voids were characterized by TEM. For specimens irradiated to 7 dpa at 290 °C, a high number density of dislocation loops were found and the void swelling was observed. At 380 °C, most of the dislocation loops were unfaulted and tangled at 7 dpa, and void swelling was observed at damage doses of 5 dpa and above.

3.3 Effects on microchemistry: radiation-induced segregation

The diffusion of vacancies and self-interstitial atoms to grain boundaries, dislocations or precipitate surfaces can cause local changes of alloy chemistry. Elements such as Cr, Mo and Fe that exchange more rapidly with vacancies are depleted near these sinks, while other elements such as Si, P and Ni that migrate by interstitial mechanisms are enriched near regions that act as sinks for point defects (Bruemmer et al., 1999; Scott 1994). Irradiation temperature and dose rate can affect these segregations.

RIS is temperature dependent. At low temperatures, reduced mobility limits the extent of RIS. At very high temperature, back diffusion can smooth out the concentration gradient. Therefore RIS tends to be most severe at intermediate temperatures. At 275–300 °C, significant RIS is observed at irradiation doses as low as 0.1 to 5 dpa (Scott, 1994; Bruemmer, 2001). The relative dominance of segregation, back diffusion and recombination of defects in various temperature zones is schematically shown in Fig. 3.7 (Was and Andresen, 2012). It can be seen that for the SCWR in-core temperature range, 375–850 °C, RIS can occur over a wide range of dose rates.

The role of RIS in IASCC has attracted significant research interest. Busby et al. (2002) reported results of an extensive study using the measurement of grain boundary chemistry and postirradiation annealing, which returns the grain boundary chemistry to the preirradiation condition, in a series of proton-irradiated alloys. They found that

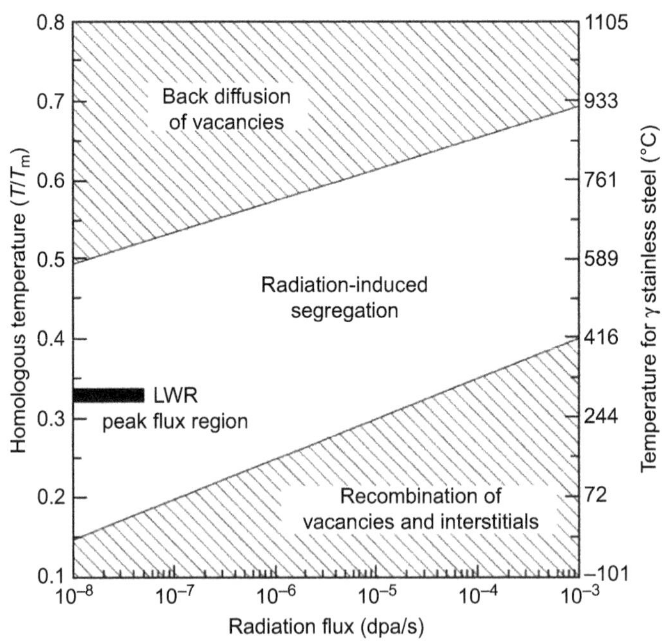

Figure 3.7 Dependence of radiation-induced segregation on homologous temperature and dose rate for austenitic stainless steels (Was and Andresen, 2012). LWR, light-water reactor.

dislocation microstructure introduced by irradiation was removed preferentially by annealing. Even when the bulk hardness was recovered only slightly after annealing at 500 °C for 45 min, the cracking had been completely mitigated, suggesting the irradiation-induced deformation structure in the bulk matrix is not a sensitive correlation parameter. They were able to isolate the chemistry change due to RIS from hardening by annealing at 600 °C for 90 min, as this annealing procedure completely recovered irradiation-induced hardening but the RIS profile had not been changed. Based on comparison of as-irradiated and annealed specimens, they found that RIS of Cr, Ni, Si or P per se is not sufficient to cause IASCC. Ni level remained constant during postirradiation annealing while cracking was mitigated rapidly by annealing, indicating that Ni enrichment at the grain boundary is not a mitigating factor for IASCC. The enrichment of Si at grain boundaries by RIS was shown to be a noncontrolling factor. Finally, their results showed that the enrichment of P has a minimal impact on IASCC.

In a follow-up study, Jiao and Was (2011) further ruled out RIS as correlating factor, at least in the case they studied where the grain boundary Cr concentration range was 12–19 wt%. Austenitic stainless steels with Ni contents ranging from 8 to 32 wt% and Cr contents from 15 to 22 wt% were irradiated to 1 and 5 dpa at 360 °C using 2–3.2 MeV protons, and SCC tests were based on slow straining test in simulated BWR water at 288 °C with straining up to 3%. The correlation strength with IASCC of localized deformation, which was characterized using the weighted average channel height, was found to be the strongest of the factors considered, which included the stacking fault energy, hardness and RIS. It was speculated that localized deformation promoted cracking through intensive interaction of dislocations in slip channels with grain boundaries.

Gupta et al. (2016) used Fe irradiation at 450 °C to study the SCC of 304 L austenitic stainless steel. After irradiation with 10 MeV Fe to 5 dpa, irradiation-induced damage in the microstructure was quantified by slip-line spacing, which was 1.8 times that of unirradiated samples. Fe-irradiated material strained to 4% in a PWR environment exhibited much more severe SCC than an unirradiated counterpart in terms of crack size and density. Fe-irradiated material subjected to 4% plastic strain in an inert Ar environment did not exhibit any cracking, confirming that localized deformation is not, by itself, sufficient for initiating cracking, and SCC requires the presence of a corrosive environment.

3.4 Creep

3.4.1 Introduction

Creep properties are essential mechanical properties for design of any high-temperature system, and creep rupture strength is a key criterion for selecting candidate SCWR fuel cladding material. Design requirements are usually given in codes, standards and specifications. For examples, the ASME code, Section III Subsection NH (ASME, 2006), outlines the requirements of the design stress for class 1 components in elevated temperature service. ASME Section III Subsection NH (ASME, 2006) requires the design stress to be less than 1/3 of ultimate tensile strength or 2/3 of yield strength, except for stainless steels and Ni-based alloys, which can be up to 90% of the yield strength. The development of creep-resistant alloys has been a major focus of supercritical fossil power plant R&D programs as these plants move to operation under USC conditions (Viswanathan and Bakker, 2000). Under neutron irradiation, creep can occur earlier and at a higher rate than pure thermal creep. One distinct difference between irradiation creep and pure thermal creep is that irradiation creep is 'nondamaging' on the microstructural level in the sense that its occurrence does not lead to triple-point crack formation or grain boundary voids and does not produce martensites (Garner, 2012). In that sense irradiation creep can be beneficial as it relaxes both externally applied and internally generated stresses in a way similar to low-temperature annealing.

Pure thermal creep occurs when the application temperature, T, is higher than 0.3 T_m. A typical creep failure exhibits three stages (Fig. 3.8): primary (Stage I), a steady secondary stage (Stage II) and a fast, final tertiary stage (Stage III). In the primary stage, primary creep occurs with a relatively high strain rate, but the deformation slows down quickly with time as a result of work hardening. The deformation rate drops to a minimum and eventually becomes nearly constant, and this is where the secondary stage (Stage II) begins. In this stage, there is a balance between work hardening and recovery. Most of the service life of a high-temperature component is spent in Stage II. At the late stage of service, Stage III, accumulation of microscopic voids (which can often be observed using TEM) and the occurrence of necking cause an exponential increase in strain rate with time. The strength of the alloy can then be lost rather quickly.

It should be noted that for some metals (e.g., lead, magnesium, tin) creep can occur even at room temperature, although for most metals and alloys creep is considered a high-temperature phenomenon.

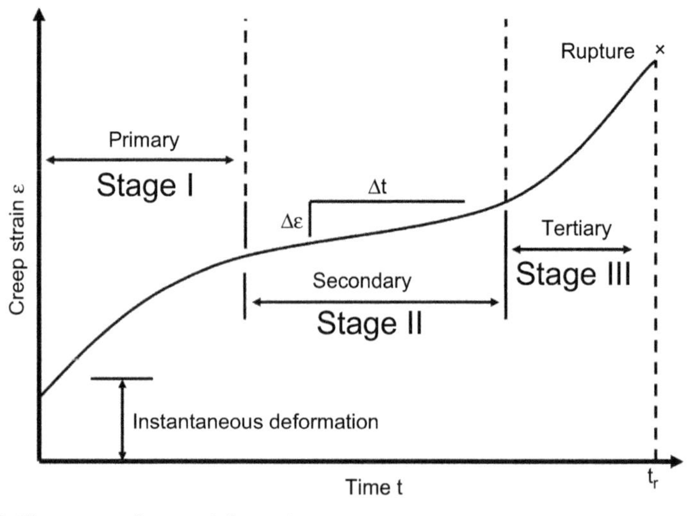

Figure 3.8 The stages of creep deformation.

Depending on the temperature range and the stress level, different mechanisms are involved in the creep process. They include the following:

1. Dislocation creep. This is the primary process for low-temperature and high-stress situations. Dislocation gliding on slip planes readily occurs at low temperature, but the rate-determining step is a dislocation climb process, which is temperature dependent as diffusional motion of vacancies is involved.
2. Grain boundary sliding. Under the action of shear stress, the grain boundaries can slide on one another creating wedge cracks and voids. Above 0.6 T_m, the grain boundary regions are believed to have lower shear strength than the grain interior.
3. Diffusional flow. With the thermally activated diffusion of atoms along the diffusion pathways of an alloy, the grains along the tensile axis become longer as atoms move from the sides to the top and the bottom of the grains. The process involving atoms at the grain boundary region is called Coble creep (Coble, 1963). When the diffusional paths are within the grains, the process is referred to as Nabarro-Herring creep. The latter is typically found in alloys of small grain size under low stress but at high temperature.

Fig. 3.9 is an example of creep strain measurement with time for a P91 steel at 625 °C. Note that the rate of strain increases rapidly with stress when the applied stress is greater than 120 MPa. This alloy is typically used for temperatures lower than 650 °C.

3.4.2 Predicting creep

The Larson–Miller parameter (LMP) method (Larson and Miller, 1952) is a conventional method for creep data extrapolation that assumes the rate of creep follows an

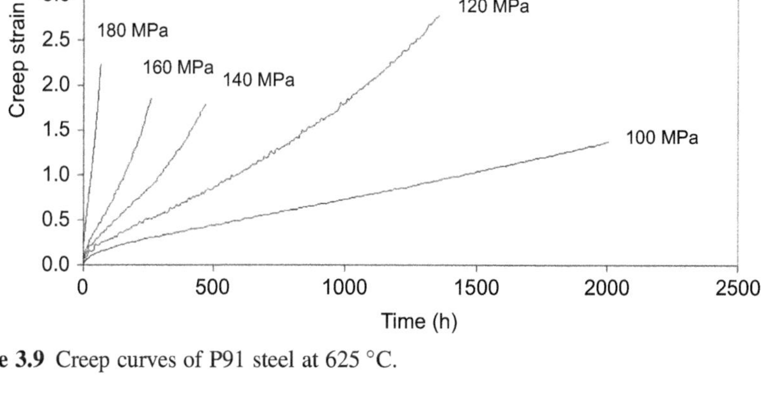

Figure 3.9 Creep curves of P91 steel at 625 °C.

Arrhenius temperature relationship at a given stress. The LMP at a given stress is defined as the following function of temperature and time to rupture:

$$LMP = T \cdot [\log(t_r) + C] \quad (3.5)$$

where T is temperature in Kelvin, t_r is the creep rupture life in hours and C is the LMP constant. It has been shown that t_r of the candidate stainless steels considered in the Canadian SCWR program can be well described by the LMP method (Xu et al., 2016b).

Xu and Amirkhiz (2016) used a fitting procedure described by Conway (1969) to obtain expressions for the LMP in terms of stress for five alloys being considered for use in the Canadian SCWR concept. Experimental data were fit with a polynomial function of log σ:

$$LMP = C_1 + C_2 \cdot \log \sigma + C_3 \cdot (\log \sigma)^2 \quad (3.6)$$

where σ is the stress in Pa and C1, C2 and C3 are fitting constants, which allows the stress to be determined as a function of LMP or, by combining Eqs. (3.5) and (3.6), as a function of temperature for a specific in-service time. The constants of the LMP creep prediction model (Table 3.4) were obtained by fitting the data collected in the database and were based on the least-squares regression method assuming normal or logarithm-normal distributions. The goodness of fit to the experimental data was assessed visually with an emphasis on the long-term test data. Fig. 3.10 shows resulting LMP plots for Alloy 800H and Alloy 625. For a given stress level, the value of the LMP parameter can be determined from the curve. Rupture life can easily be determined for any given temperature (in K).

Using the LMP approach, in-reactor creep rupture life of a 20% cold-worked 316 SS was plotted by Ukai and shown in Fig. 3.11 (Ukai et al., 2000). As shown in the figure, the in-reactor creep rupture strengths were lower than those tested in air (approximately 70%–80%) and in sodium.

For a pressure vessel SCWR concept with a fuel cladding outer diameter of 8 mm and wall thickness of ~0.6 mm, Zhang et al. (2012) estimated the hoop stress on the tube wall in SCW at 25 MPa to be about 160 MPa if the fuel rod is not internally pressurized. This stress is rather high compared with the yield and creep strengths of

Table 3.4 Larson–Miller parameters of selected alloys (Xu and Amirkhiz, 2016)

	Alloy 800H	Alloy 625	SS 310	SS 347H	Alloy 214
C	14.99	23.9	9.74	11.05	22.71
C1	2589	3270	1252	1407	100,273
C2	10,004	13,402	7190	7699	−12,939
C3	−1011	−1271	−727	−763	483

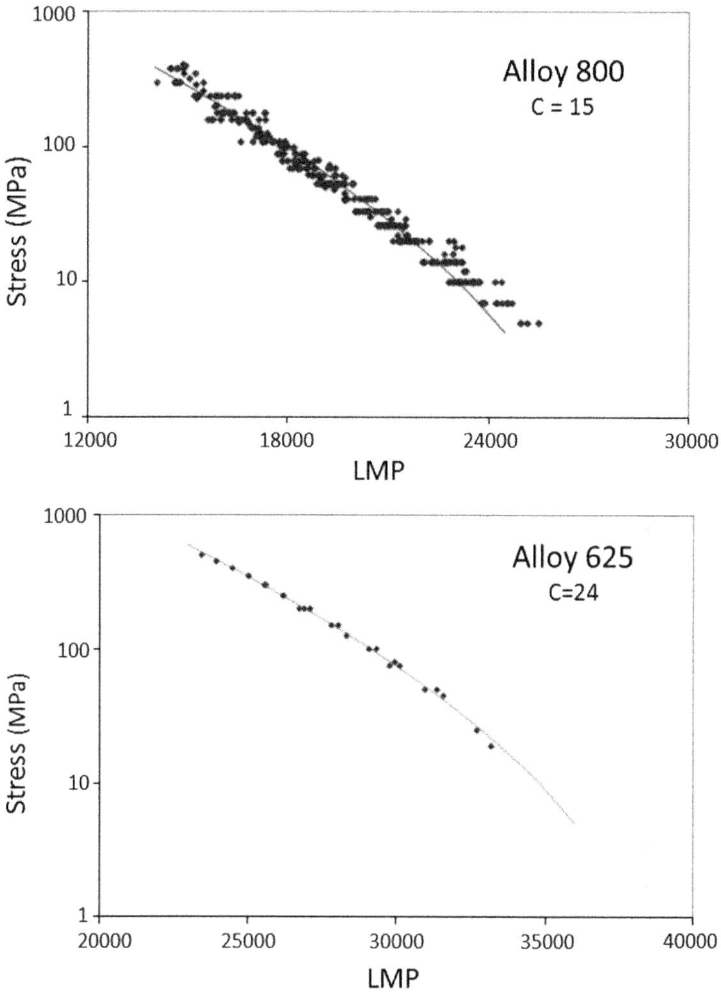

Figure 3.10 Larson−Miller parameter (LMP) plots for Alloy 800H and 625. C is the LMP constant (see Eq. 3.5).
Adapted from Xu, S., Zheng, W., Yang, L., 2016b. A review of irradiation effects on mechanical properties of candidate SCWR fuel cladding alloys for design considerations. CNL Nucl. Rev. 5 (2), 309−331. https://doi.org/10.12943/CNR.2016.00036.

candidate SCWR alloys at 650 °C. Internal pressurization of the fuel rod by He significantly reduces the cladding hoop stress, but during a loss of coolant accident that results in total depressurization of the core the fuel cladding must withstand the stresses from the internal He pressure plus that of gaseous fission products released from fuel. Zhang et al. (2012) suggested a compromise, proposing an initial He pressure of ~4 MPa which reduces the hoop stress on cladding to ~85 MPa at normal working conditions and at full depressurization (assuming a maximum fuel rod temperature not exceeding 650 °C).

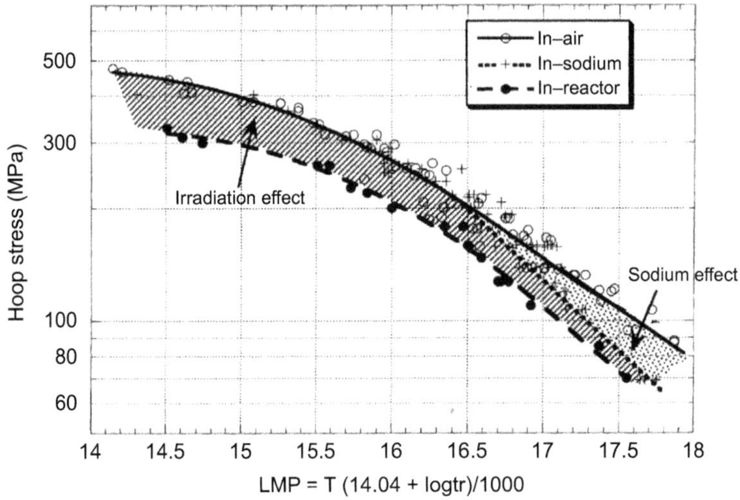

Figure 3.11 Creep rupture curves for in-air, in-sodium and in-reactor conditions (20 × 10^{26} n·m^{-2} (E > 0:1 MeV) as a function of Larson–Miller parameter (LMP) (Ukai et al., 2000).

A TEM study of 347H coupons that had undergone creep testing at 850 °C showed that NbC precipitates in the alloy can play a role against creep deformation (Shalchi Amirkhiz and Xu, 2014). It was found that NbC particles were oriented along [111] slip lines in a sample that had undergone creep for 1244 h. Moreover, as seen in Fig. 3.12, by pinning dislocations they prevent dislocation recovery and formation of subgrain boundaries, which could otherwise facilitate deformation. Through this mechanism, these particles can slow down creep deformation and strengthen material for longer creep life. No sigma or $M_{23}C_6$ carbide phases were observed in a sample crept for 1244 h. At the lowest creep stress level tested (20.4 MPa) precipitation of a Cr_2N phase with Widmanstatten morphology within the grain and large equiaxed Cr_2N particles at grain boundaries was observed. The Cr_2N phase nucleated on NbC precipitates.

Because of the difficulty in carrying out creep measurement of loaded test samples placed inside a reactor, many irradiation creep studies were performed on preirradiation samples (i.e., stress was not being applied during irradiation) in air postirradiation. The creep rates measured by the two different methods tend to differ considerably. The irradiation creep data from the literature often represent the overall effects of irradiation on time-dependent deformation but are suitable for examination of irradiation effects for design purposes. It should be noted that many irradiation data in the literature are postirradiation test data. Postirradiation test results may be different from in-reactor irradiation creep test results (i.e., creep strain measured during actual irradiation, stress and temperature conditions) but postirradiation tests offer an advantage over in-reactor tests because the loads and strain can be applied or measured accurately and the testing is more convenient and available than the in-reactor testing.

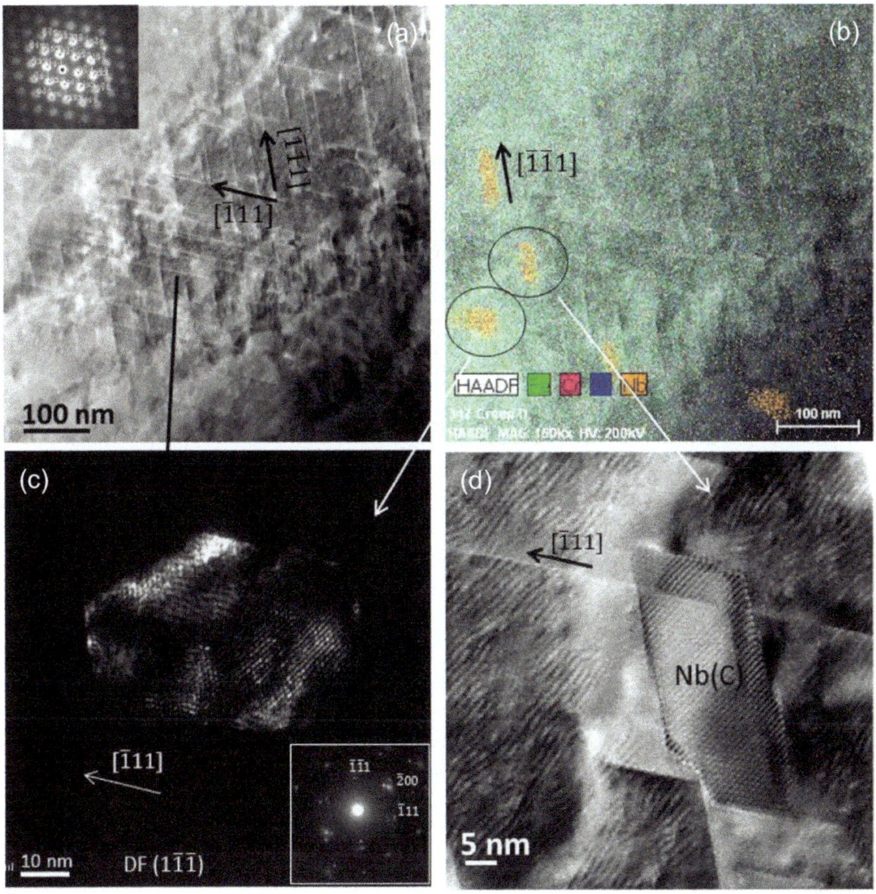

Figure 3.12 (a) High-angle annular dark-field imaging (HAADF) image and convergent beam electron diffraction (CBED) inset ([0 1 1] zone axis) showing microstructure of 1244 h crept sample of 347H SS highlighting slip lines in <111> direction (b) EDX map highlighting NbC precipitates oriented towards <111>, (c) dark field TEM image from (−1 1 1) reflection of cubic NbC, (d) high-resolution STEM-HAADF image showing one NbC particle oriented along slip lines; Moiré fringes are indicative of deformation within the lattice (Shalchi Amirkhiz and Xu, 2014).

3.4.3 Irradiation-assisted creep

Many irradiation creep mechanisms have been proposed and examined experimentally and theoretically and have been extensively reviewed (Simonen, 1980; Garner, 2006; Was, 2007). Matthews and Finnis (1988) summarized irradiation creep mechanisms, which include simple irradiation-enhanced diffusion models, stress-induced preferred nucleation of interstitial loops, stress-induced preferential absorption of point defects at dislocations, climb-enhanced glide mechanisms and enhanced recovery mechanisms.

For the steady-state irradiation creep, the deformation may be expressed as the sum of irradiation creep and swelling and the following equation is a generalized formulation (Anderoglu et al., 2012, originally from Foster et al., 1972):

$$\frac{\dot{\varepsilon}_{creep}}{\sigma^n} = B_0 + D\dot{S} \tag{3.7}$$

where $\dot{\varepsilon}_{creep}$ is the creep rate as a function of displacement damage (dpa), σ is the stress, n is a stress exponent, B_0 is called the creep compliance, $\dot{S}$ is the swelling rate as a function of dpa, and D is called the creep-swelling coupling coefficient. The creep compliance is sometimes used to compare irradiation creep dependence.

In in-reactor creep tests, especially when the accumulated dpa reaches a certain level, void swelling, creep, and precipitation-related strains can start to be coupled together. The onset of swelling in the test sample can be detected by a jump in the creep rate even before measurable void swelling produces any detectable dimensional changes (Garner, 2012). In the case of a thermalized neutron spectrum, additional creep deformation can probably be related to the large amounts of transmuted helium and hydrogen, which could cause the formation of bubbles and voids, accelerating the creep deformation, as suggested by Garner (2012).

Hazel et al. (1965) reported that the combination of fuel and reactor power levels imposed high stresses on the annular-type fuel elements used in the ESH-1 and ESH-1.1 fuel tests in the ESADE-VBWR steam superheat loop, resulting in measureable creep for almost all test claddings. A major conclusion was that none of the elements could be considered to have had a free-standing cladding under the test conditions (~540–700 °C, 106–135 days irradiation time). Fuel cladding wrinkling or ridging was found. Alloy 800 was found to be more dimensionally stable than Alloy 600.

Brinkman et al. (1973) reported creep-fatigue studies of irradiated solution-annealed 304 SS, 304L(Ti-modified) SS, 316 SS and Alloy 800 irradiated at fluences from 0.4 to 5 × 10^{22} $n\cdot cm^{-2}$, E > 0.1 MeV at 700–750 °C and tested at 700 °C. The irradiated alloys showed a marked decrease in tensile ductility and reductions in fatigue life ranging from factors of 1.5–2.5 for 304, 304L and 316 to as much as 35 for Alloy 800. The Japanese GIF Consortium examined a number of candidate alloys (316 SS, several modified versions of 310 SS, and Alloy 690) after irradiation in the Japanese Materials Testing Reactor and the Joyo Experimental Fast Reactor at temperatures up to 700 °C (Higuchi et al., 2007). They noted that creep deformation was dominated by thermal effects rather than irradiation at 700 °C. Void formation was observed in materials irradiated at the Joyo reactor at 600 °C, but its effect on component deformation was considered to be tolerable.

3.5 Microstructural instability

Testing for SCWR research has mostly been conducted using as-received (typically mill-annealed) material in as-machined condition without much consideration of the microstructure instability that can occur from simultaneous exposure to both high temperature and radiation during the in-service life. The effect of thermal ageing damage has only been recently recognized as a possible influencing factor in determining the corrosion resistance (Li et al., 2014, Jiao et al., 2015) of alloys being considered for SCWR use (Section 5.3.6). Mo et al. (2010) discussed high-temperature ageing effects of Alloy 617 and Alloy 230 in the context of the development of the Very High Temperature Reactor. Wang et al. (2017) examined long-term ageing effects for a series of Fe−18Ni−12Cr-based alumina-forming alloys being considered for SCWR use. Samples were aged at 700 °C for periods of up to 1000 h, and the microstructural stability and creep properties were measured. Increased tensile strength with increasing ageing time was observed, attributed to precipitate strengthening.

Various radiation-induced precipitates may form in this temperature range as a result of RIS at defect sinks. For example, some slowly diffusing elements (Ni, Si), which move inside the metals by interstitial mechanisms, can become enriched at sinks such as grain boundaries, whereas faster-moving elements such as Cr and Mn, as well as Mo, can be depleted with irradiation at or near sinks such as grain boundaries. Irradiation temperature, dose and dose rate are the primary variables that can affect the degree of RIS. There have been some reviews of the microstructure instability of austenitic Fe−Cr−Ni alloys at high temperature (for example, Weiss and Stickler, 1972; Sourmail, 2001), and with radiation exposure, the latter mostly in the context of LWR applications (Kenik and Busby, 2012; Maziasz, 1993). Jin et al. (2013) recently reported the results of studies of irradiation-induced precipitation of $Cr_{23}C_6$ in an austenitic stainless steel after ion irradiation.

3.5.1 Microstructural instability due to high temperature exposure

The microstructure instability at high temperatures includes the temperature/time-dependent nucleation and growth of secondary phases, typically occurring first on grain boundaries and then within the grains. The sequence is typically presented in the form of a time−temperature precipitation diagram. In general, short-term exposure is well known to 'sensitize' the microstructure due to formation of Cr-rich carbides such as $M_{23}C_6$. These carbides can decompose with prolonged exposure to form M_6C carbides, a phase with a variable composition and stability strongly linked to other constituents. Preferential formation of Ti, Nb or Zr carbides over formation of Cr-rich carbides can take place when Ti, Nb or Zr is added as a minor alloying element to 'stabilize' the microstructure against formation of chromium-rich carbides, which are linked to intergranular corrosion. More complex intermetallic phases enriched with the solute alloying elements (Cr, Ni and Mo) such as the chi (χ), laves (η) and sigma (σ) phases require a longer time to form. The χ and η phases contain Mo and therefore are of more concern for Mo-containing austenitic Fe−Cr−Ni alloys. The χ phase requires the shortest amount of time to form, whereas the σ phase requires the longest. At 700 °C, for instance, χ phase was reported to form after about 5 h in solution-annealed 316L SS, whereas σ phase precipitation takes more than 100 h (Weiss and Stickler, 1972). Formation of the σ phase is usually associated with dissolution of carbides, but it can also form independently. The η phase forms predominantly within the grains and only occasionally on grain boundaries in competition with χ and σ phases (Weiss and Stickler, 1972).

3.5.2 Formation of precipitates due to radiation exposure

RIS, introduced in Section 3.3, is linked with the formation of precipitates in materials after irradiation. With continuing irradiation, RIS at defect sinks can accumulate with dose so that the solubility level is reached and, with sufficient driving force, a precipitate phase can form. The precipitates are generally classified as either (1) radiation-enhanced/impeded; (2) radiation-modified or (3) radiation-induced (Lee et al., 1981). The first two types involve precipitates common with high temperature exposure alone, with radiation affecting the nucleation and growth rate as well as the chemistry. The third type involves precipitates not common to high temperature exposure alone (Fig. 3.13).

Kenik and Busby (2012) summarized the type of precipitates that can form in neutron irradiated 316 stainless steel. Cold working of the stainless steels can affect the precipitation process. For example, the dislocations generated by cold work serve initially as the major defect sink, which attracts both Ni and Si coming from the RIS process and subsequent pipe diffusion of these elements lead to the formation of γ' phase.

3.6 Modelling

As irradiation experiments are both time-consuming and expensive, modelling provides a valuable tool that can provide guidance to experimenters and designers. Multi-scale models are being developed that consider phenomena occurring over a range of time and length scales using techniques such as ab initio calculations and MD simulations to understand defect production, Monte Carlo simulations to study the kinetics of defect evolution and dislocation dynamics to calculate the parameters needed for continuum and finite element analysis (Samaras et al., 2009), the latter providing component performance data. MD displacement cascade damage studies on α-Fe have focused on the influence of PKA energy with temperature, interatomic potential and solute additions on defect statistics and clustering for alloys such as Fe—Cr (Virtler et al., 2008; Deo et al., 2007) and Fe—Ni (Masterakos et al., 2010; Beamish et al., 2010) on defect statistics and clustering. Malerba et al. (2008) has reviewed recent progress in the multiscale modelling of Fe—Cr alloys. The authors note that irradiation response and thermodynamic properties are particularly difficult to model in a concentrated alloy and identified some of the outstanding issues. Such modelling is cross-cutting as many of the irradiation and temperature conditions are common to all GEN IV reactor concepts (Fazio et al., 2009).

There has been little modelling work directed at alloys of interest specifically for SCWR development. Boyle and Shabib (2011) reported displacement cascade defect

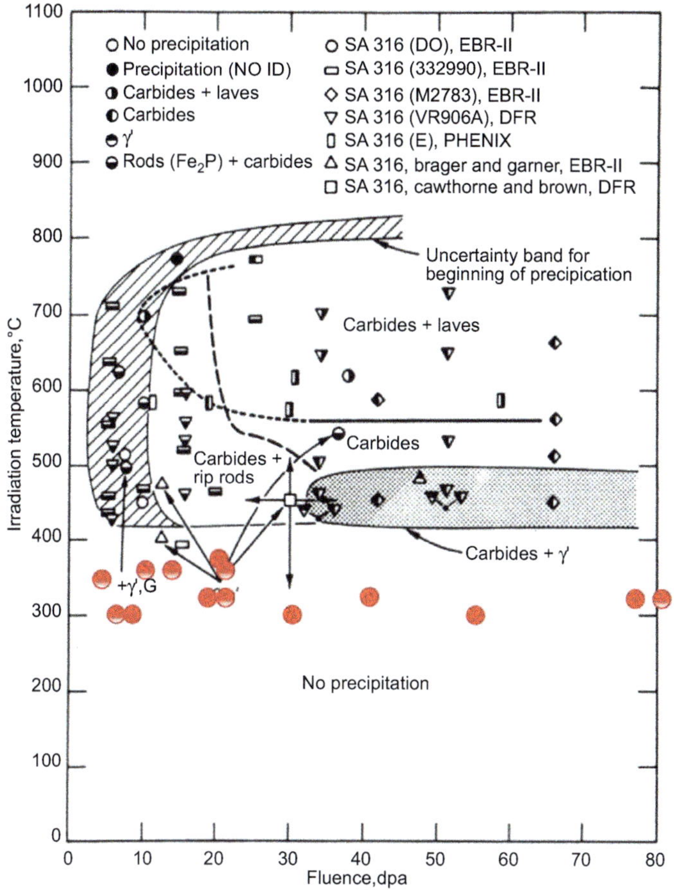

Figure 3.13 Precipitate phases observed after irradiation as a function of temperature and dose. Partially shaded data points are for γ′ phase and solid data points are for either G and related phases or an unidentified phase.
From Kenik E.A., Busby, J.T., 2012. Radiation-induced degradation of stainless steel light water reactor internals. Mater. Sci. Eng. R. 73, 67 who added the *red dots* to the black-and-white plot originally made by Maziasz, P.J., McHargue C.J., 1987. Microstructural evolution in annealed austenitic steels by neutron irradiation. Int. Mater. Rev. 2, 190.

statistics generated for α-Fe over a wide range of PKA energies (0.1 and 20 keV) and temperatures (15 and 1200K) using MD simulations, analysing the data in terms of defect production efficiency η. The results showed η to be temperature dependent, attributed to thermally activated defect recombination. The authors proposed an extended power-law representation for η that accounted for the influence of both temperature and PKA energy. Zheng et al. (2017) studied the effect of retained hydrogen on the defect microstructure of specimens of AL-6XN SS, a candidate SCWR alloy, irradiated with low energy (100 keV) H^{2+} to high concentrations of injected hydrogen and displacement damage dose up to 7 dpa. MD simulations were performed to elucidate the observed division of large dislocation loops under irradiation. As no potential for the fcc austenitic phase was available and potentials developed for bcc Fe would not accurately describe a system containing 1/3 < 111> faulted dislocation loops, the authors used a potential for fcc Ni, recognizing that it would not be ideal. In spite of the choice of potential, it was found that after 10 ps, the configuration of 1/3 < 111> loop divided into small dislocation loops in agreement with measurements. In a parallel study, Yu et al. (2017) developed a rate theory to calculate the average size and density evolution of these dislocation loops. The size and density of defect clusters were calculated under irradiation temperatures between 550 and 900 K and irradiation doses up to 15 dpa, and the simulation results were in relatively good agreement with the experimental results at 563K.

References

Allen, T.R., Busby, J.T., Gan, J., Kenik, E.A., Was, G.S., 2000. Correlation between swelling and radiation-induced segregation in iron-chromium-nickel alloys. In: Hamilton, M.L., Kumar, A.S., Rosinski, S.T., Grossbeck, M.L. (Eds.), 19th International Symposium: Effects of Radiation on Materials, No. 1366. ASTM Special Technical Publication, pp. 739–755.

Allen, T.R., Tsai, H., Cole, J.I., Ohta, J., Dohi, K., Kusanagi, H., 2002. Properties of 20% cold-worked 316 stainless steel irradiated at low dose rate. In: ASTM Radiation Effects on Materials Conferences, Tucson Arizona, June 18–20, 2002.

Allen, T.R., Cole, J.I., Trybus, C.L., Porter, D.L., Tsai, H., Garner, F., Kenik, E.A., Yoshitake, T., Ohta, J., 2006. The effect of dose rate on the response of austenitic stainless steels to neutron radiation. J. Nucl. Mater. 348 (1–2), 148–164.

Anderoglu, O., Byun, T.-S., Toloczko, M., Maloy, S.A., 2012. Mechanical performance of ferritic martensitic steels for high dose applications in advanced nuclear reactors. Metall. Mater. Trans. 44A, S70–S83.

ASME, 2006. Class 1 Components in Elevated Temperature Service. ASME Boiler & Pressure Vessel Code, Section III, Subsection NH, New York, NY, USA).

ASTM, 2009. Standard Practice for Neutron Radiation Damage Simulation by Charged-Particle Irradiation. American Society for Testing and Materials, Philadelphia, 1996, reapproved 2009.

Averback, R.S., de la Rubia, T., 1997. Displacement damage in irradiated metals and semiconductors. Solid State Phys. 51, 281–402.

Ayrault, G., Hoff, H.A., Nolfi Jr., F.V., Turner, A.P.L., 1981. Influence of helium injection rate on the microstructure of dual-ion irradiated type 316 stainless steel. J. Nucl. Mater. 104, 1035–1039.

Beamish, E., Campa, C., Woo, T.K., 2010. Grain boundary sliding in irradiated stressed Fe-Ni bicrystals: a molecular dynamics study. J. Phys. Condens. Matter 22 (34), 345006.

Boothby, R.M., 2012. Radiation effects in nickel-based alloys. In: Konings, R.J.M. (Ed.), Comprehensive Nuclear Materials. Elsevier, Oxford, ISBN 9780080560335, pp. 123–150. https://doi.org/10.1016/B978-0-08-056033-5.00090-2.

Boyle, K.P., Shabib, I., 2011. Temperature dependence of defect production efficiency in α-Fe. In: 5th International Symposium on Supercritical Water Cooled Reactors, Canadian Nuclear Society. Paper P116.

Brinkman, C., Korth, G.E., Beeston, J.M., 1973. Influence of irradiation on the creep/fatigue behavior of several austenitic stainless steels and Incoloy 800 at 700 C. In: Effects of Radiation on Substructure and Mechanical Properties of Metals and Alloys, ASTM STP 529, American Society for Testing and Materials, pp. 473–492.

Broeders, C.H.M., Konobeyev, A.Y., 2004. Defect production efficiency in metals under neutron irradiation. J. Nucl. Mater. 328, 197–214.

Bruemmer, S.M., Simonen, E.P., Scott, P.M., Andresen, P.L., Was, G.S., Nelson, J.L., 1999. Radiation-induced material changes and susceptibility to intergranular failure of light-water-reactor core internals. J. Nucl. Mater. 274, 299–314.

Bruemmer, S.M., 2001. In: Proc. 10th Intl. Symp. on Environmental Degradation of Materials in Nuclear Power Systems – Water Reactor. NACE, Houston, TX. Paper No. 0008V.

Busby, J.T., Was, G.S., Kenik, E.A., 2002. Isolating the effect of radiation-induced segregation in irradiation-assisted stress corrosion cracking of austenitic stainless steels. J. Nucl. Mater. 302, 20–40.

Bysice, S., Pencer, J., Walters, L., Bromley, B., 2016. Comparison of dpa and helium production in candidate fule cladding materials for the Canadian SCWR. CNL Nuc. Rev. 5, 269–275.

Byun, T., Hashimoto, N., 2006. Strain localization in irradiated materials. Nucl. Eng. Technol. 38 (7), 619–638.

Chow, C.-K., Khartabil, H.F., 2008. Conceptual fuel channel designs for CANDU-SCWR. Nucl. Eng. Technol. 40, 139–146.

Chung, H.M., 2006. Assessment of Void Swelling in Austenitic Stainless Steel PWR Core Internals. Argonne National Laboratory Technical report ANL-04/28 TRN: US1006107.

Coble, R.L., 1963. A model for boundary diffusion controlled creep in polycrystalline materials. J. Appl. Phys. 34, 1679.

Comprelli, F.A., Busboom, H.J., Spalrais, C.N., 1969. Comparison of radiation damage studies and fuel cladding performance for Incoly-800. In: Irradiation Effects in Structural Alloys for Thermal and Fast Reactors, ASTM STP 457, American Society for Testing and Materials, pp. 400–413.

Conway, J.B., 1969. Stress-Rupture Parameters: Origin, Calculation and Use. Gordon and Breach, New York.

Dai, Y., Odette, G., Yamamoto, T., 2012. The effects of helium in irradiated structural alloys. In: Konings, R.J.M. (Ed.), Comprehensive Nuclear Materials. Elsevier, Oxford, pp. 141–193.

Deo, C.S., Okuniewski, M.A., Srivilliputhur, S.G., Maloy, S.A., Baskes, M.I., James, M.R., Stubbins, J.F., 2007. The effects of helium on irradiation damage in single crystal iron. J. Nucl. Mater. 367–370A, 451–456.

Erlich, K., Konys, J., Heikinheimo, L., 2004. Materials for high performance light water reactors. J. Nucl. Mater. 327, 140.

Fazio, C., Alamo, A., Almazouzi, A., De Grandis, S., Gomez-Briceno, D., Henry, J., Malerba, L., Rieth, M., 2009. European cross-cutting research on structural materials for generation IV and transmutation systems. J. Nucl. Mater. 392, 316–323.

Foster, J.P., Wolfer, W.G., Biancheria, A., Boltax, A., 1972. Analysis of irradiation-induced creep of stainless steel in fast spectrum reactors. In: Proc. European Conference on Irradiation Embrittlement and Creep in Fuel Cladding, British Nuclear Society, London, pp. 273–281.

Freyer, P.D., Mager, T.R., Burke, M.A., 2007. Hot cell crack initiation testing of various heats of highly irradiated 315 stainless steel component obtained from three commercial PWRs. In: Proceedings of the 13th International Conference on Environmental Degradation of Materials in Nuclear Power Systems, Whistler, British Columbia, Canada, April 19–23, 2007.

Garner, F.A., Brager, H.R., 1985. Dependence of neutron-induced swelling on composition in iron-based austenitic alloys. In: Garner, F.A., Gelles, D.S., Witten, F.W. (Eds.), Optimizing Materials for Nuclear Applications. The Metallurgical Society of AIME, pp. 87–109.

Garner, F.A., Brager, H.R., 1988. The influence of Mo, Si, P, C, Ti, Cr, Zr and various trace elements on the neutron-induced swelling of AISI 316 stainless steel. J. Nucl. Mater. 155–157, 833–837.

Garner, F.A., Kumar, A.S., 1987. The influence of both major and minor element composition on void swelling in austenitic steels. In: Garner, F.A., Packan, N.H., Kumar, A.S. (Eds.), Radiation-Induced Changes in Microstructure: 13th International Symposium (Part I). ASTM Special Technical Publication, pp. 289–314.

Garner, F.A., Brager, H.R., Gelles, D.S., McCarthy, J.M., 1987. Neutron irradiation of Fe-Mn, Fe-Cr-Mn and Fe-Cr-Ni alloys and an explanation of their differences in swelling behavior. J. Nucl. Mater. 148, 294–301.

Garner, F.A., 1984. Recent insights on the swelling and creep of irradiated austenitic alloys. J. Nucl. Mater. 122 (1–3), 459–471.

Garner, F.A., Gelles, D.S., 1990. Proceedings of Symposium on Effects of Radiation on Materials: 14th International Symposium, ASTM STP 1046, vol. II, pp. 673–683.

Garner, F.A., 2006. Irradiation performance of cladding and structural steels in liquid metal reactors. In: Frost, B.R.T. (Ed.), Materials Science and Technology: A Comprehensive Treatment. VCH Publishers, pp. 419–543 (Chapter: 10A).

Garner, F.A., 2012. Radiation damage in austenitic steels. In: Konings, R.J.M. (Ed.), Comprehensive Nuclear Materials. Elsevier, Oxford, ISBN 9780080560335, pp. 33–95. https://doi.org/10.1016/B978-0-08-056033-5.00065-3.

Gessel, G.R., Rowcliffe, A.F., 1977. The effect of solute additions on the swelling of an Fe-7.5Cr-20Ni alloy. In: Bleiberg, L., Bennett, J.W. (Eds.), Radiation Effects in Breeder Reactor Structural Materials. Metallurgical Society of AIME, pp. 431–442.

Gittus, J., 1978. Irradiation Effects in Crystalline Solids. Elsevier Applied Science Publishers, Limited.

Greenwood, L., Oliver, B.M., 2006. Comparison of predicted and measured helium production in U.S. BWR reactors. J. ASTM Int. 3 (3), 1–9.

Greenwood, L.R., 1983. A new calculation of thermal neutron damage and helium production in nickel. J. Nucl. Mater. 115, 137–142.

Gupta, J., Hure, J., Tanguy, B., Laffont, L., Lafont, M.-C., Andrieu, E., 2016. Evaluation of stress corrosion cracking of irradiated 304L stainless steel in PWR environment using heavy ion irradiation. J. Nucl. Mater. 476, 82–92.

Hardie, C.D., Williams, C.A., Xu, S., Roberts, S.G., 2013. Effects of irradiation temperature and dose rate on the mechanical properties of self-ion implanted Fe and Fe–Cr alloys. J. Nucl. Mater. 439 (1–3), 33–40.

Harries, D.R., 1979. Neutron irradiation-induced embrittlement in type 316 and other austenitic steels and alloys. J. Nucl. Mater. 82, 2–21.

Hazel, V.E., Boyle, R.F., Busboom, H.J., Murdock, T.B., Skarpelos, J.M., Spalaris, C.N., 1965. Fuel Irradiation in the ESADA-VBWR Nuclear Superheat Loop. General Electric Atomic Power Report GEAP-4775.

Herschbach, K., Schneider, W., Ehrlich, K., 1993. Effects of minor alloying elements upon swelling and in-pile creep in model plain Fe-15Cr-15Ni stainless steels and in commercial DIN 1.4970 alloys. J. Nucl. Mater. 203, 233–248.

Higuchi, S., Sakurai, S., Ishida, T., 2007. A study of fuel behavior in an SCWR core with high power density. In: Proc. ICAPP'07, Nice, France, May 13–18, 2007. Paper 7206.

Hojná, A., 2013. Irradiation-assisted stress corrosion cracking and impact on life extension. Corrosion 69 (10), 964–974.

IAEA, 2011. Stress Corrosion Cracking in Light Water Reactors: Good Practices and Lessons Learned. IAEA Nuclear Energy Series No. NP-T-3.13. International Atomic Energy Agency, Vienna.
Idrees, Y., Yao, Z., Sattari, M., Kirk, M.A., Daymond, M.R., 2013. Irradiation induced microstructural changes in Zr-Excel alloy. J. Nucl. Mater. 441, 138–151.
Jiao, Z., Was, G.S., 2011. Impact of localized deformation on IASCC in austenitic stainless steels. J. Nucl. Mater. 408 (3), 246–256.
Jiao, Y., Zheng, W., Guzonas, D.A., Cook, W.G., Kish, J.R., 2015. Effect of thermal treatment on the corrosion resistance of Type 316L stainless steel exposed in supercritical water. J. Nucl. Mater. 464 (2), 356.
Jin, S., Guo, L., Luo, F., Yao, Z., Ma, S., Tang, R., 2013. Ion irradiation-induced precipitation of $Cr_{23}C_6$ at dislocation loops in austenitic steel. Scr. Mater. 68 (2), 138–141.
Johnston, W.G., Rosolowski, J.H., Turkalo, A.M., Lauritzen, T., 1974. An experimental survey of swelling in commercial Fe-Cr-Ni alloys bombarded with 5 MeV Ni ions. J. Nucl. Mater. 54 (1), 24–40.
Johnston, W.G., Lauritzen, T., Rosolowski, J.H., Turkalo, A.M., 1976. Void swelling in fast reactor materials: a metallurgical problem. J. Met. 28 (6), 20–24.
Kenik, E.A., Busby, J.T., 2012. Radiation-induced degradation of stainless steel light water reactor internals. Mater. Sci. Eng. R. 73, 67.
Kinchin, G.H., Pease, R.S., 1955. The displacement of atoms in solids by radiation. Rep. Prog. Phys. 18 (1), 1.
Kiritani, M., 1989. Radiation rate dependence of microstructure evolution. J. Nucl. Mater. 169, 89–94.
Kittel, 1976. Introduction to Solid State Physics. John Wiley and Sons, New York.
Klassen, R.J., Rajakumar, H., 2015. Combined effect of irradiation and temperature on the mechanical strength of Inconel 800H and AISI 310 alloys for in-core components of a GEN-IV SCWR. In: 7th International Symposium on Supercritical Water-cooled Reactors (ISSCWR7), Helsinki, Finland, March 15–18, 2015. Paper ISSCWR7-2006.
Kozlov, A.V., Portnykh, I.A., Bryushkova, S.V., Kinev, E.A., 2004. Dependence of maximum swelling temperature on damage dose in cold worked 16Cr-15Ni-2Mo-1Mn cladding irradiated in BN-600. In: Grossbeck, M.L., Allen, T.R., Lott, R.G., Kumar, A.S. (Eds.), Radiation-induced Changes in Microstructure: 21th International Symposium, No. 1447, pp. 446–453.
Larson, F.R., Miller, J., 1952. A time-temperature relationship for rupture and creep stresses. Trans. ASME 74, 765–775.
Lauritzen, T., Vaidyanathan, S., Bell, W.L., Yang, W.J.S., 1987. Irradiation-induced swelling in AISI 316 steel: effect of tensile and compressive stresses. In: Garner, F.A., Packan, N.H., Kumar, A.S. (Eds.), Radiation-Induced Changes in Microstructure: 13th International Symposium (Part I). ASTM Special Technical Publication, pp. 101–113.
Lee, E.H., Maziasz, P.J., Rowcliffe, A.F., Holland, J.R., et al., 1981. The Metallurgical Society of AIME, Warrendale, PA, pp. 191–218.
Lee, E.H., Byun, T.S., Hunn, J.D., Yoo, M.H., Farrell, K., Mansur, L.K., 2001a. On the origin of deformation microstructures in austenitic stainless steel: Part I- Microstructures. Acta Mater. 49, 3269.
Lee, E.H., Yoo, M.H., Byun, T.S., Hunn, J.D., Farrell, K., Mansur, L.K., 2001b. On the origin of deformation microstructures in austenitic stainless steel: Part II-Mechanisms. Acta Mater. 49, 3277.
Li, J., Zheng, W., Penttilä, S., Liu, P., Woo, O.T., Guzonas, D., 2014. Microstructure stability of candidate stainless steels for GEN-IV SCWR fuel cladding application. J. Nucl. Mater. 454, 7.

Malerba, L., Caro, A., Wallnius, J., 2008. Multiscale modelling of radiation damage and phase transformations: the challenge of FeCr alloys. J. Nucl. Mater. 382, 112–125.

Mastorakos, I.N., Le, N., Zeine, M., Zbib, H.M., Khaleel, M., 2010. Multiscale modeling of irradiation induced hardening in α-Fe, Fe-Cr and Fe-Ni systems. Mater. Res. Soc. Symp. Proc. 1264, 259–264.

Matsui, H., Sato, Y., Saito, N., Kano, F., Ooshima, K., Kaneda, J., Moriya, K., Ohtsuka, S., Oka, Y., 2007. Material development for supercritical water-cooled reactors. In: Proc. ICAPP 2007, Nice, France, May 13–18, 2007. Paper 7447.

Matthews, J.R., Finnis, M.W., 1988. Irradiation creep models-an overview. J. Nucl. Mater. 159, 257–285.

Maziasz, P.J., McHargue, C.J., 1987. Microstructural evolution in annealed austenitic steels by neutron irradiation. Int. Mater. Rev. 2, 190.

Maziasz, P.J., 1993. Overview of microstructural evolution in neutron-irradiated austenitic stainless steels. J. Nucl. Mater. 205, 118–145.

Maziasz, P.J., 1993. Swelling and swelling resistance possibilities of austenitic stainless steels in fusion reactors. J. Nucl. Mater. 122 (1–3), 472–486.

Mo, K., Lovicu, G., Tung, H.-M., Chen, X., Stubbins, J.F., 2010. High temperature aging and corrosion study on alloy 617 and alloy 230. In: Proc. Int. Conf. on Nuclear Engineering (ICONE18), May 17–21, 2010, Xi'an, China.

Murphy, S.M., 1987. The influence of helium trapping by vacancies on the behavior of metals under irradiation. In: Garner, F.A., Packan, N.H., Kumar, A.S. (Eds.), Radiation-Induced Changes in Microstructure: 13th International Symposium (Part I). ASTM Special Technical Publication, pp. 330–344.

Nava-Domínguez, A., Onder, N., Rao, Y., Leung, L., 2016. Evolution of the Canadian SCWR fuel-assembly concept and assessment of the 64 element assembly for thermalhydraulic performance. CNL Nuc. Rev. 56, 221–238.

Norgett, M.J., Robinson, M.T., Torrens, I.M., 1975. A proposed method of calculating displacement dose rates. Nucl. Eng. Des. 33, 50–54.

OECD, 2015. Primary Radiation Damage in Materials. NEA/NSC/DOC(2015)9. https://www.oecd-nea.org/science/docs/2015/nsc-doc2015-9.pdf.

Ohnuki, S., Takahashi, H., Nagasaki, R., 1988. Effect of helium on radiation-induced segregation in stainless steel. J. Nucl. Mater. 155, 823–827.

Okita, T., Kamada, T., Sekimura, N., 2000. Effects of dose rate on microstructural evolution and swelling in austenitic steels under irradiation. J. Nucl. Mater. 283–287, 220–223.

Okita, T., Sato, T., Sekimura, N., Garner, F.A., Greenwood, L.R., 2002. The primary origin of dose rate effects on microstructural evolution of austenitic alloys during neutron irradiation. J. Nucl. Mater. 307–311, 322–326.

Pawel, J.P., Ioka, I., Rowcliffe, A.F., Grossbeck, M.L., Jitsukawa, S., 1999. In: In Effects of Radiation on Materials: 18th International Symposium; ASTM STP 1325, pp. 671–688.

Porter, D.L., Takata, M.L., Wood, E.L., 1983. Direct evidence for stress-enhanced swelling in type 316 stainless steel. J. Nucl. Mater. 116, 272–276.

Rowcliffe, A.F., Mansur, L.K., Hoelzer, D.T., Nanstad, R.K., 2009. Perspectives on radiation effects in nickel-base alloys for applications in advanced reactors. J. Nucl. Mater. 392, 341–352.

Samaras, M., Victoria, M., Hoffelner, W., 2009. Advanced materials modeling – E.U. perspectives. J. Nucl. Mater. 392, 286–291.

Schroeder, H., 1988. High temperature helium embrittlement in austenitic stainless steels – correlations between microstructure and mechanical properties. J. Nucl. Mater. 155–157, 1032–1037.

Scott, P., 1994. A review of irradiation assisted stress corrosion cracking. J. Nucl. Mater. 211, 101–122.
Shalchi Amirkhiz, B., Xu, S., 2014. NbC precipitation and deformation of SS 347H crept at 850 °C. Microsc. Microanal. 20 (Suppl. 3). Paper 1494 (Proceeding of Microscopy and Microanalysis Conference, August 3–7, 2014, Hartford, CT, USA).
Simonen, E.P., 1980. Irradiation creep in transient irradiation environments. J. Nucl. Mater 90, 282–289.
Singh, B.N., Leffers, T., Makin, M.J., Walters, G.P., Foreman, A.J.E., 1981. Effects of implanted helium on void nucleation during Hvem irradiation of stainless steel containing silicon. J. Nucl. Mater. 104, 1041–1045.
Sourmail, T., 2001. Precipitation in creep resistant austenitic stainless steels. Mater. Sci. Technol. 17, 1–14.
Stephenson, K., Was, G.S., 2016. The role of dislocation channeling in IASCC initiation of neutron irradiated stainless steel. J. Nucl. Mater 481, 214–225.
Stoller, R.E., 2012. Primary radiation damage formation. In: Konings, R.J.M. (Ed.), Comprehensive Nuclear Materials. Elsevier, Oxford, ISBN 9780080560335, pp. 293–332.
Strasser, A., Santucci, J., Lindquist, K., Yario, W., Stern, G., Goldstein, L., Joseph, L., 1982. An Evaluation of Stainless Steel Cladding for Use in Current Design LWRs. Electric Power Research Institute Report NP-2642.
Thiele, B.A., Diehl, H., Ohly, W., Weber, H., 1984. Investigation into the irradiation behavior of high temperature alloys for high-temperature gas-cooled reactor applications. Nucl. Technol. 66, 597–606.
Ukai, S., Mizuta, S., Kaito, T., Okada, H., 2000. In-reactor creep rupture properties of 20% CW modified 316 stainless steel. J. Nucl. Mater. 278, 320–327.
Virtler, K., Björkas, C., Terentyev, D., Malerba, L., Nordlund, K., 2008. The effect of Cr concentration on radiation damage in Fe-Cr alloys. J. Nucl. Mater. 382, 24–30.
Viswanathan, R., Bakker, W.T., 2000. Materials for boilers in ultra supercritical power plants. In: Proc. of 2000 Int. Joint Power Generation Conference, Miami Beach, Florida, July 23–26, 2000. Paper JPGC2000-15049.
Walters, L., Donohue, S., 2016. Development of high performance pressure tube material for Canadian SCWR concept. JOM 68 (2), 490–495.
Walters, L., Wright, M., Guzonas, D., 2017. Irradiation issues and material selection for Canadian SCWR components. In: 8th Int. Symp. on Super-critical Water-cooled Reactors, March 13–15, 2017, Chengdu, China.
Wang, M., Zhou, Z., Jang, J., Sun, S., 2017. Microstructural stability and mechanical properties of AFA steels for SCWR core application. In: 8th Int. Symp. on Super-critical Water-cooled Reactors, March 13–15, 2017, Chengdu, China.
Was, G.S., 2007. Irradiation creep and growth. Fundam. Radiat. Mater. Sci. 71, 17–63.
Was, G.S., Andresen, P.L., 2012. Irradiation Assisted Corrosion and Stress Corrosion Cracking (IAC/IASCC) in Nuclear Reactor Systems and Components. Woodhead Publishing.
Weiss, B., Stickler, R., 1972. Phase instabilities during high temperature exposure of 316 austenitic stainless steel. Metall. Trans. 3, 851–866.
Wolfer, W.G., Garner, F.A., Thomas, L.E., 1982. On radiation-induced segregation and the compositional dependence of swelling in Fe-Ni-Cr alloys. In: Brager, H.R., Perrin, J.S. (Eds.), 11th International Symposium: Effects of Radiation on Materials, 782. ASTM Special Technical Publication, pp. 1023–1041.
Xu, R., Yetisir, M., Hamilton, H., 2014. Thermal-mechanical behavior of fuel element in SCWR design. In: Canada-China Conference on Advanced Reactor Development, Niagara Falls, Ontario Canada, April 27–30, 2014. Paper CCCARD2014-005.

Xu, S., Shalchi Amirkhiz, B., 2016. Mechanical properties of fuel cladding candidate alloys for Canadian SCWR concept. J. Met. 68 (2), 469–474.

Xu, C., Zhang, L., Qian, W., Mei, X.L.J., 2016a. The studies of irradiation hardening of stainless steel reactor internals under proton and xenon irradiation. Nucl. Eng. Technol. 48 (3), 758–764.

Xu, S., Zheng, W., Yang, L., 2016b. A review of irradiation effects on mechanical properties of candidate SCWR fuel cladding alloys for design considerations. CNL Nucl. Rev. 5 (2), 309–331. https://doi.org/10.12943/CNR.2016.00036.

Yetisir, M., Hamilton, H., Xu, R., Gaudet, M., Rhodes, D., King, M., Kittmer, A., Benson, B., 2017. Fuel assembly concept of the Canadian SCWR. In: 8th Int. Symp. on Super-critical Water-cooled Reactors, March 13–15, 2017, Chengdu, China.

Yu, Y., Guo, L., Zheng, Z., Tang, R., Yi, W., 2017. Rate theory studies on dislocation loops evolution in AL-6XN austenitic stainless steel under proton irradiation. In: 8th Int. Symp. on Super-critical Water-cooled Reactors, March 13–15, 2017, Chengdu, China.

Zhang, L., Bao, Y., Tang, R., 2012. Selection and corrosion evaluation tests of candidate SCWR fuel cladding materials. Nucl. Eng. Des. 249, 180–187.

Zheng, Z., Tang, R., Gao, N., Yu, Y., Zhang, W., Shen, Z., Long, Y., Wei, Y., Guo, L., 2017. Evolution of dislocation loops in austenitic stainless steels implanted with high concentration of hydrogen. In: 8th Int. Symp. on Super-critical Water-cooled Reactors, March 13–15, 2017, Chengdu, China.

Zheng, W., Guzonas, D., Boyle, K.P., Li, J., Xu, S., 2016. Materials assessment for the Canadian SCWR core concept. J. Met. 68 (2), 456–462.

Zhou, R., West, E.A., Jiao, Z., Was, G.S., 2009. Irradiation-assisted stress corrosion cracking of austenitic alloys in supercritical water. J. Nucl. Mater. 395, 11–22.

Zinkle, S.J., 2012. Radiation-induced effects on microstructure. In: Konings, R.J.M. (Ed.), Comprehensive Nuclear Materials. Elsevier, Oxford, ISBN 9780080560335, pp. 65–98.

Water chemistry

4

4.1 Introduction

Thermal power plants require:

1. a medium to transfer the heat from the source (fossil fuels, nuclear fission);
2. pipework to contain the heat transfer medium.

Water is a commonly used heat transfer medium because it is abundant, cheap, nontoxic, stable to high temperature and has good thermal properties. However, it is reactive with many materials and the critical transition occurs within the optimum operating temperature range for high thermal efficiency. Therefore, in any power cycle using high-temperature water as the heat transfer medium, water chemistry has a major influence on materials degradation, and it is necessary to understand those water properties that influence material degradation processes. In particular, for any supercritical water-cooled reactor (SCWR) concept the properties of supercritical water (SCW) directly affect general and localized corrosion (such as environmentally assisted cracking (EAC)) of system materials and the transport of corrosion products to and from the core. The deposition of corrosion products on surfaces can affect heat transfer to the coolant leading to higher metal temperatures, which can also affect bulk degradation mechanisms such as creep. Cohen (1985) and Kritsky (1999) are excellent references on the basics of water chemistry practices in current generation water-cooled reactors (WCRs).

4.1.1 *What is supercritical water?*

All fluids[1] exhibit a critical point in the P–V–T diagram defined by T_c, P_c and V_c above which it is not possible to liquefy the gas phase by application of pressure. For water, these values are $T_c = 647.096$ K, $P_c = 22.064$ MPa, $\rho_c = 322\ \text{kg}\cdot\text{m}^{-3}$. In the supercritical region, there is no phase segregation leading to liquid and vapour coexistence (no boiling). The physical chemistry of the critical transition has been the subject of great interest, the thermodynamic behaviour of fluids in the near-critical region being described in terms of critical power laws and critical scaling laws using dimensionless thermodynamic quantities defined in terms of T_c, P_c and ρ_c. A detailed exposition of this topic is beyond the scope of this book; for the most part, such a detailed understanding of the critical transition is not required for SCWR development[2] as the critical scaling laws are valid only over a small range of temperatures and densities near the critical point. Readers interested in exploring this topic will find the review by Anisimov et al. (2004) an excellent introduction.

[1] In this context fluid refers to a state of a substance that does not retain its shape if not confined.

[2] The exception being in the area of model development under conditions in the vicinity of the critical point.

Materials and Water Chemistry for Supercritical Water-cooled Reactors
https://doi.org/10.1016/B978-0-08-102049-4.00004-0

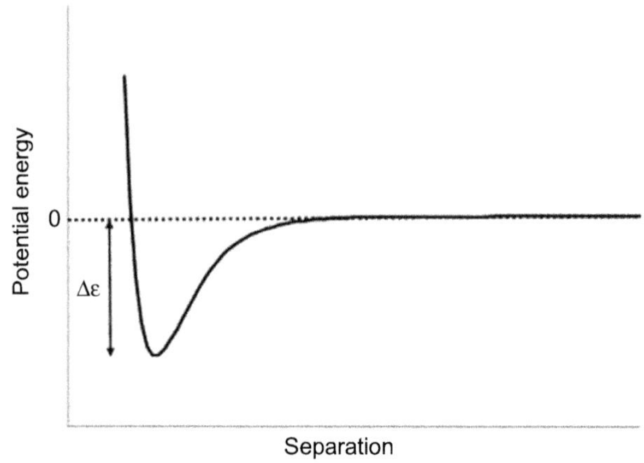

Figure 4.1 Simplified schematic of the dependence of potential on separation distance for two interacting molecules in the liquid state. When k_BT exceeds $\Delta\varepsilon$, molecules will always escape from the potential well and no permanent liquid phase can form.
Adapted from Tabor, D., 1969. Gases, Liquids and Solids: And Other States of Matter. Cambridge University Press.

Some insight into the molecular-scale meaning of the critical transition can be obtained by considering the interaction energy ($\Delta\varepsilon$) between fluid molecules (Fig. 4.1). When $T > \Delta\varepsilon/k_B$, the average thermal energy of fluid molecules (k_BT) is larger than the interaction energy between molecules ($\Delta\varepsilon$), and a persistent liquid phase cannot be formed; the temperature at which $T = \Delta\varepsilon/k_B$ is T_c. Short-lived molecular clusters will still form and break up, and a supercritical fluid consists of regions of high-density clusters of molecules and low-density regions of individual molecules (Fig. 4.2). This inhomogeneity in molecular distribution is a characteristic property of supercritical fluids. In the vicinity of the critical point the fluid undergoes significant density fluctuations.

While the term SCW is sometimes restricted to the area of the T-P phase diagram where both $T > T_c$ and $P > P_c$ are satisfied, considering the entire region where $T > T_c$ as SCW (as in the shaded area in Fig. 1.1) better highlights the connection between corrosion phenomena in superheated steam ($T > T_c$, $P < P_c$) and in SCW at higher pressures ($T > T_c$, $P > P_c$) and this convention has been adopted in this book. It should be noted here that various physical properties (e.g., isothermal compressibility, heat capacity, speed of sound, thermal diffusivity) pass through extrema in the vicinity of the critical point, and this behaviour extends into the single-phase region as a 'continuation' of the liquid–gas equilibrium[3]. These Widom lines[4] (Fig. 1.1) decrease in magnitude and broaden as the distance from the critical

[3] For example, the temperature at which the isobaric heat capacity of a fluid reaches a maximum at pressures greater than P_c is referred to as the pseudocritical point (384.87 °C at 25 MPa).

[4] The locus of the extrema of a thermodynamic response function (or related quantity) in P–T space (Imre et al., 2012). These extrema are close in P–T space but do not coincide, and diverge as the distance from the critical point increases; there is thus a set of Widom lines.

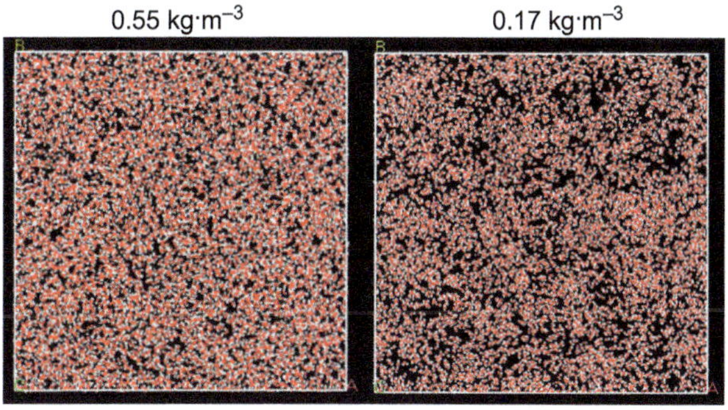

Figure 4.2 Molecular dynamics simulations of the structure of supercritical water at densities of 55 and 17 $kg \cdot m^{-3}$, showing 3-D distribution of oxygen (red) and hydrogen (white) atoms after 20 ns equilibration, T = 400 °C.
Adapted from Metatla, N., Jay-Gerin, J-P., Soldera, A., 2011. Molecular dynamics simulation of sub-critical and supercritical water at different densities. In: 5th International Symposium on Supercritical Water-cooled Reactors (ISSCWR-5), Vancouver, British Columbia, Canada, March 13–16, 2011.

point increases (Brazhkin et al., 2011; Imre et al., 2012) and are therefore only important near the critical point. The Widom lines serve to divide the SCW fluid region into a 'liquidlike' and a 'gaslike' region (Simeoni et al., 2010; Imre et al., 2012), and as noted by Imre et al. (2012), this division has more physical significance than divisions based on only the critical pressure.

Water is a unique substance when compared with other chemical compounds of similar molecular weight, and to compounds of hydrogen with oxygen's neighbours in the periodic table, due to its ability to form an extensive three-dimensional network of intermolecular hydrogen bonds that give rise to differences in properties (boiling point, dielectric constant) compared with nonhydrogen-bonded liquids. This is illustrated in Fig. 4.3, which shows the boiling points of hydrogen compounds with period 2 p-block elements; as the ability of the molecule to hydrogen bond increases the boiling point passes through a maximum.

The hydrogen bond energy is not large, and as the temperature increases thermal motion decreases the extent of hydrogen bonding (Fig. 4.4). In parallel, the dielectric constant drops from 80 at room temperature to ~10 at T_c and to ~2 at 425 °C. The low dielectric constant significantly changes the ionic product (K_w) and pH (Fig. 4.5) and SCW is only slightly dissociated compared with subcritical water. As K_w decreases SCW behaves like a nonpolar solvent. While water at low temperatures is a very incompressible fluid (isothermal compressibility[5] $\kappa_T = 0.000453\ MPa^{-1}$ at

[5] Isothermal compressibility is a measure of the relative volume change of a fluid or solid as a response to a pressure (or mean stress) change,

$$k_T = \rho^{-1}\left(\frac{\delta\rho}{\delta p}\right)_T \equiv -V^{-1}\left(\frac{\delta V}{\delta p}\right)_T$$

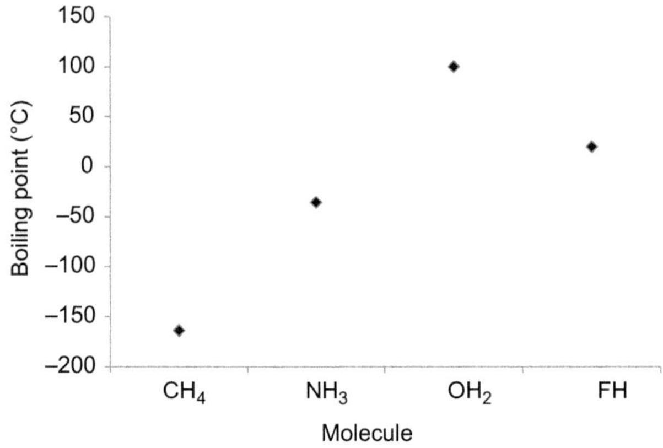

Figure 4.3 Boiling points of compounds of hydrogen and period 2 p-block elements (methane, ammonia, water and hydrogen fluoride).

25 °C, saturated liquid (Anisimov et al., 2004)), κ_T increases rapidly as the critical point is approached and diverges at the critical point. These large variations in properties make SCW an ideal medium for chemistry, as the properties (e.g., dielectric constant) can be 'tuned' over a wide range of values by only small changes in temperature (Akiya and Savage, 2002; Galkin and Lunin, 2005).

The changing water properties, combined with the high temperatures and pressure and the intense radiation fields, result in an 'extreme' in-core environment in an SCWR (Guzonas, 2014) that must be understood to advance materials development. Referring to Fig. 2.4, at the core inlet ('evaporator' region) the density is 'liquidlike' while in the 'superheater' region the density is more 'gaslike' and relatively constant. The rapid changes in properties in the near-critical region make modelling of phenomena such as corrosion product deposition and water radiolysis difficult in the vicinity of the critical point.

The three major water chemistry concerns in any direct cycle SCWR concepts are illustrated in Fig. 4.6:

1. Transport of impurities into the reactor core; these impurities can be corrosion products released from feedtrain surfaces, or impurities released from purification media and introduced by make-up water.
2. The in-core environment, in particular, the effects of water radiolysis and corrosion product deposition.
3. Transport of impurities from the core to the turbines and low pressure regions. Of major concern in an SCWR are the transport of oxidizing radiolysis products and radioactive isotopes (activity transport).

Similar concerns will exist in the primary circuit of indirect cycle SCWRs, where the feedtrain is replaced by the main primary loop(s), the turbines are replaced by steam generators, and the purification system is a side-stream system attached to the main loop(s).

This chapter roughly follows the water chemistry of the SCWR around this 'circuit', recognizing that these three concerns are not independent but overlap from

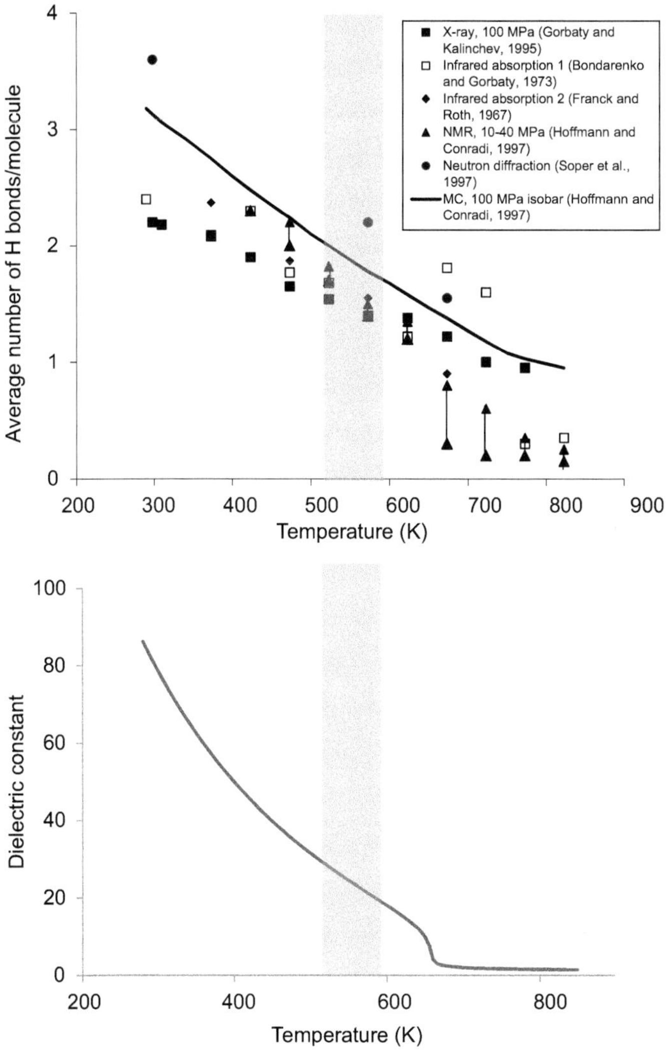

Figure 4.4 Temperature dependencies of the average number of hydrogen bonds per molecule (top) and dielectric constant (bottom) of water. The *grey bands* indicate the primary coolant operating temperature range of current water-cooled reactors.
Top figure adapted from Kalinichev, A.G., Bass, J.D., 1997. Hydrogen bonding in supercritical water. 2. Computer simulations. J. Phys. Chem. A 101, 9720–9727. Lower figure adapted from Guzonas, D., 2014. Extreme water chemistry-how Gen IV water chemistry research improves Gen III water-cooled reactors. In: 19th Pacific Basin Nuclear Conference (PBNC 2014), Vancouver, B.C., Canada, August 24–28, 2014.

one region to the next. For example, corrosion products originating in the feedtrain, whose production is affected by feedwater chemistry, can deposit in the core where their behaviour is affected by in-core chemistry, e.g., water radiolysis. The corrosion products can become radioactive by neutron activation and later be transported out of the core, depositing on downstream surfaces such as the turbine blades.

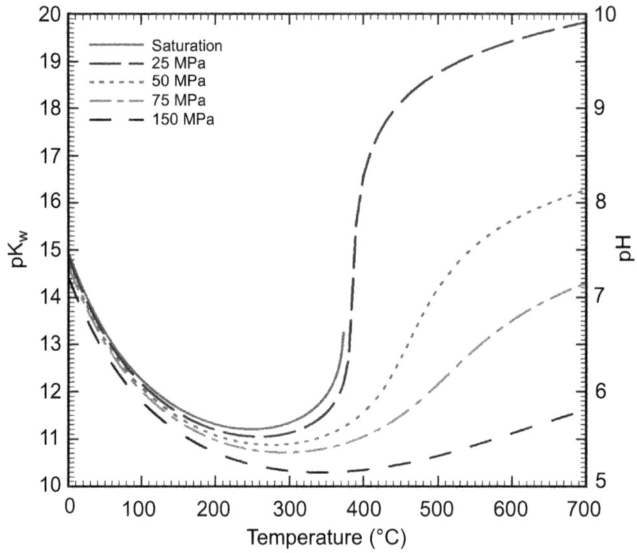

Figure 4.5 Ionic product and pH of neutral water as a function of temperature (Guzonas and Cook, 2012).

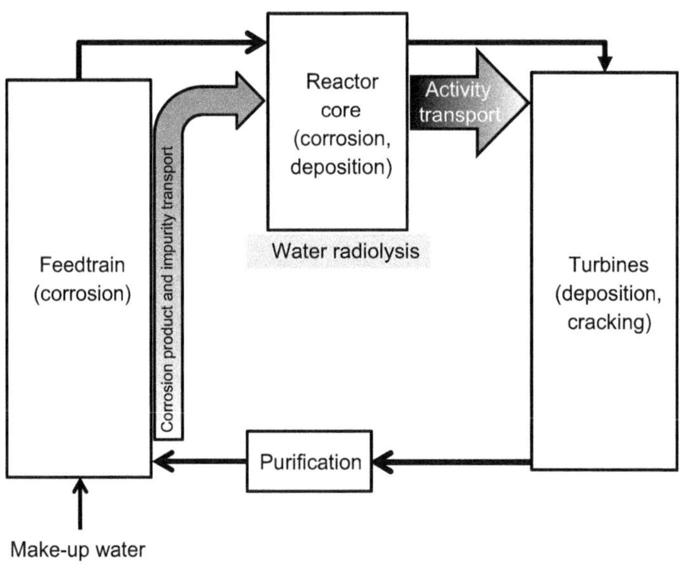

Figure 4.6 Areas of concern for water chemistry in a direct cycle supercritical water-cooled reactor.
Adapted from Guzonas, S., Cook, W., 2015. Water chemistry specifications for the Canadian supercritical water-cooled reactor concept. In: 7th International Symposium on Supercritical Water-cooled Reactors (ISSCWR-7), March 15–18, 2015, Helsinki, Finland, Paper ISSCWR7-2089.

4.2 Feedwater chemistry

The balance-of-plant (BOP) of any water-cooled thermal power plant is a massive and complex chemical/mechanical engineering system whose design and operation have been optimized over many decades of operating experience. A large fraction of maintenance and repair costs for nuclear power plants (NPPs) have been associated with the BOP. For example, the costs of pressurized water reactor (PWR) steam generator replacements due to materials degradation have been estimated at about $350M USD (Ru and Staehle, 2013). The wide range of temperature and pressure conditions in the BOP, coupled with cost considerations that preclude the use of more corrosion resistant but higher-cost alloys for all systems, can lead to environmentally assisted material degradation issues (e.g., flow accelerated corrosion (FAC) and fretting corrosion) and ultimately to pipe failures.

As a result, control of the feedwater chemistry[6] is a major requirement for plant operators to 'preserve the assets'. Similar to current SCWR concepts, supercritical fossil power plant (SCFPP) boilers usually operate in 'once-through' mode (no recirculation), which determines which water chemistry regimes are feasible. The use of organic amines such as morpholine or cyclohexylamine is not possible, and solid reagents such as sodium hydroxide or sodium phosphate are not added to the feedwater. Full flow purification (condensate polishing) is usually used (for a review of technology see Shields et al. (2006)).

It is important to recognize that feedwater chemistry is not always the same as boiler (core) chemistry. The feedtrain is at temperatures below T_c and SCFPP water chemistry is optimized to minimize feedtrain corrosion and the transport of impurities to the turbines. Concentrations of feedwater chemistry control additives may change through the system (e.g., concentration of dissolved oxygen (DO)), and this may in fact be desirable, e.g., to minimize the DO concentration in the boiler. Woolsey (1989) discusses issues and experiences with the use of oxygenated feedwater chemistries in SCFPPs using austenitic stainless steels in the superheater region (discussed further in Chapter 6).

As with plant design, feedwater chemistry control has been carefully optimized as operating experience has accumulated. Organizations such as EPRI, VGB and IAPWS provide guidance in the form of recommended operating specifications or best practices, which are continually updated as operating experience evolves. Reviews of water chemistry practices by Woolsey (1989), Dooley et al. (2004), Gabrielli and Schwevers (2008) and others also provide excellent overviews of chemistry control practices in SCFFPs. The reader is encouraged to consult such documents to gain a deeper understanding of the rationale behind current best practices. This section focuses only on those aspects of feedwater chemistry that affect SCWR in-core materials degradation processes so that the rationale behind proposed feedwater chemistry specifications for SCWR design concepts can be understood.

[6] Feedwater chemistry in SCFPPs and BWRs is similar to secondary side chemistry in PWRs and PHWRs. These latter two WCRs designs also strictly control the primary side chemistry, but as the primary circuit is a closed, pressurized loop the chemistry considerations are somewhat different.

4.2.1 Transport of corrosion products and other impurities

It is inevitable in any operating direct cycle power plant that corrosion of BOP piping will occur. During the corrosion process, some fraction of the metal oxides formed will dissolve into the flowing water. In addition, ingress of trace concentrations of impurities such as chloride, sulphate and silica can occur due to condenser tube leaks, leaching from filters and ion-exchange resins in the condensate polisher and make-up water additions. While their concentrations are usually low ($\mu g \cdot kg^{-1}$ or $ng \cdot kg^{-1}$), various concentration mechanisms exist that can result in locally high concentrations known to be problematic. In addition, even short periods of off-specification operation (tube leaks, unavailability of purification) can lead to parts-per-million concentrations of impurities known to be harmful to materials. Therefore, it is crucial to understand the behaviour and effects of these impurities in the core so that limits on their concentrations at the core inlet can be developed. Once these impurities enter the core, there is little that can be done to mitigate their effects.

The importance of corrosion product transport and deposition in an SCWR was recognized in 1960 (Machaterre and Petrick, 1960) who noted in their conclusions that 'The major gap in SCW technology pertaining to a reactor system is the lack of information on the magnitude of the problems of deposition of radioactivity in the external system and of the buildup of internal crud[7] under irradiation'. Corrosion product deposition in the primary circuit has been an issue for WCRs since their development (Cohen, 1985) and continues to cause operational issues, especially when operating conditions change (e.g., power uprating, changes in water chemistry). Recent issues include axial offset anomaly (AOA) in PWRs (Henshaw et al., 2006) and crud-induced localized corrosion in boiling water reactors (BWRs) (Cowan et al., 2011, 2012). Deposition of corrosion products transport in the core can result in (1) fouling of fuel cladding surfaces leading to reduced heat transfer and the increased possibility of fuel failures, (2) flow restrictions leading to insufficient fuel cooling and (3) increased production of radioactive species by neutron activation resulting in increased out-of-core radiation fields and worker dose. The formation of deposits in system piping has also been an issue for supercritical water oxidation process development (Hodes et al., 2004).

The effects of corrosion product layers on the thermal hydraulics in an SCWR core have not yet been assessed, although Pencer et al. (2011) modelled the effects of in-core deposits on core physics. Edwards et al. (2014) estimated the rise in cladding surface temperature due to the presence of a surface oxide as a function of the oxide thickness for oxide porosities ranging from 0 to 75%, assuming an oxide thermal conductivity of 3 $Wm^{-1} \cdot K^{-1}$ and a 40 $kW \cdot m^{-1}$ heat flux. While the presence of a compact (nonporous) 20 μm thick oxide layer resulted in a modest ~5K increase in the cladding temperature, a similar thickness of 75% porosity oxide was found to result in a roughly 60K increase in cladding temperature. The oxide layers formed in SCW are often porous and the porosity can vary significantly with test conditions

[7] Used to denote deposited and/or circulating particulate corrosion products. Said to be an acronym for Chalk River Unidentified Deposits.

(Chapter 5). Solberg and Brown (1968) compared heat transfer through ferritic and austenitic steels exposed to superheated steam (SHS) for long periods with heat transfer through new tubes of identical size and composition. The ferritic steels (2.25Cr1Mo, 3Cr1Mo, 5Cr0.5Mo, 5Cr0.5Ti, 5Cr1.25Si, 9Cr and an alloy identified as Ge CrMoVa) were exposed at 593 or 649 °C for periods up to 30 months at a pressure of 13.8 MPa. The austenitic steels (304, 310, 316, 321, 347, 16-25-6, 17-14CuMo and 15-15N) were exposed at 649, 732 or 816 °C for periods up to 18 months. The authors reported that for the ferritic alloys the oxide reduced heat transfer, attributed to the very porous nature of the outer layer of the oxide formed. The reduction in heat transfer was lower for the higher Cr steels. The austenitic alloys showed improved heat transfer with oxide formation, attributed to an increase in surface roughness that compensated for the increased thermal resistance of the oxide. These results highlight the need for studies of heat transfer in SCW using oxide-covered samples, as well as the need to minimize formation of porous oxide layers in an operating SCWR core.

Very little is known about how corrosion products will behave in an SCWR, but examination of data from SCFPPs suggests a significant risk of deposition of corrosion products released from feedtrain surfaces onto fuel cladding surfaces. SCFPPs typically use lower cost alloys (carbon steel, low alloy steels) in their feedtrains, but modelling suggests that deposition can occur in an SCWR core even when feedtrain materials with low general corrosion rates are used. Plant data for iron-oxide deposits inside boiler tubes operated using SCW show that the heaviest deposits occur around T_c with a significant effect of surface heat flux on the location of the deposition peak. Dickinson et al. (1958) studied deposition and tube overheating in SCW in the fossil-fired boiler of a pilot plant. The boiler tube was carbon-moly (A-209) steel (1.43 cm OD, 0.80 cm ID), 150 m long with a 33 m long type A-321 stainless steel superheater tube (all other dimensions identical to the boiler tube). The water contained 5 μg $Cu \cdot kg^{-1}$, 10 μg $Fe \cdot kg^{-1}$ and 20 μg $Si \cdot kg^{-1}$ and was slightly alkaline with ammonia at pH 8.5–9.5. The coolant temperature appeared to rise roughly linearly with distance over the temperature range 379–482 °C. Deposition occurred over a temperature range of 304–416 °C with maximum deposition (4.84 $mg \cdot cm^{-2}$) at 416 °C and 34 MPa. When the boiler pressure was 23.9 MPa, deposition at T_c was inferred from a change in the wall temperature. Based on data from Russian supercritical boilers described by Zenkevitch and Sekretar (1976), Woolsey (1989) concluded that the oxide deposition profile was largely determined by the pseudophase change at T_c rather than by the decrease in oxide solubility.

Fig. 4.7 shows data on internal boiler tube deposits as a function of temperature in the lower radiant section of a boiler operated under SCW conditions after 8400 h exposure at pH_{25} 8–8.5 adjusted with hydrazine (Chudnovskaya et al., 1988). The outer layer of the three-layer deposit was brown to black depending on whether the surface faced fire side or the rear water wall. The inner deposit layer[8] was black and fine grained. Deposits were strongly adherent to piping surfaces when only hydrazine was added to the coolant, but not when hydrazine and ammonia were added. The outer

[8] Probably under the loose outer deposit but distinguished from the corrosion film.

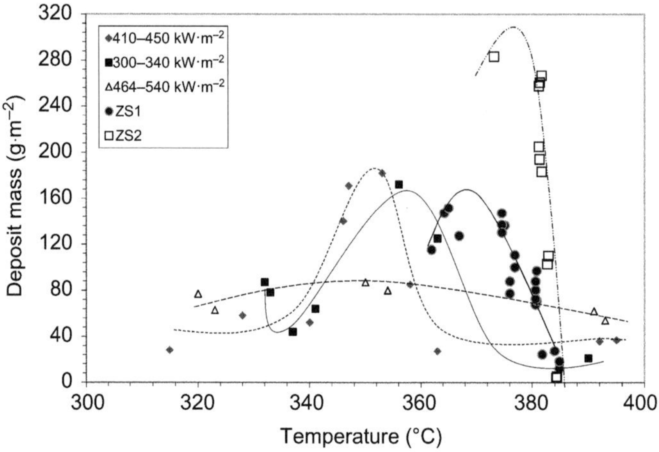

Figure 4.7 Boiler tube deposits in supercritical fossil power plants. The *diamonds*, *filled squares* and *triangles* are after 8400 h exposure at $pH_{25\ °C}$ 8–8.5 with hydrazine (adapted from Chudnovskaya et al. (1988)). The data labelled ZS1 and ZS2 are for Russian plants (Zenkevitch and Sekretar (1976), adapted from Woolsey (1989)) for two different boilers; Woolsey noted that the pressure was believed to be 25.5 MPa. The *lines* are guides to the eye only.

deposit weight peaked at 350 °C at 180 $g \cdot m^{-2}$ and the inner deposit layer peaked at 320–340 °C at 80–150 $g \cdot m^{-2}$. The deposit weight appeared to be a complex function of the surface heat flux, water temperature and tube material. Data from Russian supercritical boilers described by Zenkevitch and Sekretar (1976) (taken from Woolsey (1989)) are also included in Fig. 4.7 to further illustrate the range of deposit weights and peak temperatures possible. Deposit data reported by Vasilenko (1976) from a Russian SCFPP operated at pH_{25} 8.8–9.8 with ammonia plus hydrazine for 10,000 h exposure show peaks in the deposit weight between 370 and 390 °C. In spite of the range of temperatures over which the maximum in deposition is observed, the maximum deposition is always within ~20 °C of T_c.

Corrosion product deposition using ammonia or ammonia–hydrazine water chemistries is typically higher than when oxygenated water chemistries are used. Kontorovich and Rogal'skaya (1987) noted that in SCFPPs using hydrazine–ammonia chemistry in the feedtrain, deposit formation lead to the need for chemical cleaning of the boiler heating surfaces at intervals of between 4000 and 15,000 h of operation depending on the heat load. While chemical cleaning[9] of SCWR in-core surfaces would be possible, the additional difficulties arising from the activation of the deposits significantly increases the technical and financial challenges. Ultrasonic fuel cleaning to remove deposits from fuel assemblies could be performed during refuelling outages.

[9] Chemical cleaning of nuclear reactor primary circuits is called chemical decontamination by the nuclear industry (IAEA, 1994), and involves the injection of chemicals able to dissolve metal oxides without damaging materials of construction.

There have been few laboratory studies of corrosion product deposition under SCWR core conditions and none under irradiation. Karakama et al. (2012) reported deposit thickness data for iron oxide on 316L SS at 23 MPa and temperatures around T_c at pH_{25} values of about 3 and 9. The Fe^{2+} concentration in the test solution was 5 $mmol \cdot L^{-1}$. Deposit thickness and oxide adhesion were measured after the test by destructive analysis. On-line temperature measurements during the tests enabled the evolution of deposition to be monitored during the test. The temperature was constant along the test section length. An increase in wall temperature was observed above T_c, at both pH values studied, being highest at pH 9. It was noted that at pH 9, the on-line temperature data suggested oxide exfoliation occurred readily. The authors also reported that the deposits formed above T_c were more difficult to remove than those formed below T_c. Attempts to model deposition were unsuccessful, attributed to neglect of surface attachment and removal rates in the model used.

Using SCFPP data, available (but limited) data on magnetite solubility in SCW and extrapolation of data measured in subcritical water, Burrill (2000) predicted that peak deposit masses in an SCWR core would be about 10 times greater than the deposit loadings seen in Russian SCFPPs (Fig. 4.7), and 10^4 times higher than values typically observed on fuel removed from a PHWR. Saito et al. (2006) constructed potential-pH (Pourbaix) diagrams for relevant Fe, Cr and Ni oxides at 400 and 450 °C and at 50 and 100 MPa and also concluded that corrosion products could potentially deposit on the fuel cladding in an SCWR. Woolsey (1989) noted that if orifices are used to control boiler tube flow, corrosion product deposition can lead to blockage or flow restrictions, resulting in insufficient coolant flow; this could be an issue for SCWR fuel assemblies with complex geometries or tight lattice spacings. He noted that this has largely been an issue in SCFPPs using deoxygenated feedwater chemistry regimes.

Much of the oxide solubility data in SCW has been generated by geochemists interested in hydrothermal geological processes and mineral formation. As a result the data are typically measured at much higher pressure (e.g., 100 MPa) than those to be used in an SCWR. Under SCW conditions, solubilities show a strong pressure (density) dependence, and therefore, from an SCWR development perspective, the geochemical data are only useful for qualitative purpose. Walther (1986) has summarized some of the available geochemical data, and Qiu and Guzonas (2010) have given a brief summary of data relevant to SCWR development. Of greatest interest for SCWR development are the solubilities of Fe, Ni and Cr oxides and hydroxides as a function of temperature above T_c. In spite of their importance, there are limited experimental data (Fig. 4.8) for these values due to the extremely low concentrations involved and the experimental challenges in measuring solubilities at high temperatures and pressures (Wesolowski et al., 2004). Adschiri et al. (2001) gave estimates of these solubilities as a function of pH at 400 °C and 30 and 60 MPa. While metal oxide solubilities generally decrease sharply in the vicinity of the critical point, some do not; for example, the solubility of WO_3 increases with increasing T.

There are few reported measurements of the solubility of magnetite in SCW. An early study by Holster and Schneer (1961) reported that the solubility of magnetite in pure water was below their method detection limit (MDL) of 20 $\mu g \cdot kg^{-1}$ at temperatures from 400 to 500 °C and a pressure of 48.5 MPa. Martynova (1973) reported the

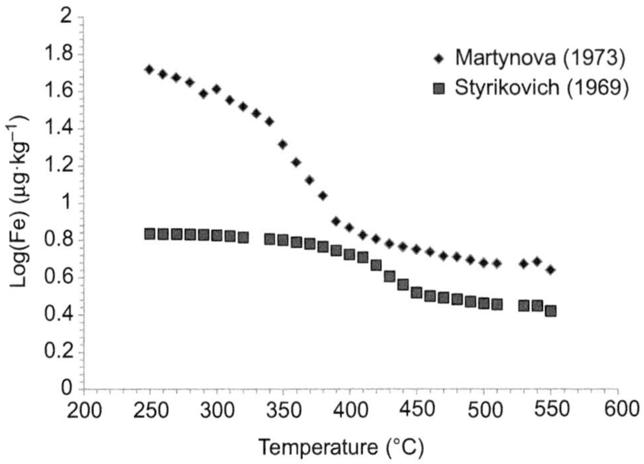

Figure 4.8 Measured solubility of magnetite as a function of temperature at T > T_c.

solubility of magnetite as a function of temperature from 250 to 550 °C in pure water at a pressure of 25 MPa (Fig. 4.8), although the paper did not report details of the measurement technique. The solubility decreased rapidly with increasing temperature near the critical point; at T > 400 °C the rate of decrease slowed. Magnetite solubility was ~5.4 $\mu g \cdot kg^{-1}$ at 550 °C. The solubility of magnetite in steam as a function of temperature in the superheat region was reported by Styrikovich (1969) (Fig. 4.8), the value at 550 °C (2.1 $\mu g \cdot kg^{-1}$) being in good agreement with the value reported by Martynova (1973) at the same temperature.

Until appropriate experiments can be performed, modelling of corrosion product deposition processes in an SCWR core is the only route to predicting possible challenges. Most of the impurities of concern in an SCWR are electrolytes (e.g., NaCl) although nonelectrolytes such as gases are also of some interest. The modelling of the behaviour of nonelectrolytes in the vicinity of the critical point is much better developed than that of electrolytes. This is a consequence of how the standard theories, developed to treat aqueous electrolyte solutions at temperatures below T_c, behave in the vicinity of the critical point. Electrolyte solutions contain charge carriers, and as the temperature increases, water becomes highly compressible and the decrease in dielectric constant leads to ion association. Majer et al. (2004) present a useful summary of various models for standard thermodynamic properties as a function of temperature and pressure. Anisimov et al. (2004) discuss the near-critical behaviour of both nonelectrolyte and electrolyte systems, including the difficulties in modelling charged systems in the vicinity of the critical point. Tremaine et al. (2004) and Tremaine and Arcis (2014) summarize approaches taken to model thermodynamic properties in high-temperature water.

Important considerations in modelling of aqueous solutions are choice of standard states and concentration variables. While several semiempirical models now exist, to date modelling of corrosion product transport in an SCWR has only been carried out using the Helgeson–Kirkham–Flowers (HKF) equation and its refinements. The use

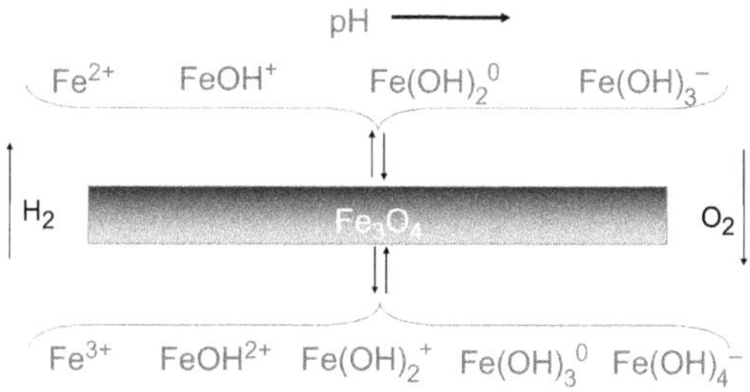

Figure 4.9 Illustration of the various reactions associated with magnetite dissolution in water and the dependences on pH and redox conditions.
Adapted from Brosseau, F.J.R., 2010. Complexation and Hydrolysis of Aqueous Iron and Nickel With Chloride and Ammonia Under Hydrothermal Conditions (MSc. thesis). Faculty of Graduate Studies, University of Guelph.

of this model for SCWR corrosion product deposition is described in Guzonas et al. (2012), Brosseau (2010) and Olive (2012). A major limitation of all semiempirical approaches is the limited amount of reliable high-temperature experimental data available for use in developing the correlations required by the models and to validate model predictions. Of significant concern is the possible existence of high-temperature species that are unstable, and hence undetected, at lower temperatures. Such species would not be predicted by correlations developed solely from low-temperature data. Guzonas et al. (2012) highlighted the importance of ensuring that any database used for model development is internally self-consistent.

Modelling the solubility of magnetite is particularly difficult because magnetite is a mixed-valence oxide containing both Fe^{2+} and Fe^{3+} ions. Its dissolution therefore involves a number of redox and hydrolysis reactions (Fig. 4.9) whose importance depends on the pH and concentrations of oxidant and reductants. While these parameters can be carefully controlled in a laboratory setting, their values are often poorly known in an operating power plant. Thus while there are several published data sets for magnetite solubility as a function of pH and temperature in subcritical water (Sweeton and Baes, 1970; Tremaine and LeBlanc, 1980; Bohnsack, 1987; Ziemniak et al., 1995; Wesolowski et al., 2000) that are in reasonable agreement considering the experimental uncertainties, there is still controversy regarding the relative importance of Fe(II) and Fe(III) hydrolysis species at high temperatures and the appropriate model for use in alkaline solutions, required for prediction of corrosion product deposition in PWRs and PHWRs (e.g., Guzonas and Qiu, 2004). As the redox conditions expected in an SCWR core are currently essentially completely unknown due to the limited state of knowledge of water radiolysis in SCW (Section 4.4), predictions of corrosion product deposition on an SCWR core are necessarily only qualitative at this time. Wesolowski et al. (2004), Brosseau (2010), Brosseau et al. (2010) and Olive (2012) provide recent discussions of the issues around the measurement and modelling of magnetite solubility at high temperatures.

Fig. 4.10 shows the calculated solubilities of a number of oxides of metals used in alloys proposed for use in SCWR concepts obtained using the revised HKF model

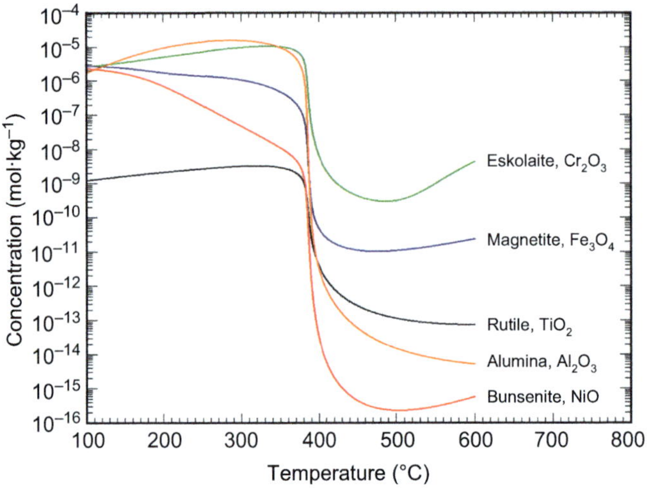

Figure 4.10 Calculated metal oxide solubilities in neutral water obtained using the revised Helgeson–Kirkham–Flowers model (Guzonas and Cook, 2012; Olive, 2012). For magnetite and eskolaite, the hydrogen activity was set at 1 bar.

(Guzonas and Cook, 2012; Olive, 2012). The solubilities exhibit a rapid drop in the vicinity of the critical point; some species (NiO, Cr_2O_3) show a small increase above ~500 °C. The data at 400 °C are in reasonable agreement with the estimates of Adschiri et al. (2001), but the calculated solubilities of magnetite at $T > T_c$ are several orders of magnitude lower than the measured values shown in Fig. 4.8. This may be a result of the experimental techniques used, and/or the choice of redox conditions or Fe species used in the model. Burrill (2000) concluded that the solubility values reported by Martynova (1973) were ~100 times too high based on an assessment of the reported oxide deposition rates in SCFPP boilers. In the absence of experimental data on magnetite solubility in SCW obtained under carefully controlled conditions, it is impossible to further refine the models. Finally, it is important to note that, while the corrosion product concentrations predicted by Fig. 4.10 may be low, the large coolant volumes circulated in a WCR (and SCWR) can still result in the deposition of significant masses of material if the deposition is localized.

Using the most recent data on magnetite solubility, Cook and Olive (2012a, 2013) and Olive (2012) modelled iron and nickel deposition in the core of the Canadian SCWR using the revised HKF model for two scenarios: (1) coolant saturated in the species of interest at the core inlet and (2) coolant unsaturated in the species of interest at the core inlet (Fig. 4.11). The Canadian SCWR concept has a central flow tube down which the coolant passes with minimal heating before reversing direction at the channel bottom and flow up through the fuel where it is heated (Yetisir et al., 2013). When the coolant was saturated, deposition started at the core inlet, reached a maximum about 1 m into the core and continued until the core outlet. For unsaturated coolant (1 $\mu g \cdot kg^{-1}$ dissolved Fe), deposition started roughly 1 m into the core and continued until the core outlet. The effect of the deposition rate constant (which has never been

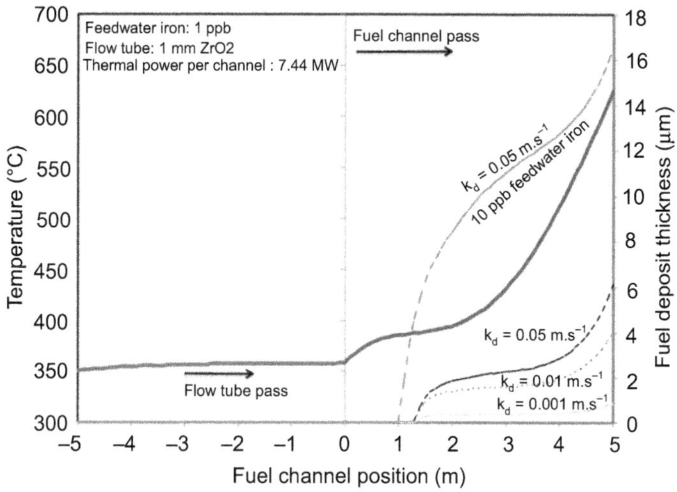

Figure 4.11 Oxide deposit profiles for an insulated flow tube in the Canadian supercritical water-cooled reactor (SCWR) concept reentrant fuel channel design over a 1-year period of operation (Cook and Olive, 2013). The Canadian SCWR concept has a central flow tube down which the coolant passes with minimal heating (negative values of the fuel channel position) before reversing direction at the channel bottom and flowing upwards through the fuel (positive fuel channel positions) where it is heated.

measured in SCW) on iron loading was also studied by parametric analysis. Very little difference was noted when the deposition rate constant was varied from 0.05 to 0.1 $m \cdot s^{-1}$, suggesting that at values of 0.05 $m \cdot s^{-1}$ and higher, the rate-limiting step affecting deposition is actually the drop in solubility as the fluid is heated when passing through the core. Comparing deposition rate constants of 0.01 and 0.001–0.05 $m \cdot s^{-1}$, Olive (2012) noted that the maximum iron loading decreased and the loading further up the channel increased; at 0.001 $m \cdot s^{-1}$, the loading was nearly uniform with distance past the onset of deposition. The study found that, for unsaturated coolant at 1 $\mu g \cdot kg^{-1}$ dissolved iron at the core inlet, peak magnetite thickness could be anywhere between 10 and 65 μm after 1 year depending on the deposition rate constant and corrosion rate.

A comparison of Figs 4.11 and 4.7 show that the model fails to predict the observed peaks in the SCFPP data. However, the model of Cook and Olive did not include a heat flux dependence, which, as evident from Fig. 4.7, has a significant effect on both the magnitude of deposition and the peak deposition temperature in operating SCFPPs, and this may account for the differences. There is also considerable uncertainty in the value of the deposition rate constant. It is also possible that the use of the HKF model does not properly describe the transition through the near-critical region, for the reasons touched on earlier. A further shortcoming of current modelling efforts is their neglect of possible radiolysis effects on the redox state of iron; Mayanovic et al. (2012) found that, under irradiation, iron species could be oxidized, reduced or precipitated depending on the pressure, temperature and irradiation conditions.

The modelling work suggests that corrosion product deposition could be maintained at an acceptable level if the iron concentration at the core inlet can be minimized. An early prediction by Guzonas et al. (2007) was that use of feedtrain materials with corrosion rates of $\sim 0.1\ mg \cdot dm^2 \cdot d^{-1}$ (mdd) would minimize in-core deposition. A simple extrapolation of the predictions of Cook and Olive (2012a) suggests that for an optimized feedtrain with $0.1\ \mu g \cdot kg^{-1}$ dissolved Fe at the core inlet, the deposit thickness after one fuel cycle (425 d) would be about 15 μm, not significantly higher than the average oxide thickness of 5.4 μm reported for PWR spent fuel by Hazelton (1987). This suggests that in-core deposition in an SCWR could be managed if the feedtrain materials and chemistry are optimized such that the dissolved iron concentration at the core inlet is $< 0.1\ \mu g \cdot kg^{-1}$. Currently, plants that cascade feedwater heater drains back to the turbine condenser have the lowest final feedwater concentrations ($0.1-0.5\ \mu g \cdot kg^{-1}$; Stellwag et al., 2011), and this should be the design option used by SCWRs. To minimize deposit build-up during subsequent fuel cycles, fuel bundles being reloaded could be ultrasonically cleaned, as is the practice with some PWRs to address the issue of AOA (EPRI, 2004). This would also reduce activity transport due to activation and release of the deposits. More precise predictions require additional experimental data on deposition under SCWR conditions, especially on deposition rate constants and the effects of water radiolysis and heat flux. The latter two aspects can be studied in in-reactor loop tests.

The dissolved iron and nickel concentrations in the coolant of current generation of WCRs are primarily a function of the corrosion rates of the materials of construction and the solubilities of the oxides formed on the alloy surfaces. It has long been recognized that magnetite and mixed ferrite solubilities decrease with increasing temperature under acid conditions. Raising the pH to slightly alkaline values causes the solubilities of these oxides to increase with temperature due to the presence of hydrolysed metal species. On low alloy steels, redox potential also plays an important role, with reducing conditions favouring magnetite and mixed ferrites, while oxidizing conditions favour hematite, which has a lower solubility. A number of chemistry control strategies have been used in the SCFPP industry to control the corrosion rate and the solubilities of corrosion products, including the use of pH control additives such as NH_3, and through redox control additives, such as N_2H_4 or O_2 (Table 4.1). However, neutral species such as $Fe(OH)_2^0(aq)$ and $Fe(OH)_3^0(aq)$ will dominate in the supercritical region, and pH control eventually becomes ineffective at higher temperatures.

LiOH (or KOH) is commonly used in PWRs and PHWRs to control the pH of the primary circuit. It is not used in BWRs or fossil power plants because it is nonvolatile and therefore does not partition into the steam phase to any useful (from a pH control perspective) extent. Some pH control agent must be present in the steam to mitigate corrosion of BOP components downstream of the core or boiler. However, at supercritical pressures, the solubility of LiOH is higher than at BWR pressures, and Guzonas et al. (2012), Brosseau et al. (2010), Plugatyr et al. (2011a), Carvajal-Ortiz et al. (2012a,b) and Cook and Olive (2012a) have investigated the use of LiOH for pH control in an SCWR. All these authors agree that at LiOH concentrations above $\sim 1\ mmol \cdot kg^{-1}$, LiOH would ensure a positive solubility gradient of magnetite in the SCWR core up to a temperature of about 400 °C. Above 400 °C, LiOH is not

Table 4.1 **Summary of all-volatile and oxygen water treatments used in supercritical water fossil-fired power plants (Guzonas et al., 2012).**

Water Chemistry	pH_{25}	Comments	Reference
Ammonia + Hydrazine	–	0.7–1 mg amine kg^{-1}	Larsen et al. (1996)
	9.1–9.4	$NH_3 + N_2H_4$	Chudnovskaya et al. (1988)
	8.5–9.5	NH_3	Dickinson et al. (1958)
	9.1–9.6	–	Margulova (1978)
	9.1	$NH_3 < 0.8$ mg kg^{-1} to avoid copper transport	Vasilenko (1976)
	>9	NH_3 or cyclohexylamine + hydrazine	Cialone et al. (1986)
Hydrazine only	8–8.5	N_2H_4 at 60–100 μg kg^{-1}	Chudnovskaya et al. (1988)
	7.7	N_2H_4 at 60–100 μg kg^{-1}	Deeva et al. (1986)
Chelate + NH_3 + N_2H_4	–	80 μg kg^{-1} chelate, 0.8 mg kg^{-1} ammonia, 0.2 mg kg^{-1} hydrazine	Oliker and Armor (1992)
	–	Increases thermal conductivity of iron oxide deposits	Vasilenko (1978)
pH 7 with oxygen	–	50–200 μg O_2 kg^{-1}, conductivity <0.1 μS cm^{-1}	Larsen et al. (1996)
	6.5–7.3	–	Margulova (1978)
	–	Reaction of $Fe(OH)_2 + O_2$ slow compared to reaction with H_2O_2	Kontorovich et al. (1982)
	–	<0.15 μS cm^{-1}, 0.2–0.4 mg O_2 kg^{-1} at boiler inlet	
Combined mode	8–8.5	$NH_3 + O_2$–NH_3 provides slight buffering of pH so that impurities do not cause wide swings in the pH, especially to the acid side where corrosion rates can increase	Kontorovich and Rogal'skava (1987)

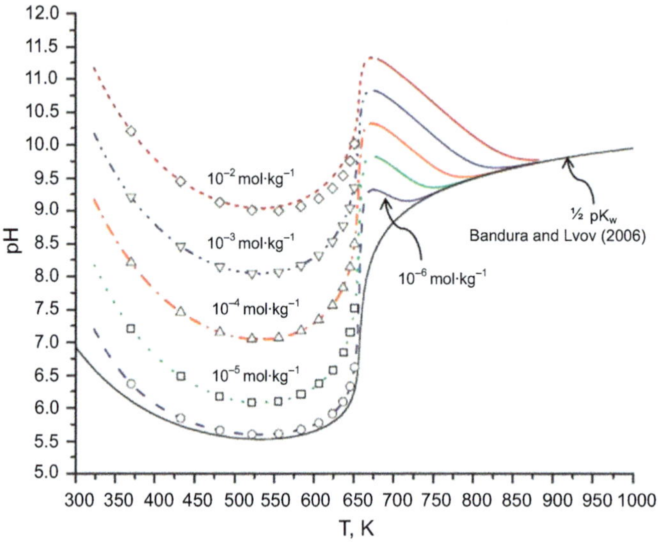

Figure 4.12 The pH values for dilute $LiOH_{(aq)}$ solutions along the 25 MPa isobar. *Lines* represent the pH values calculated using ion association constants of LiOH obtained by molecular dynamics simulations (Plugatyr et al., 2011a). *Open circles* are the pH values calculated using experimental K_a(LiOH) data (Ho et al., 2000).

dissociated to a significant extent and pH control is gradually lost (Fig. 4.12) as LiOH ion pairs become the dominant solution species (Carvajal-Ortiz et al., 2012b). Carvajal-Ortiz et al. (2012b) suggest that pH control using LiOH would be impracticable at SCW densities less than $86\ kg \cdot m^{-3}$ ($T > 514\ °C$ at 25 MPa). This was elegantly demonstrated by Svishchev et al. (2013a) who measured metal release (an indirect measure of corrosion rate) from 316 stainless steel over the temperature range 350–650 °C under neutral ($pH_{25} = 7 \pm 0.3$) and alkaline ($pH_{25} = 10 \pm 0.3$) water chemistries using LiOH as the pH control agent. As seen in Fig. 4.13 while below ~500–550 °C the two release rates are different, above this temperature range the two release rates are the same. This observation is consistent with the decrease in LiOH dissociation observed in Fig. 4.12. As this temperature is reached at the fuel cladding surface relatively close to the core inlet, LiOH would not be effective for pH control through most of the core. In addition, the reduction in LiOH solubility as the coolant density decreases along the core would likely result in LiOH precipitation on the fuel cladding and in the main steam piping and high-pressure turbine making its use impractical.

Ammonia was used for pH control in Unit 2 at the Beloyarsk NPP. Because ammonia is a much weaker base than LiOH at room temperature, higher concentrations ($>30\ mmol \cdot kg^{-1}$) are required to achieve the negative solubility gradient for magnetite solubility required to minimize in-core deposition (Cook and Olive, 2012a; Brosseau, 2010). As thermodynamic data for these species are scarce or nonexistent at high temperatures, Brosseau et al. (2010) used room-temperature data for the $Fe(NH_3)_4^{2+}$ complex to show that this complex is the dominant species in neutral to

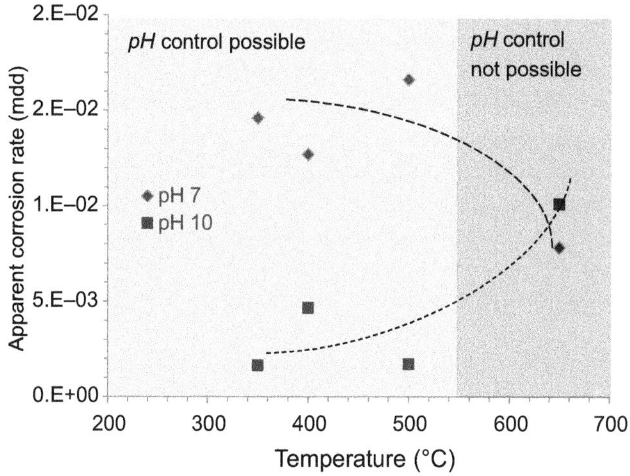

Figure 4.13 Metal release rate from 316 stainless steel as a function of temperature in supercritical water at neutral and alkaline pH.
Adapted from Svishchev, I.M., Carvajal-Ortiz, R.A., Choudhry, K.I., Guzonas, D.A., 2013a. Corrosion behavior of stainless steel 316 in sub- and supercritical aqueous environments: effect of LiOH additions. Corros. Sci. 72, 20–25.

alkaline conditions at high temperature, and results in a significant increase in the solubility of magnetite under SCWR core conditions. However, the degree of dissociation of ammonia drops with increasing temperature, so that a positive solubility gradient cannot be maintained above ~350 °C. The use of high concentrations of ammonia in the SCWR coolant raises concerns about high ^{16}N radiation fields around downstream piping, but Yurmanov et al. (2010) noted that at Beloyarsk Unit 2, no issues with the transport of activated nitrogen were reported. An additional concern with the use of ammonia is the radiolytic production of nitrates and nitrites; at Beloyarsk Unit 2 the concentration of nitrate ion at a reactor power of 130 MW_{th} was 1.24 $mg \cdot kg^{-1}$ when ammonia was used to adjust the feedtrain pH. Tests using static Hastelloy C and Alloy 625 autoclaves have shown that metal release, particularly nickel, can be very high in SCW containing ammonia (Guzonas et al., 2010). The concern is confirmed by operating experience from the Beloyarsk NPP that showed that corrosion product deposition was higher in ammonia than with an oxygenated feedwater treatment (Yurmanov et al., 2009).

Hydrazine is used as an oxygen scavenger in the all-volatile treatment but has also been used for in situ hydrogen generation to suppress water radiolysis in VVERs (Vodo-Vodyanoi Energetichesky Reaktor (Water-Water Power Reactor); Russian PWR reactor design). Plugatyr et al. (2011b) have reported the thermal decomposition of hydrazine as a function of temperature from room temperature to SCWR temperatures. In an SCWR, most of the hydrazine added in the feedtrain would be consumed by reaction with oxygen or thermally decomposed before reaching the core. This option is further disadvantaged by environmental restrictions placed on hydrazine use in many jurisdictions.

The most common feedwater chemistry used in SCFPPs is the oxygenated treatment (OT) (neutral pH plus added O_2) or variations denoted combined treatment or combined mode (injection of O_2 or H_2O_2 plus addition of ammonia to give a pH_{25} between 8.0 and 8.5) (Table 4.1), which was developed in the 1970s in the former West Germany and subsequently adopted in Russia. In the 1990s, it started to be adopted in the United States as well. A detailed exposition of the OT and guidelines for its use can be found in Bursik et al. (1994). The implementation details vary from country to country, but in general, oxygen is added either as air, gaseous oxygen or hydrogen peroxide, to achieve an oxygen concentration of between 50 and 300 $\mu g \cdot kg^{-1}$. The presence of oxygen reduces the corrosion rate of the carbon steel feedtrain piping, reducing corrosion product transport to the boilers. This reduction in corrosion rate has been attributed to a change in the nature of the oxide formed (from magnetite to either hematite or an iron oxyhydroxide). The OT cannot be used in plants containing copper alloys in the feedtrain because the oxide formed on copper alloys dissolves in high-temperature oxygenated water. A key prerequisite for the use of the OT is the ability to produce high purity feedwater to prevent SCC (Chapter 6), which requires the use of full-flow condensate polishing.

4.2.2 Transport of other impurities to the core

Direct cycle NPPs such as the SCWR risk exposing reactor core components to impurities that could enhance corrosion rates or lead to SCC (Woolsey, 1989). Ru and Staehle (2013) highlighted the importance of impurities such as chloride on the degradation of materials for an SCWR, noting that materials testing during the US nuclear reheat development program in the 1960s found that chloride deposition during testing eventually led to failure of the stainless steels tested by SCC, even with the best available efforts to remove chloride.

An early study of chloride deposition in the BONUS reactor showed that wet steam containing chlorides and oxygen caused chloride-induced SCC failure of 304 and 347 stainless steels. No detectable deposition of chlorides was found on the test cartridge heaters except in an area where condensation had occurred. The absence of chloride deposition was attributed to the addition of 1−3 °C of superheat to the steam, which essentially removed all chloride-bearing moisture before the steam passed over the test cartridge heaters (Bevilacqua and Brown, 1963a). A series of 13 follow-up experiments by Bevilacqua and Brown (1963b) in which ^{36}Cl was added to the steam (which contained ~50 $\mu g \cdot kg^{-1}$ chloride) entering the test loop was used to improve the detection limit for chloride (1×10^{-6} g of ^{36}Cl) and quantify the mass of chloride deposited under various conditions onto superheater cladding surfaces. Chloride deposition from drying of moist steam resulted in heavy, adherent localized deposits. Dry steam gave the least deposition. These heavy deposits together with the presence of oxygen and water were conducive to severe chloride-induced SCC of austenitic steels. A strong effect of surface preparation was noted with mechanically produced surface defects collecting far less ^{36}Cl activity than defects produced by chemical attack. It should be noted, however, that Unit 2 at the Beloyarsk NPP operated for many years

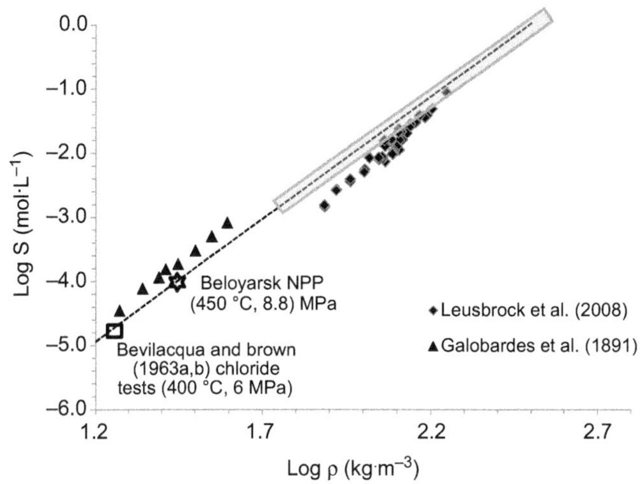

Figure 4.14 Logarithm of the measured solubility of NaCl as a function of the logarithm of water density at 400 °C. The shaded area indicates the range of supercritical water densities expected in a 25 MPa supercritical water-cooled reactor core, up to a peak cladding temperature of 850 °C. The open star indicates the operating water density of the Beloyarsk NPP, and the open square indicates the lowest water density studied in the corrosion tests of Bevilacqua and Brown (1963a,b).
Adapted from Guzonas, S., Cook, W., 2015. Water chemistry specifications for the Canadian supercritical water-cooled reactor concept. 7th International Symposium on Supercritical Water-cooled Reactors (ISSCWR-7), March 15–18, 2015, Helsinki, Finland, Paper ISSCWR7-2089.

with a typical chloride concentration of 25 $\mu g \cdot kg^{-1}$ with no reported negative effects (Yurmanov et al., 2010).

No testing in which representative concentrations of chloride (or other impurities known to promote cracking) were added to the test solution have been carried out in recent materials tests for the various SCWR concepts (but see discussion in Chapter 6). Current BWRs operate with feedwater chloride concentrations as low as 0.25 $\mu g \cdot kg^{-1}$ (typically the method detection limit) and sulphate concentrations as low as 2 $\mu g \cdot kg^{-1}$ (Stellwag et al., 2011). Tests should be conducted with impurities in this range as well as at concentrations representative of upset conditions.

The solubility of NaCl in high-temperature water as a function of pressure has been reported by many groups (Galobardes et al., 1981; Leusbrock et al., 2008). A reasonable approximation for the dependence of solubility (S) on water density (ρ) is given by $\log(S) \approx \log(\rho)$ (Fig. 4.14), while the temperature dependence is relatively weak. As highlighted by Palmer et al. (2004), care must be taken in the assessment of the experimental data used in such plots. These authors give a concise but thorough assessment of the available data and models for NaCl solubility in steam. Chialvo et al. (2010) discuss the challenges of modelling NaCl association and transport at ultrasupercritical temperatures and pressures, particularly at lower SCW densities. Bearing these caveats in mind, Fig. 4.14 shows that the solubility is an order of magnitude higher at 25 MPa than at 8.8 MPa (Beloyarsk NPP) and more than an order of magnitude higher than at 6 MPa, the steam pressure used in the chloride deposition tests of Bevilacqua and Brown (1963a,b). Since there is no phase change in the SCWR core, the solubility of NaCl should be sufficiently high under in-core conditions that operation with a feedwater chloride concentration of 0.25 $\mu g \cdot kg^{-1}$ should result in low chloride deposition, even at the peak cladding temperature. The major risk could be hideout within the oxide layers on the fuel cladding surfaces. However, while the increased solubilities of impurities should minimize the risk of their in-core deposition, it can also result in increased carryover to the turbines, a phenomenon well known and studied by the SCFPP industry (Zhou and Turnbull, 2002).

Essentially nothing is known about hideout[10] phenomena in SCW. The large changes in solubility that occur in the vicinity of T_c suggest that many species could potentially precipitate within the deposits and redissolve during shutdown, resulting in large spikes in impurity concentrations in the coolant. The effects of the precipitated impurities on localized corrosion are also an issue, as noted above for chloride.

Bevilacqua and Brown (1963a) noted that loop transients, even those of brief duration that resulted in only slight cooling of loop piping, resulted in large increases in ^{36}Cl concentration. In some instances the ^{36}Cl concentration in the water increased to a value 100 times higher than the value prior to shutdown. Such observations highlight the importance (well known in existing WCR experience) of transient periods of reactor operation. Degradation issues in current WCRs are often the result of such transient (upset) conditions, and these will need to be studied in future SCWR R&D programs.

4.3 Activity transport

An aspect of corrosion product transport unique to an SCWR when compared with an SCFPP is the transport of radioactive material out of the reactor core, a process denoted activity transport. This phenomenon has been known and studied since the advent of nuclear reactor technology, as it can lead to the production of hazardous radiation fields external to the reactor core, and the uptake of radiation dose by plant staff. In addition, the leakage of radioactive coolant from the system piping can lead to a significant contamination hazard that has the potential to be spread around the plant. Cohen (1985) provides a good introduction to this topic.

Radioactive elements are unstable and decay to a more stable state, accompanied by the emission of either an α-particle (He nucleus) or a β-particle (election), often in conjunction with emission of a high-energy photon (γ-ray). In some instances the photon is of lower energy (X-ray). These various forms of radiation interact strongly

[10] Hideout and hideout return are well-known phenomena in WCRs. Hideout is the result of the increase in concentration of water contaminants and chemistry control additives by boiling and precipitation in local hot spots under deposits and in crevices. In these regions, very high thermal concentration gradients are possible when the system is at full power operation. When the power is reduced and the temperature drops the contaminants can dissolve and 'return' to the coolant.

with matter, including human tissue, and therefore pose a significant hazard to human health. The ranges of these three forms of radiation differ substantially. The range of a 4–5 MeV α-particle in air is ~3 cm, while the range of a 3.0 MeV β-particle in air is ~1 m. The range of γ-rays in air is, for practical purposes, essentially infinite. These differences in range strongly affect the type of hazard to human health the various types of radiation pose.

The rate of radioactive decay is given by:

$$\frac{dN}{dt} = -\lambda N \tag{4.1}$$

where N is the number of atoms of the element present at time t and λ is the decay constant. A commonly used characteristic time for the radioactive decay of an element is the half-life, $t^1/_2$, which is the time required for the number of atoms N, to decay to $^1/_2$ N, and is given by:

$$t_{1/2} = \ln 2 \cdot \lambda^{-1} = 0.691 \cdot \lambda^{-1} \tag{4.2}$$

The differences in ranges of the various forms of radiation leads to two distinct hazards: internal dose in which the radionuclide is ingested into the body by eating, drinking or breathing and external dose, which occurs when the radionuclide remains outside of the body and the radiation penetrates the skin. Except during an accident, most dose to NPP workers during operation and maintenance is external dose.

The radionuclides released from the core can be broadly classed into two groups:

1. Activation products formed by the reaction of a neutron (or proton) with the nucleus of an atom in one of the materials used in the core (fuel cladding, support structure) or in any grown-on or deposited oxide. An example is the production of ^{60}Co from the naturally occurring nonradioactive ^{59}Co:

$$^{59}Co(n, \gamma)^{60}Co$$

2. Fission products and actinides formed by the nuclear reactions of the fissile fuel isotopes (e.g., ^{235}U). Actinides result from neutron absorption, while fission products result from the radioactive decay of the fragments produced by the nuclear fission process.

Radionuclides in both groups have the potential to be released from the reactor core, although the mechanisms of release differ. Fig. 4.15 depicts some of the processes associated with activity transport. Impurities transported into the core can be in the form of dissolved species or particles. While some activation will occur during the brief period that the flowing coolant resides in the core, a bigger concern is the deposition of the impurities on in-core surfaces either by precipitation, exchange of ions in the solution with ions in the in-core oxide, or by deposition of particles from solution. Once the impurities are deposited activation can occur until the impurity is released from the surface by dissolution, ion-exchange back into solution or particle release.

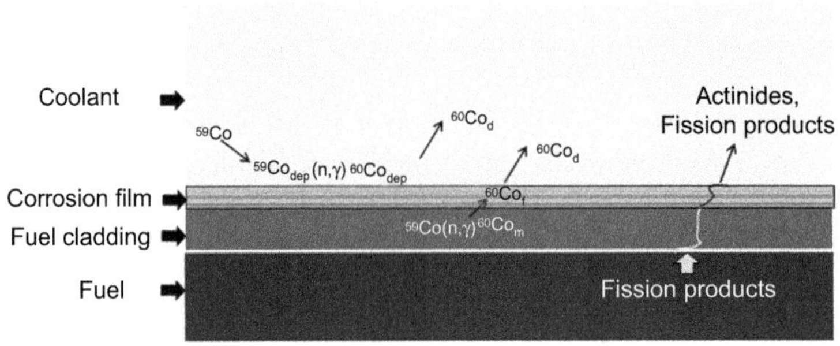

Figure 4.15 The three source terms for activity transport in a supercritical water-cooled reactor. From left to right the processes are (1) deposition of corrosion products (in this example ^{59}Co) transported to the core from the feedtrain, neutron activation and then release of ^{60}Co to the coolant; (2) neutron activation of ^{59}Co in the alloy and subsequent diffusion of the resultant ^{60}Co into the corrosion film and release into the coolant and (3) release of fission products into the coolant through a fuel defect. Subscripts indicate location: 'dep' for within deposit, 'd' for dissolved in coolant, 'm' for within metal and 'f' for within film.
Adapted from Guzonas, D.A., Qiu, L., 2013. Activity transport in a supercritical water-cooled reactor. In: 6th International Symposium on Supercritical Water-cooled Reactors, March 03–07, 2013, Shenzhen, Guangdong, China.

If, as predicted by the modelling work discussed in the previous section, the SCWR coolant remains supersaturated in dissolved metals relative to the solubility of the relevant metal oxide and there is no oxide dissolution, little release of metal ions is anticipated during normal operation[11]. In that case the most likely mechanism for activity release from these deposits is release during reactor shutdown as the solubility increases by several orders of magnitude as the temperature decreases below the critical temperature. Other possible release mechanisms are mechanical wear (e.g., due to vibration of the fuel rods), erosion by the flowing coolant or spalling of the oxide from surfaces.

4.3.1 Activation of in-core materials

While corrosion products transported from the out-of-core sources are typically deposited and released over relatively short periods[12], some in-core materials can reside in the core for long periods. During this time they are subjected to nearly continuous irradiation leading to changes in alloy composition due to transmutation and various forms of irradiation damage as discussed in Chapter 3. Some of the transmutation products are potentially hazardous radionuclides (Table 4.2) that can be released into the coolant by corrosion or mechanical wear processes. The concentrations of these radionuclides in the alloy can be quite high due to the long irradiation time. While generally not

[11] The rates of dynamic exchange of isotopes in the coolant with those in the oxide are expected to be low but this process may release a (probably small) amount of activity.

[12] For example, in a typical batch-fuelled SCWR the entire fuel load will be replaced every three fuel cycles (~3–4 years).

Table 4.2 Parent isotopes, production reactions and half-lives for key activation products

Isotope	Reaction	Half-life
^{51}Cr	$^{50}Cr(n,\gamma)^{51}Cr$	27.8 d
^{54}Mn	$^{54}Fe(n,p)^{54}Mn$	312.2 d
^{55}Fe	$^{54}Fe(n,\gamma)^{55}Fe$	2.7 y
^{56}Mn	$^{55}Mn(n,\gamma)^{56}Mn$	2.58 h
^{59}Fe	$^{58}Fe(n,\gamma)^{59}Fe$	44.5 d
^{58}Co	$^{58}Ni(n,p)^{58}Co$	71 d
^{60}Co	$^{59}Co(n,\gamma)^{60}Co$	5.27 y
^{64}Cu	$^{63}Cu(n,\gamma)^{64}Cu$	12.7 h
^{65}Zn	$^{64}Zn(n,\gamma)^{65}Zn$	244 d
^{95}Zr	$^{94}Zr(n,\gamma)^{95}Zr$	64 d
^{95}Nb	^{95}Nb by beta decay of ^{95}Zr	35 d
^{94}Nb	$^{93}Nb(n,\gamma)^{94}Nb$	20,000 y
^{122}Sb	$^{121}Sb(n,\gamma)^{122}Sb$	2.7 d
^{124}Sb	$^{123}Sb(n,\gamma)^{124}Sb$	60.2 d

significant from a metallurgical standpoint, these concentrations can be high enough that even low corrosion rates can release hazardous amounts of material[13]. In addition, handling of the alloys on their removal from the core requires significant precautions and adds to the eventual cost of plant decommissioning.

Of the radionuclides listed in Table 4.2, ^{60}Co (from cobalt activation) and ^{58}Co (from nickel activation) have been the most troublesome in existing WCRs. Cobalt is typically present in steels as an impurity at $\mu g \cdot kg^{-1}$ concentrations but can be a major component of some specialty alloys. ^{60}Co is particularly hazardous as it emits two high-energy (~ 1 MeV) gamma rays and has a relatively long half-life that allows it to build up to significant concentrations on out-of-core piping. Short-lived radionuclides do not build up on out-of-core surfaces but can result in high doses during outages, especially if they are mobilized by chemistry transients during reactor shutdown. The behaviour of ^{58}Co during outages in PWRs is a well-known example. Antimony is also a trace impurity in many steels and is also used in some bearing and seal materials, and radioantimony has been an issue in many WCRs. Zr and Nb have been

[13] As an illustration, a 1 mg particle of Co neutron activated to its equilibrium activity of ^{60}Co (rate of production of ^{60}Co and rate of radioactive decay reach a steady state) has an activity of about 1 Ci (37 GBq).

included in the table because both elements are sometimes added to steels to improve their properties, typically at concentrations on the order of 1%.

There have been only a few studies of the metal release into the SCWR coolant. Guzonas and Cook (2012) reported data on the release of iron, manganese, nickel and chromium from 403 stainless steel in static autoclaves, as well as data on the release of various elements into SCW from Hastelloy C and Alloy 625 autoclaves. Han and Muroya (2009) quantified the release of ^{60}Co radiotracer from irradiated 304 SS samples as a function of time and temperature in SCW by collecting the material released on ion-exchange resins and measuring the amount collected by gamma spectrometry. They noted that while the corrosion rate of the 304 SS test specimens increased with increasing temperature (being about 10 times higher at 550 °C than at 300 °C) the ^{60}Co release rate at 550 °C was less than 1% of the 300 °C value. They attributed this to the decrease in oxide solubility above T_c. Choudhry et al. (2016) reported data on the release of Fe, Cr, Ni, Ti, Mn, Cu and Al from Alloy 800H as a function of time in oxygenated ($\sim$20 $\mu g \cdot kg^{-1}$) feedwater at a volumetric flow rate of 0.1 $mL \cdot min^{-1}$ at 550, 650 and 700 °C and a pressure of 25 MPa (Chapter 5). The release of each element was a complex function of time over several hundred hours, in some cases changing significantly over the course of just a few hours (Fig. 5.36).

Saito et al. (2006) modelled the rate of ^{60}Co transport to the turbines of an SCWR concept (1000 MW_e, $T_{in} = 290$ °C, $T_{out} = 550$ °C, P = 25 MPa) relative to that of a BWR of similar output. In the absence of data on corrosion rates and concentrations of dissolved ^{59}Co in the coolant, they examined a range of relevant input parameters. The model included both ^{60}Co release from fuel cladding corrosion and ^{60}Co release from deposits on the cladding surface resulting from deposition of ^{59}Co carried to the core from feedtrain alloy corrosion. The corrosion release was calculated using:

$$AR = C_R \cdot C_{RR} \cdot C_{Co} \cdot SA \tag{4.3}$$

where C_R is the corrosion rate of the cladding alloy, C_{RR} the release fraction (amount of corroded element released to the coolant, C_{Co} the concentration of Co in the alloy, and SA the cladding surface area[14]. The corrosion rate was varied from 0.0001 to 10 times the rate for the same alloy in a BWR. ^{60}Co transport rates (relative to a BWR) were found to vary from 0.1 to greater than 100 depending on the input parameters chosen. This study highlighted the need for more data on corrosion rates and, more importantly, on the fraction of the corroded alloy moving into the coolant rather than remaining in the surface oxide layer.

With the benefit of recent data on corrosion and metal release rates in SCW, Guzonas and Qiu (2013) used the metal release data of Carvajal-Ortiz et al. (2012b) to estimate the ^{60}Co activity release from the stainless steel fuel cladding expected to be used in an SCWR for two limiting cases. Care must be taken when using experimental release data, since, as noted above, the release rates may take hundreds of

[14] Not the original notation of Saito et al. (2006).

hours to reach steady state. Cohen (1985) discussed the activation of 304 SS fuel cladding under conditions relevant to the Yankee PWR; after 3 years irradiation (roughly the expected SCWR fuel cladding in-service life) in a typical PWR flux activation of 304 SS fuel cladding resulted in a ^{60}Co activity in the alloy of about 2 $MBq \cdot mg^{-1}$. A reasonable estimate for the activity release rate AR (in $MBq \cdot d^{-1}$) for ^{60}Co can be obtained using:

$$AR = 2MBq \cdot mg^{-1} \times S_{ac} \times R \tag{4.4}$$

where S_{ac} is the surface area of the cladding and R is the metal release rate from the surface (equal to $C_R \cdot C_{RR}$ in Eq. 4.3). The value of the release rate, R, will vary with location in the core, as it depends on corrosion rate and on the solubility of the resulting surface oxide film. The model assumed that an element was released from an alloy in proportion to its concentration in the alloy, which, while known to be incorrect (Chapter 5), is acceptable for such an estimate given all the other uncertainties. Two limiting cases were examined:

Case 1, 'Evaporator' region: Coolant temperature is low and coolant density is high, consequently oxide solubility is relatively high. The peak ^{60}Co release rate was estimated to be ~50,000 $MBq \cdot d^{-1}$ (fuel cladding surface area ~2000 m^2). In an operating SCWR the concentration of metal species in the coolant is likely to be nonzero and dissolution of the oxide film may be very low if the coolant is already saturated with the relevant dissolved metal species (although ion exchange between inactive dissolved species and activated oxide species is still possible). Thus the presence of a **very low** concentration of dissolved metal species in the coolant at the core inlet may be beneficial from an activity transport perspective.

Case 2, 'Superheater' region: Coolant temperature is high, the coolant density is low, oxide solubility is very low and most of the metal species released by corrosion will remain on the surface. An average release rate of 27,500 $MBq \cdot d^{-1}$ was estimated; based on a coolant flow rate of 1000 $kg \cdot s^{-1}$, the coolant concentration of ^{60}Co was estimated at 320 $Bq \cdot kg^{-1}$, on the same order of magnitude as the ^{60}Co coolant concentration in current PWRs and PHWRs. This activity corresponds to a dissolved Co concentration of ~10^{13} $mol \cdot kg^{-1}$. More detailed predictions are not possible at this time because the required data on oxide solubilities are not currently available and the in-core chemistry conditions (particularly water radiolysis) cannot yet be adequately specified.

Essentially the only data on the transport of activation products at supercritical temperatures come from Beloyarsk NPP. ^{51}Cr, ^{54}Mn, ^{58}Co, ^{60}Co, ^{65}Zn and ^{124}Sb (the same activation products found in current WCRs) were reported in deposits and coolant samples (Dollezhal et al., 1969; Aleksandrova et al., 1968; Veselkin et al., 1968). The surface activities measured on the blades of different turbine stages at the Beloyarsk AMB-1NPP (Fig. 4.16) showed that ^{60}Co, and to a slightly lesser extent ^{124}Sb, appeared to behave similarly to ^{54}Mn, an activation product of iron and therefore a marker for iron transport (Veselkin et al., 1968). A minimum in deposited ^{60}Co, ^{124}Sb and ^{54}Mn activity was observed around stage 11 or 12 of the turbine; however, the pattern of ^{60}Co deposition exhibited a time dependence that made detailed

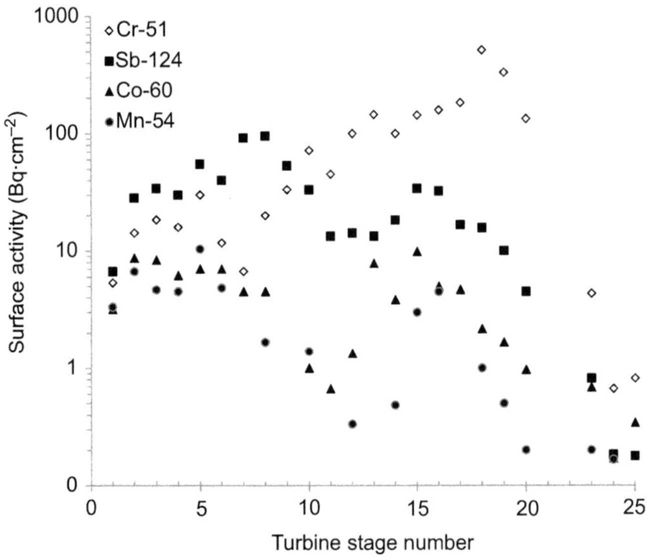

Figure 4.16 Surface activity of radioactive deposits on the blades of different turbine stages at the Beloyarsk NPP.
Adapted from Veselkin, A.P., Lyutov, M.A., Khandamirov, Yu.E., 1968. Radioactive deposits on the surfaces of the technological equipment of the I.V. Kurchatov nuclear power station at Beloyarsk. At. Energ. 24, 219

interpretation of these data impossible (Veselkin et al., 1971). Dollezhal et al. (1958) reported that the turbine was two-stage, which could explain the observed minimum near stages 11 and 12. Although turbine designs are complex, in general, the temperature and pressure decrease as the steam passes through the turbine stages; using realistic assumptions, it is estimated that the steam density decreased from about 24 $kg \cdot m^{-3}$ to about 3.5 $kg \cdot m^{-3}$ at the turbine outlet. As a result, precipitation of dissolved metal species throughout the turbine is expected. In addition, transport and deposition of particles is likely; coolant samples suggested most of the activity was associated with particles.

Interestingly, ^{51}Cr seemed to behave very differently than the other three radionuclides shown in Fig. 4.16 in that there is no obvious minimum near turbine stages 11 and 12; no explanation for this behaviour was given, but this observation suggests differences in release or transport mechanisms (e.g., particulate iron vs. dissolved chromium species).

In general, the data on radioactive corrosion product deposition from Beloyarsk agree well with data reported for corrosion product deposition in SCFPPs. An initial decrease in solubility leads to an increase in deposition, an increase in solubility at the reheater leads to a decrease in deposition, and a final further decrease in solubility leads to increased deposition.

It was reported that the radiation dose close to the secondary circuit equipment (i.e., outside the active zone of the reactor) did not exceed $\sim 3\ mR \cdot h^{-1}$

(Veselkin et al., 1968). The radiation fields around the turbines were lower than those at other direct cycle NPPs, and (as with other WCRs) most of the dose to workers occurred during maintenance activities. ^{51}Cr, ^{54}Mn, ^{58}Co, ^{60}Co and ^{124}Sb were detected in samples of SHS collected after the reactor (condensate) and of the steam condensate after the turbine (Aleksandrova et al., 1968). The ^{60}Co specific activity of the SHS was reported to be 50–150 $pCi \cdot kg^{-1}$; assuming that the one order of magnitude difference in NaCl solubility between an SCWR and Beloyarsk in Fig. 4.14 is applicable to cobalt oxides, the estimated ^{60}Co activity in an SCWR is about 0.5–1.5 $nCi \cdot kg^{-1}$. It was concluded that a considerable part of the radioactivity in the SHS condensate was present in the form of suspended matter (the majority of the particles had sizes in the range 3–10 μm); however, the samples were cooled and some or all of the particulates detected could have precipitated in the sample lines during cooling. ^{54}Mn was reported to have been present as dissolved species (50%–70% cationic and <3% anionic) (Yurmanov et al., 2010). The activity of the impurities in the steam condensate after the turbine was considerably less (but not zero) than in the SHS, confirming that significant amounts of activated corrosion products deposited on the piping between the reactor core and the turbines and in the turbines themselves (Fig. 4.16). The presence of activity in condensate streams means that in any SCWR design *all* condensate lines must return to the system upstream of the condensate polisher to minimize contamination of the feedtrain.

One of the only other reported measurements of activated corrosion products was by Mravca and Simpson (1961), who found that 4 months after the failure of fuel assembly SH-4B in the SADE loop[15] (by which time fission product activity had largely decayed) the main source of loop piping activity were the corrosion products ^{51}Cr, ^{58}Co, ^{60}Co, ^{65}Zn and ^{110m}Ag.

4.3.2 Defected fuel in a supercritical water-cooled reactor

A fuel defect is a breach in the fuel cladding that allows the primary coolant to enter the fuel-to-sheath gap and allows fuel particles and fission products to escape. Defected fuel phenomena in current generation WCRs are reasonably well understood, and various detection, location, diagnostic and management techniques have been developed (e.g., IAEA, 2002).

Fuel release from defected UO_2 elements in WCRs is caused by erosion and enhanced by fuel oxidation (Lewis et al., 1993). The erosion rate is affected by defect geometry (i.e., hole, large axial split, burst hydride) and the fuel oxidation rate is mainly driven by the fuel temperature. As SCWR development proceeds, the interaction of the SCW and fuel must be investigated, as well as the solubility and transport of fuel material and fission products in SCW. The interaction between SCW and the inner surfaces of the fuel cladding must also be studied to determine defect degradation behaviour (e.g., internal cladding oxidation).

[15] Maximum cladding temperature = 621 °C, 617 h irradiation time under superheat condition. Cladding material was 28 mil thick 304 SS. Cladding failure was attributed to chloride-induced SCC.

Once fission products are released into the coolant they can be detected in the system, the presence of a fuel defect identified and defect diagnostics can be carried out. The measured release rate to the coolant and coolant inventory can be used to infer the number, size, location, power, fuel burn-up, etc. using various techniques (Locke, 1972; Likhanskii et al., 2009). Some of these techniques may be applicable to an SCWR; the basic understanding of FP release and transport developed for current WCRs will also facilitate development of new techniques.

To date there have been few systematic measurements of the solubilities of fission products and actinides in SCW. Qiu (2014) reviewed the limited data on the solubilities of U, Th and some possible fission products in SCW. Zimmerman et al. (2014) measured the association constants of strontium with hydroxide and chloride ions at temperatures up to 350 °C using high-pressure flow AC conductance. The systems $[Sr(OH)_2 + H_2O]$ and $[SrCl_2 + H_2O]$ were chosen because they are soluble in near-critical water allowing them to be used as model systems for other M^{2+} species. Under SCWR conditions the formation of $SrOH^+$ and $Sr(OH)_2^0$ ion pairs was established to be greater than $SrCl^+$ and $SrCl_2^0$ ion pairs, so that hydroxyl complexes will be the dominant strontium species in solution. Concentrations of neutral species were found to be substantial at 350 °C at concentrations above 10^{-3} mol·kg^{-1} for both salts.

4.3.2.1 *Plant experience and experimental data*

Noble gases and iodines

During reactor operation radioactive noble gases (e.g., ^{133}Xe, ^{88}Kr) and iodines (e.g., ^{131}I) are produced by the nuclear fission and diffuse through the fuel matrix to the gap between the fuel and fuel cladding. If the cladding is breached, these gases can escape into the coolant; the releases are often found to increase during operational transients (Cohen, 1985). Fission gas release and transport in WCRs has been the subject of much study, and the on-line detection of released fission gases is used to detect and monitor fuel defects. There is a significant amount of data on the behaviour of noble gases and iodines in SHS from the US nuclear steam superheat program carried out in the 1960s. Because gases are infinitely soluble in SCW, data on noble gas transport obtained at lower pressures are directly relevant to the transport of these gases in an SCWR.

During the US program to study nuclear steam superheat, a principal goal of the chemistry program of the ESADE-VBWR test program was to detect the failure of fuel cladding and measure fission product release rates. Radiochemistry measurements during the ESH-1 irradiation did not indicate the presence of defective fuel, although a small defect was later found by sipping[16] and subsequent metallographic sectioning. However, the presence of a known fuel defect in the VBWR itself severely limited the sensitivity of the measurements (Hazel et al., 1965). After failure of the SH-4B fuel element in the SADE loop (Spalaris et al., 1961), the loop piping was found to be contaminated with fission products (mainly radioiodines) with radiation fields up

[16] Sipping is the detection of defects in spent fuel stored in a water-filled fuel bay by sampling the water locally around the fuel bundle (Deng and Deng., 2012).

to 20 $mR \cdot h^{-1}$ (Mravca and Simpson, 1961). Several fuel defects occurred during the experimental program at the EVESR, which was deliberately operated for several months with a grossly defected fuel bundle (KB-41) to quantify release of fission products (Murray, 1965). The KB-41 bundle was found to be defective about 3 h after the reactor power was increased from 15 MW_{th} to 17 MW_{th}, as detected by an increase in the plant off-gas noble gas monitor from 1.5 to 15 $mR \cdot h^{-1}$ in a 15-min period (Busboom et al., 1966). It was found that the release of fission products was highly dependent on the reactor power and the reactor exit steam temperature. Operation at the fuel full-rated temperature led to the deposition of tellurium and other fission products on the superheat steam piping, resulting in surface dose rates of up to 300 $mR \cdot h^{-1}$ on piping external to the containment building and up to 5 $R \cdot h^{-1}$ on one valve in the reactor pipe gallery. Radiation fields were highest close to the reactor and at piping irregularities such as fittings or valves. The radiation levels decreased with a 78-h half-life following reactor shutdown.

A test (KB-41-2) was performed with a purposely defected fuel rod (a 0.005–0.013 inch wide by 1 inch long slit in the 0.016 inch thick Alloy 800 cladding). The purpose of the test was to place a known defect in a known position in the core to study the release of fission products and the consequences of such operation on a steam superheat system. Data on the release of 13 noble gases and five iodine isotopes were reported (Roof, 1966).

A thorough examination of these data in the context of SCWR development has yet to be performed but as in-reactor test facilities become available such an assessment could assist in the design of experimental programs to study fuel defects in an SCWR.

Fuel leaching tests

Guzonas et al. (2016) reported the results of leaching tests of a simulated fuel (SIMFUEL) in SCW performed to identify those fission product species capable of dissolving from the fuel matrix and determine if the fuel itself might dissolve in SCW. While the composition of the current Canadian SCWR reference fuel is 13 wt% PuO_2 in ThO_2 (Bergeron and Hamilton, 2013), the radiotoxicity of plutonium complicates handling of fresh fuel and irradiated fuel presents additional challenges. SIMFUEL, containing natural UO_2 as a surrogate for PuO_2 and nonradioactive additives to simulate fission products present in irradiated fuel, was therefore used. A 60 $GWd \cdot tonne^{-1}$ burn-up SIMFUEL was chosen for the tests because it had the highest concentrations of fission product surrogates, increasing the likelihood of their release and their detection in solution. The tests were performed in a Hastelloy C-276 static autoclave at 400 and 500 °C, 25 MPa. A temperature of 400 °C was chosen because the solubility of the SIMFUEL constituents was expected to be relatively high at this temperature, while 500 °C was the maximum autoclave operating temperature.

Significant releases of corrosion products (Fe, Ni, Cr, Mo) from the autoclave were detected. Of the simulated fission products, only Sr and Ba were detected in all tests at significant concentrations (Fig. 4.17); both species appeared to have relatively high solubilities under the test conditions, although equilibrium solubility may not have been reached as the dissolution rate may have been limited by diffusion through the sintered SIMFUEL pellets. Mo release from SIMFUEL could not be quantified because Mo is also a major constituent of Hastelloy C-276, and the wetted surface area of the autoclave was much higher than that of the test specimens.

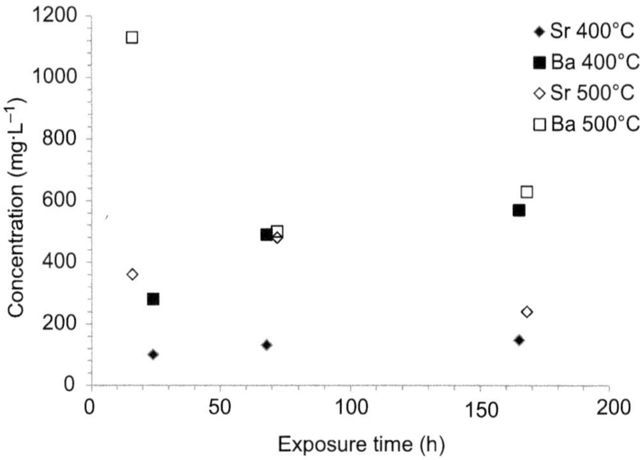

Figure 4.17 Concentrations of Sr and Ba as a function of exposure time at 400 and 500 °C at 25 MPa.
Adapted from Guzonas, D., Qiu, L., Livingstone, S., Rousseau, S., 2016. Fission product release under supercritical water-cooled reactor conditions. J. Nucl. Eng. Rad. Sci. 2, 021010-021011–021010–021016.

At 400 °C, Zr, Ru, Pd, Ag and Sn were detected in one or more tests at close to the MDL of ICP-MS, suggesting these elements have low solubilities under the test conditions. Similar results were obtained at 500 °C, and, with the exception of the first data point measured for Ba, the concentrations were not significantly higher than those measured at 400 °C. At 500 °C, Rh, Ce, La and Nd were detected in one or more tests at close to the MDL of ICP-MS. Rhodium was the only element other than Sr, Ba and the autoclave corrosion products detected at both 400 and 500 °C, the concentration at 500 °C being about one order of magnitude lower than that at 400 °C.

A measureable concentration of U in solution was reported in two tests: 0.031 $mg \cdot kg^{-1}$ ($\log_{10}[U/mol \cdot kg^{-1}] = -9.9$) at 400 °C and 0.064 $mg \cdot kg^{-1}$ ($\log_{10}[U/mol \cdot kg^{-1}] = -9.5$). These values are in reasonable agreement with literature values; for example, Red'kin et al. (1989, 1990) reported an average value of $\log_{10}[U/mol \cdot kg^{-1}] = -9.0 \pm 0.5$ at 100 MPa over the temperature range 300–600 °C. Guzonas et al. (2016) also reported results of a simple solubility measurement of thoria powder at 400 °C and 25 MPa. The concentration of Th in solution after cooling and filtering was less than 0.2 $\mu g \cdot L^{-1}$.

4.4 Water radiolysis

One of the most significant water chemistry challenges for all SCWR designs is to predict the effects of water radiolysis on materials performance and corrosion product transport, and if possible, develop mitigating strategies. The SCWR coolant will be subjected to an intense radiation field as it passes through the reactor core. The radiation chemistry of water in the core of a nuclear reactor is initiated by the passage of γ- and fast-neutron radiation, which ionize the water. The radiation chemistry may also be influenced by α-radiolysis close to fuel cladding surfaces and fission fragment radiolysis from tramp uranium. Typical radiation doses in a reactor core are on

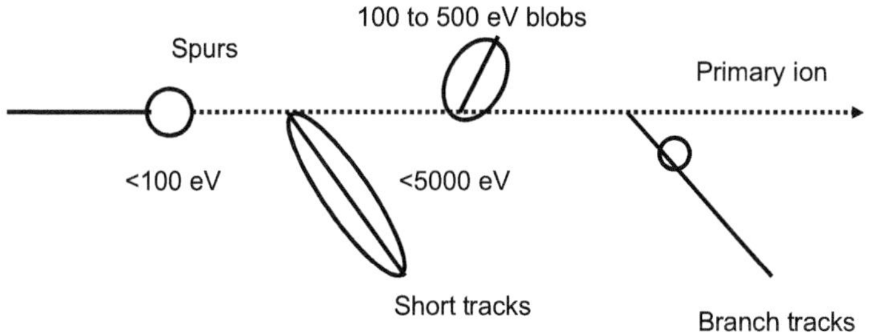

Figure 4.18 Diagram showing the track structure of an incident primary ionizing particle in a liquid.
Adapted from Spinks, J.W.T., Woods, R.J., 1990. An Introduction to Radiation Chemistry, third ed. Wiley, New York; original figure by Burton, M., 1969. Chem. Eng. News 47, 86.

the order of 1000 $kGy \cdot h^{-1}$. The different types of radiation interact differently with water, so modelling work must eventually consider all relevant types.

When a moving charged particle is slowed down in matter it loses energy producing a trail of excited atoms and molecules in its path. This process is illustrated in Fig. 4.18 (cf. Fig. 3.1). Electromagnetic energy (γ- and X-rays) has a similar effect via the production of electrons and positrons, which act as the 'primary particle'. Since neutrons are uncharged they do not directly result in ionization of matter. Many neutron-matter interactions do result in the production of charged particles, which subsequently act as 'primary particles'. These various forms of radiation lose energy to the medium at different rates, typically expressed in terms of linear energy transfer (LET). LET is defined (Spinks and Woods, 1990) as 'the linear rate of loss of energy (locally absorbed) by an ionizing particle traversing a material medium', i.e., dE/dx for ionizing particles of energy E over a distance x. Photons are low-LET particles, while fast neutrons and α-particles are high LET particles. LET is known to affect radiolysis product yields (see for example, Burns and Sims, 1981).

As illustrated in Fig. 4.19, in the first picoseconds following the ionization event, dissociation, nonradiative decay, proton transfer, electron thermalization and hydration, dissociative electron attachment and recombination reactions occur leading to the formation of hydrated electrons (e_{aq}^-), hydroxide radicals (•OH), hydrogen atoms (•H), molecular hydrogen (H_2), hydrogen peroxide (H_2O_2), superoxide radicals ($HO_2/O_2^{\bullet -}$) and the proton (H^+) (Spinks and Woods 1990; Elliot and Bartels, 2009). These primary species then undergo additional reactions as they diffuse away from the initial ionization centre to form the stable end-species hydrogen, oxygen and hydrogen peroxide. While it is the concentrations of these stable end products that mainly determines the aggressiveness of the irradiated water downstream of the core, the interactions with surface of short-lived reactive intermediates such as hydroxyl radical (HO•), the hydrated electron (e_{aq}^-) or oxygen radicals (O_2^-) should not be neglected. Ishigure et al. (1980) reported that γ-irradiation increased the release rate of dissolved iron species from 304 SS but not the release of particulate iron oxides. Christensen (1981) proposed that O_2^- was responsible for an observed increase in the corrosion of Zircaloy-2 when irradiated in core. Moreau et al. (2014) showed that both HO• and e_{aq}^- can be scavenged by metal surfaces (316L SS and Hastelloy). Such reactions of intermediates with surface oxides or alloys are likely to be most significant in confined regions such as pores or crevices.

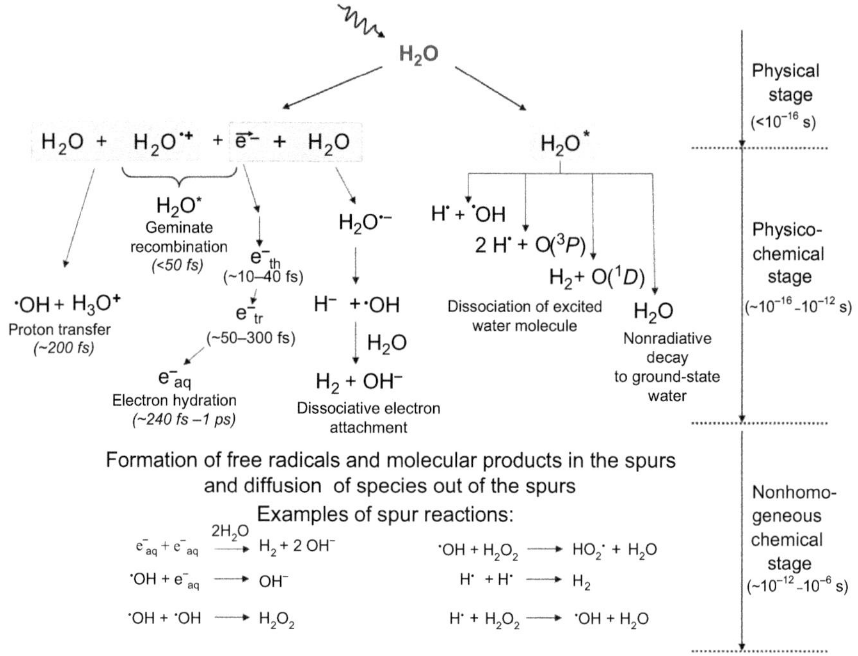

Figure 4.19 Water radiolysis pathways.
Figure courtesy Prof. J.-P. Jay-Gerin, Université de Sherbrooke.

Measurement of concentrations inside a reactor core is not possible, but the concentrations of stable species can be obtained by sampling and measurement downstream bearing in mind the difficulties in measuring reactive oxidizing species such as oxygen and hydrogen peroxide in a high-temperature aqueous system. Indirect measurements using electrochemical corrosion potential probes have been made in BWRs. The concentrations of these end-species in the core and downstream are a crucial input for defining the appropriate oxidant concentration for materials testing. Most of the recent materials testing for SCWR development has been carried out with DO concentrations of 8 $mg \cdot kg^{-1}$ or less.

Studies of water radiolysis in SCW to date have been laboratory-based (see Chapter 2 for a description of facilities) rather than reactor-based. Therefore, current understanding of the possible effects of water radiolysis in an SCWR is based on modelling. Such modelling requires knowledge of the temperature and density dependencies of the chemical yields[17] of oxidizing (•OH, H_2O_2 and HO_2/O_2) and reducing (e_{aq}^-, H•, H_2) radiolytic products and of the rate constants

[17] Yields (the number of species produced or consumed per unit absorbed energy) are expressed in terms of the g- or G-value, conventional units being molecule per 100 eV (Spinks and Woods, 1990). A distinction is made between the primary yields (normally measured at the end of spur expansion (end of nonhomogeneous chemical stage in Fig. 4.19)), for which a lower case g is used, e.g., $g(e_{aq}^-)$, and the yield of stable products, for which an upper case G is used, e.g., $G(H_2)$.

for the reactions taking place in spurs and tracks that result in these primary yields. While this information is available over the temperature range relevant to current WCRs operating with subcritical water, knowledge of these parameters in SCW is incomplete, and the results of modelling work must be considered very preliminary at this time. Lin et al. (2010) is a useful summary of recent experimental work.

The primary radiolysis products include both ionic and radical species, and therefore it seems reasonable to expect that the rapid change in water properties in the vicinity of the critical point (e.g., dielectric constant) would affect the various reactions of these species. In addition, the change in water density changes the extent to which various active species are hindered from moving away from each other. Finally, the inhomogeneous nature of SCW means that species could segregate into locally high- or low-density regions depending on whether their interactions with water are favourable or unfavourable. It is expected that modelling water radiolysis will be the most difficult in the near-critical region.

The g-values for the primary yields formed by using low-LET (γ or fast electron) radiation at temperatures from ambient to above T_c have been measured by several groups. In general, the measured temperature dependencies of these rate constants do not follow an Arrhenius relationship, and their values in SCW cannot therefore be predicted by simple extrapolations of existing subcritical water data. As with the thermodynamic quantities discussed earlier in the chapter, determination of high-temperature rate constants is exceedingly challenging, especially under SCWR conditions. In addition to the experimental challenges, the commonly used technique of using chemical 'scavengers'[18] to follow reactions of the intermediate radiolysis products may be problematic, in that little is known about the behaviour of the scavenger molecules in SCW. A further complication is that different experimental techniques may measure yields at different timescales between the initial interaction of a water molecule with radiation and the eventual end of the spur lifetime and return to homogeneous kinetics (Sanguanmith et al., 2013). For example, Haygarth and Bartels (2010) suggested that the yields for the hydrated electron (e_{aq}^-) obtained by their experiments were close to the escape yield[19] while those of Lin et al. (2005) were measured closer to the initial excitation event.

The temperature dependencies for many of the relevant reactions show a local minimum (for both light and heavy water) near the critical point (Fig. 4.20); this minimum has been reported to occur at thermodynamic conditions where the compressibility of the fluid is a maximum (Ghandi et al., 2002, 2003; Percival et al., 2007; Marin et al., 2005; Bonin et al., 2007). As discussed by Alcorn et al. (2014), such a minimum could arise from effects such as density inhomogeneity (Tucker, 1999; Ghandi et al., 2012), 'critical slowing' of reactions at

[18] Ion or radical scavengers are chemical species added to the test solution to react quantitatively with a specific primary radiolysis product so that its concentration can be measured.

[19] End of spur lifetime and return to homogeneous kinetics.

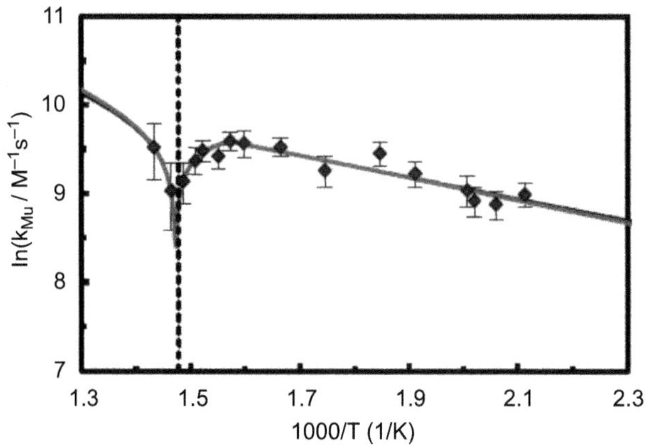

Figure 4.20 Rate constants determined for the reaction of muonium with H_2O at 30 MPa in the temperature range 35−425 °C. *Black and red* (gray in print versions) *lines* represent best fits of the model to data. The vertical dotted black line denotes the temperature of maximum compressibility (Alcorn et al., 2014).

thermodynamic critical points (Procaccia and Gitterman, 1981) and the solvent cage effect[20] (Ghandi et al., 2002, 2003; Percival et al., 2007). Liu et al. (2016) proposed a cage effect model and used it to predict the rate constants in SCW of a number of equilibrium reactions involved in water radiolysis.

It is known that density fluctuations and solvent inhomogeneity occur near the critical point where the solvent compressibility is very large (Tucker, 1999; Ghandi et al., 2012). At the critical point localized clustering of water molecules occurs producing areas of high and low density (Fig. 4.2). Hydrophobic species such as H• (Mu•) (Tse and Klein, 1983; Roduner, 2005) will find it energetically favourable to be located in the voids between water clusters, while hydrophilic species such as ions will be located within the water clusters. In a reaction such as $H\bullet + H_2O$ the low concentration of water molecules in the voids could result in a minimum in the observed rate constant by reducing the number of encounters of H• with water.

4.4.1 Approaches to modelling

From an SCWR designer's perspective, what is required are not the concentrations of primary and secondary transient radiolysis products (Fig. 4.18) but rather the steady-state concentrations of the final products (hydrogen, oxygen and hydrogen peroxide).

[20] The cage effect is the tendency for encounters between molecules within solvent clusters to last longer, providing more opportunities for reactants to collide during each encounter. Gas phase encounters usually consist of a single collision due to the low density. The cage effect can have an important influence on the temperature dependence of rate constants when fluid structure changes with temperature (Leblanc et al., 2014).

Two approaches have been pursued for SCWR development. One uses experimental data and computer simulations to develop predictive models based on a detailed microscopic knowledge of radiolysis yields and their dependence on temperature, density, SCW structure and other parameters. The other approach is semiempirical, similar to the approach used by current WCRs.

4.4.1.1 Microscopic models

Monte Carlo simulations offer a promising avenue to quantitatively describe the temperature and pressure (density) effects in SCW radiolysis, to identify key parameters, and also to help elucidate mechanisms by which radiation interacts with SCW. Molecular dynamics (MD) simulations provide a microscopic understanding of SCW and the precise role of its molecular structure in radiolysis for a wide range of densities (Metatla et al., 2016). As with corrosion product transport modelling, this type of radiolysis model requires the availability of high-quality experimental data measured under SCW conditions.

The importance of such changes in reaction rate can be illustrated using the reaction of hydrogen and hydroxyl radical to form hydrogen radical and water:

$$H_2 + HO\cdot \leftrightarrow H\cdot + H_2O \tag{4.5}$$

a key reaction in the suppression[21] of net water radiolysis by addition of molecular hydrogen to the coolant of WCRs. This is the only reaction occurring in the nonhomogeneous chemical stage whose rate can be made fast enough (by addition of a high concentration of hydrogen) to convert the oxidizing •OH radical into the reducing •H radical before the •OH radical can encounter and oxidize another species. The back reaction is relatively slow at current WCR operating temperatures. The relative concentrations of reactants and products in Eq. (4.5) can be altered by adjusting the H_2 concentration. As this is the only reaction in the reaction set that removes H_2 (Elliot and Bartels, 2009), molecular hydrogen is usually injected into PWR and PHWR coolant systems to suppress the net formation of oxidizing radiolysis products. The back reaction is known to be responsible (Swiatla-Wojcik and Buxton, 2005; Bartels, 2009; Swiatia-Wojcik and Buxton, 2010) for the increase in the concentration of H_2 observed in pulse radiolysis experiments of high-temperature water.

The rate constant for Eq. (4.5) shows Arrhenius behaviour at temperatures up to 250 °C; above this temperature, the reaction rate slows and it has been suggested that the backward reaction may become significant. The rate constant of the back reaction under hydrothermal conditions has been the subject of much discussion (Swiatla-Wojcik and Buxton, 2005; Bartels, 2009; Swiatia-Wojcik and Buxton, 2010), due to a lack of data on solvation energies of transient species in SCW. In the near-critical region changes in relative stabilities of ionic and radical species in solution with changes in temperature can alter the relative importance of various reaction pathways (e.g., diffusion vs. recombination). A sensitivity analysis using Monte Carlo simulations from room temperature

[21] Suppression of net radiolysis means undetectable concentrations of oxidants at the core outlet.

up to 350 °C (Sanguanmith et al., 2011; Sanguanmith, 2012) examined the effect of the rate constants of Eq. (4.5) on H_2 yield, concluding that the best agreement between simulated and experimental results was obtained with the rate constant $k = 10^4\ M^{-1} \cdot s^{-1}$ at 300 °C obtained by muon spin spectroscopy (Alcorn et al., 2011).

To explain the measured temperature dependence of Eq. (4.5), Alcorn et al. (2014) proposed two competing reactions, one with low activation energy at low temperature (Eq. 4.6) and another with a much higher activation energy at high temperature (the backward reaction in Eq. 4.5).

$$H\cdot + H_2O \leftrightarrow H_3O^+ + e_{aq}{}^- \tag{4.6}$$

Such a reaction scheme is consistent with the changes in water properties with temperature; at low temperature, the higher water density stabilizes the ions produced by Eq. (4.6), but as the temperature increases production of neutral species is favoured (Eq. 4.5). At much higher temperatures the reaction is expected to become gaslike with an activation energy approaching the gas phase value for H· atom abstraction. As there are no significant fluctuations in solvent properties outside of the near-critical region the Arrhenius behaviour is again expected in this region.

A major focus of radiolysis modelling in support of SCWR development has been on the effects of SCW density on radiolytic yields. Guzonas et al. (2012) suggested that in low-density SCW, water radiolysis is dominated by the gaslike component based on the considerable differences in the physical track structure processes, chemical reactions and the product yields in irradiated water vapour and irradiated liquid water. As an example (Meesungnoen et al., 2013), the recombination of a subexcitation electron with its parent cation (shaded boxes at the top left-hand side of Fig. 4.16) prior to thermalization of the electron requires close approach of opposite charges. Previous simulations (Goulet et al., 1990) showed that, on average, recombination occurs in the first steps of the subexcitation electron random walk. The electrons that recombine tend to be the ones that, in the absence of cations, would be thermalized relatively close to their starting position, a condition not favoured at low SCW density where there are relatively few water molecules. The decrease in geminate electron–cation recombination in low-density SCW will increase the initial yield of $e_{aq}{}^-$ (at a timescale of 10^{-12} s). Meesungnoen et al. (2013) assumed a simple linear density dependence for the probability of the electron–cation geminate recombination, significantly improving the calculated values of $e_{aq}{}^-$ yields at low density (Fig. 4.21). Good agreement was found with the experimental values at 1 ns, although the simulations still slightly underpredicted the experimental data at 60 ps and low density[22]. This work illustrates how a detailed understanding of the experimentally observed density dependencies of yields in irradiated SCW (especially at low density) requires detailed consideration of the effects of density on water structure.

[22] The authors note that they are comparing calculated yields of $e_{aq}{}^-$ for H_2O with values obtained experimentally in D_2O. Bartels et al. (2001) predicted that the 'initial' yield $e_{aq}{}^-$ in D_2O should be ~7% larger than in H_2O; making this correction leads to very good agreement between the measured and calculated $e_{aq}{}^-$ yields at 60 ps.

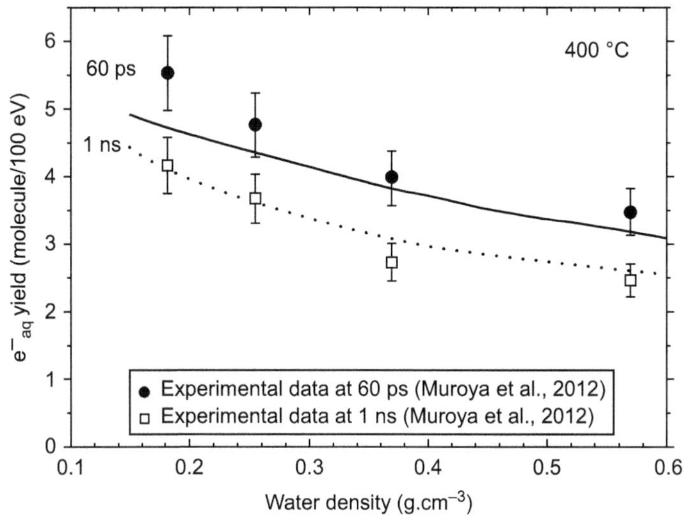

Figure 4.21 Density dependence of the yields of e^-_{aq} in supercritical water at 400 °C measured directly by picosecond pulse radiolysis experiments in D_2O (Muroya et al., 2010; Muroya et al., 2012). The solid and dotted lines show the results of Monte Carlo simulations (Meesungnoen et al., 2013) at 60 ps and 1 ns, respectively, when the density dependence of the geminate electron−cation recombination is incorporated into the calculations.

While microscopic models of SCW radiolysis are providing valuable insights into the role of parameters such as water density on radiolysis yields, they have not yet advanced to the point of being able to provide the quantitative predictions of the concentrations of oxidizing species needed to define the in-core chemistry of an SCWR.

4.4.1.2 Semiempirical modelling

Several groups have taken a more macroscopic, less first-principles approach to develop models of the steady-state concentrations of hydrogen, oxygen and hydrogen peroxide in an SCWR similar to the approach taken by current generation WCRs to predict oxidant concentrations in the coolant. These models require the choice of a set of elementary reactions for the various reactions of the transient radiolysis products and (most importantly) the rate constants for these reactions. The steady-state concentrations can then be calculated by solving the coupled time-dependent rate equations using appropriate values for parameters such as dose and in-core residence time.

The first such simulations published (Yeh et al., 2012; Yeh and Wang, 2013) were for a US pressure vessel-type SCWR concept no longer being actively developed and used g-values at 400 °C that in some cases were extrapolated from 300 °C values using an Arrhenius relationship which, as shown earlier, is not valid. Nonetheless, the final results were valuable in that they showed that very high concentrations of oxidants were possible in an SCWR core. They also modelled the effect of addition of $1\ mg \cdot kg^{-1}$ H_2 at the core inlet, finding that the concentrations of H_2O_2 and O_2 were reduced by several orders of magnitude (Yeh and Wang, 2013).

Subsequent work by Subramanian et al. (2016) led to development of models for the radiation chemistry of water in the near-critical region that attempted to better account for some of the effects of the changing properties of water in this region. Two chemical kinetics models were developed that approached the densities of the near-critical region from liquidlike and gaslike values. The liquid-like radiolysis model (LLRM) used a reaction set similar to that used in an existing radiolysis model developed and validated for liquid water at ambient temperatures (Wren and Ball, 2001; Wren and Glowa, 2000; Joseph et al., 2008; Yakabuskie et al., 2010, 2011). The vapour-like radiolysis model (VLRM) used a reaction set developed for water vapour radiolysis (Arkhipov et al., 2007) and applied to water vapour chemistry at high temperatures (900 K), although at much lower pressures (0.01−1 MPa) and vapour densities ($0.25-1\ kg \cdot m^{-3}$) than those to be used in an SCWR.

Calculations started with a set of initial species concentrations and continued for 10^4 s; this time scale bounds the expected residence time of the coolant in an SCWR core, about 1 s. Calculations were carried out as a function of temperature from 25 to 400 °C using both the VLRM and LLRM models to compare model predictions. Calculations started with pure water (neutral pH, no additives, deaerated) prior to a step function initiation of a continuous radiation flux (in-core γ-radiation fields are estimated to be on the order of a few $1000\ kGy \cdot h^{-1}$ (Janik et al., 2007)). For liquid water and vapour, the system pressure was the saturation pressure (Alexandrov and Grigoreiv, 1999) at the target temperature. In both LLRM and VLRM calculations changes in water density affect the amount of radiation energy absorbed per volume. In the VLRM model calculations, the water density also affects the pseudo-second-order rate constants used for three-body reactions involving a water molecule.

The concentrations of different radiolysis products evolve at different rates with time, and the core residence time was found to have a large effect on the concentrations of the various radiolysis products. The LLRM predicts only a small effect on the radiolysis product concentrations with changes in temperature for a 1 s core residence time because the slower, more temperature-dependent chemical reactions in the water phase only become important at longer times. The VLRM predicts that changes in temperature have a greater effect on the radiolysis product concentrations at $T < 150$ °C (Fig. 4.22). These preliminary calculations were performed at pressures and temperatures along the saturation pressure curve, and as a result the water vapour density changed considerably. The increase in water vapour density increases the rate of energy absorption per unit volume, the rates of the primary radiolysis product formation and the rate of termolecular reactions that involve water molecules. Increasing divergence between LLRM and VLRM predictions with increasing time was found, consistent with the way the two models treat radiolysis and chemistry. At short times the oxidizing species concentrations should depend more strongly on the primary radiolysis production rates than on the reactions of secondary radiolysis products.

In the near-critical region, the predictions of the two models differ by only 2−3 orders of magnitude at the short in-core residence times relevant to an SCWR. At the same time these results illustrate that, as with corrosion product deposition, the near-critical region will be the most challenging for modelling and chemistry control.

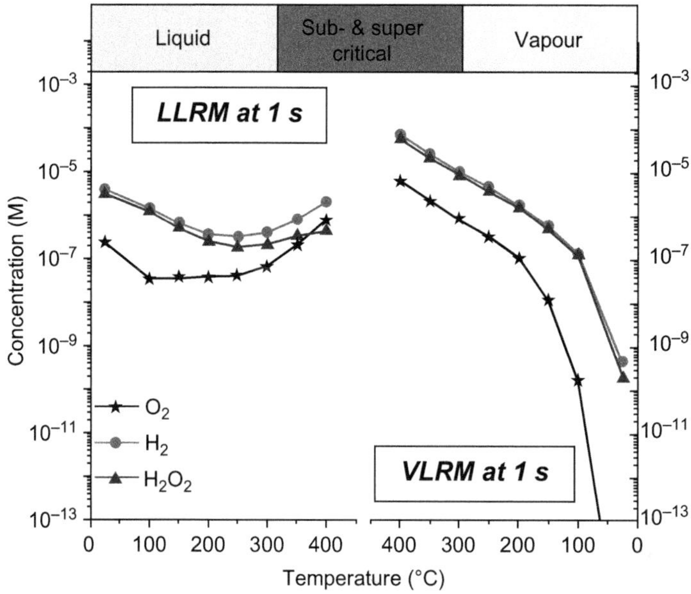

Figure 4.22 Radiolysis product concentrations as a function of temperature predicted by the liquid-like radiolysis model (LLRM) and vapour-like radiolysis model (VLRM) at 1 s and irradiation at a dose rate of 1000 $kGy \cdot h^{-1}$ (Subramanian et al., 2016).

4.4.1.3 Large-scale loop or in-reactor studies

While facilities such as the Canada-Ukraine Electron Irradiation Test Facility (CU-EITF) SCW convection loop in Ukraine and the SCW loop for materials and chemistry testing at Centrum Výzkumu Řež (both described in Chapter 2) have been constructed to study the effects of water radiolysis on material performance, to date only the Ukraine facility has been operational, providing some valuable data on water radiolysis effects. Of particular concern for SCWR materials selection is the behaviour of Cr under oxidizing conditions; the passive films formed on almost all alloys being considered for in-core use in an SCWR are Cr-oxides. Chromium oxides are soluble under oxidizing conditions (Cook and Olive, 2012b) due to the formation of soluble Cr(VI) species. Chromium oxide dissolution as H_2CrO_4 (transpassive dissolution) is observed in BWRs that operate with normal water chemistry (i.e., no hydrogen addition) because of the high concentrations of oxidizing species produced by water radiolysis, and similar behaviour is expected in an SCWR if water radiolysis cannot be controlled (Guzonas et al., 2012; Karasawa et al., 2004; Fujiwara et al., 2007).

Formation of soluble Cr(VI) species has been observed in oxidizing SCW at 25 MPa (e.g., Diago et al., 2007a,b; see also Chapter 5). The equivalent gas phase process of chromia evaporation has been well documented in mixtures of oxygen and water at temperatures above 500 °C (Halversson et al., 2006; Holcomb, 2008, 2009). Chromia evaporation is attributed to the following reaction:

$$1/2Cr_2O_3(s) + 3/4O_2(g) + H_2O \leftrightarrow CrO_2(OH)_2(g) \quad (4.7)$$

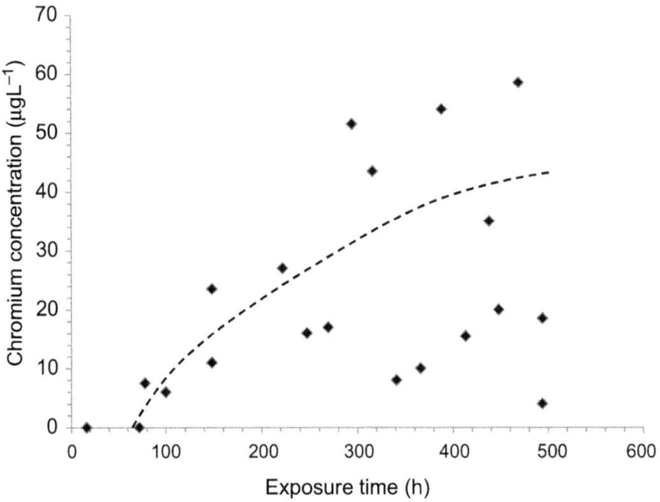

Figure 4.23 Chromium content measured in the loop water during irradiation cycle – 2 at the CU–EITF facility (Bakai et al., 2013).

$CrO_2(OH)_2$ is chemically identical to H_2CrO_4. The process has been found to have a strong dependence on flow, showing that it depends on the mass transfer of the volatile (dissolved) species away from surface.

Significantly, most of the materials testing in support of the development of nuclear steam reheat in the United States was carried out with DO concentrations of 20–50 $mg \cdot kg^{-1}$, whereas most of the recent testing for the SCWR has been carried out with concentrations of 8 $mg \cdot kg^{-1}$ or less. Initial ~500 h tests (Bakai et al., 2013) at temperatures in the vicinity of the critical point under electron irradiation using the CU–EITF facility provide clear evidence for chromium oxide dissolution under the oxidizing conditions produced by water radiolysis. Samples of the outlet water contained 3–5 $\mu g \cdot kg^{-1}$ of oxygen[23] (Fig. 4.23). The conductivity of the water at the test section outlet increased systematically at a typical rate of $\sim 0.03 \pm 0.1\ \mu S \cdot cm^{-1} \cdot h^{-1}$ up to a cumulative total of $\sim 23\ \mu S \cdot cm^{-1}$ indicating release of metals into the coolant due to corrosion. Elemental analysis by ICP-OES indicated chromium concentrations up to 54 $\mu g \cdot L^{-1}$ in the water; there was no detectable chromium in the water before exposure to the electron beam.

In addition to possible enhanced corrosion of in-core alloys, water radiolysis may also affect corrosion product transport. Yakabuskie et al. (2011b) have reported that γ-irradiation of subcritical deaerated aqueous $FeSO_4$ solutions leads to the formation of uniform-sized colloidal particles of γ-FeOOH. Mayanovic et al. (2012) found that radiolysis-induced effects may cause colloidal precipitate formation from tungsten ions in SCW, and that iron species may be either oxidized, reduced or form a

[23] Due to the long sample line length required to isolate the chemistry monitors from the irradiation zone, the actual oxygen concentration in the test sections was likely higher.

precipitate depending on the pressure, temperature and irradiation conditions. The possibility of water radiolysis in pores or crevices inside deposited or corrosion oxides should also be considered (Henshaw et al., 2006).

As discussed earlier, Eq. (4.5) is key to the control of water radiolysis in WCRs. The addition of hydrogen (or ammonia, which breaks down to form hydrogen) to the coolant at a sufficiently high concentration (which varies for different reactor designs) suppresses the net radiolytic formation of oxidizing species. The suppression of net radiolysis under high-temperature steam conditions was studied at the Beloyarsk NPP (Yurmanov et al., 2010; Gruzdev et al., 1970). Initially, Unit 1 used ammonia alone to control pH and suppress net radiolysis indirectly through the radiolysis of ammonia to produce hydrogen. Despite a hydrogen concentration of 45–88 $mL \cdot kg^{-1}$ (at NTP) in the steam entering the superheat channels, the oxygen concentration in the SHS was still 2.3 $mg \cdot kg^{-1}$. Unit 1 tested hydrogen injection to suppress net radiolysis in the superheat channels, discontinuing ammonia feedwater treatment during the studies. Hydrogen injection successfully maintained steam hydrogen concentrations at 1.2–6.2 $mL \cdot kg^{-1}$, resulting in an oxygen concentration in the SHS of less than 0.15 $mg \cdot kg^{-1}$. Further testing showed that oxygen concentration in the SHS could be maintained below 0.03 $mg \cdot kg^{-1}$ with a steam hydrogen concentration greater than 45 $mL \cdot kg^{-1}$. Bartels et al. (2010) postulated that the addition of hydrogen to an SCWR at a concentration of roughly 10 times the current value used in PWRs might be sufficient to suppress the net formation of oxidizing species in rough agreement with the results reported by Yurmanov et al. (2010). It is therefore possible that water radiolysis in the low-density SCW at the core outlet of an SCWR (~70 $kg \cdot m^{-3}$ at 625 °C, 25 MPa) could be controlled in the same manner as in the low-density steam at Beloyarsk (~26.5 $kg \cdot m^{-3}$).

Yurmanov et al. (2010) also noted that, with hydrogen addition to the superheater channels in Unit 1 at Beloyarsk NPP, the turbine ejectors produced an explosive steam–gas mixture of 62% hydrogen and 8% oxygen, requiring air dilution, and a similar issue may arise in an SCWR. The production of additional hydrogen by corrosion (Choudhry et al., 2013) must also be considered.

As part of a joint UKAEA/AECL development program on WCR systems, one of the pressurized water loops (X-4 loop) in the NRX reactor at AECL was converted to steam cooling (LeSurf et al., 1963) to examine the feasibility of using steam as a power reactor coolant. As pH control by addition of alkalis was considered unsuitable in such a reactor concept because of the risk of deposition on the fuel elements causing cladding failure, ammonia was investigated as a substitute, and the radiolysis of ammonia in both boiling water and SHS was studied with various gaseous compositions of the steam in the loop. The steam from the boiling zone of a steam-cooled reactor was expected to contain high concentrations of oxygen (10–30 $mg \cdot kg^{-1}$) and hydrogen (14 $mg \cdot kg^{-1}$). It was hoped that catalytic recombination of oxygen and hydrogen in the steam produced in the boiling zone of such a reactor could be made to recombine in the superheat zone to give steam with a low inlet oxygen content. The behaviour of oxygen and hydrogen added to the X-4 loop was investigated using hydrogen peroxide addition to the coolant as a source of oxygen. Samples were taken before and after the test section continuously and analysed for oxygen. The

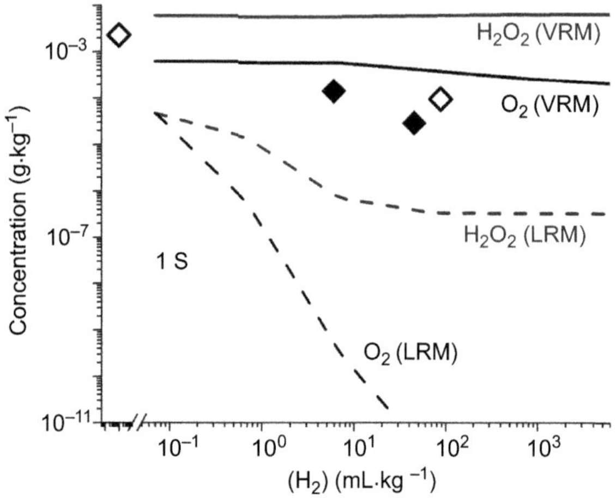

Figure 4.24 Radiolysis product concentrations as a function of initial H_2 addition predicted by the liquid-like radiolysis model (LLRM) and vapour-like radiolysis model (VLRM) at 1 s irradiation at 400 °C and 1000 kGy·h^{-1} (Subramanian et al., 2016). The symbols represent plant data for oxygen concentration measured at Beloyarsk NPP Unit 1 (◆ – direct H_2 addition and ◇ – NH_3 addition).

conditions of operation varied with different fuel charges, but the following is a typical set of conditions:

- Operating pressure (boiler) 57 kgf·cm^{-2} (~5.7 MPa)
- Steam flow through test section 320 kg·h^{-1}
- Total water flow through pumps 2100 kg·h^{-1}
- Temperature of steam entering test section 330 °C
- Temperature of steam leaving test section 430 °C
- Fuel sheath temperature 550–650 °C

Tests with very high concentrations of added hydrogen (1110 mL·kg^{-1}) were carried out with no fuel present in the loop. It was expected that equilibrium would not be established across the test section as it was believed that the presence of a hot fuel element surface was needed to catalytically assist in attainment of equilibrium. The results suggested that significant radiolytically driven recombination of oxygen with hydrogen may be achieved within the superheat zone of a reactor at the oxygen concentrations expected in the steam (20–30 mg·kg^{-1}) provided hydrogen was added at concentrations sufficient to create a considerable excess over stoichiometric proportions.

The models developed by Subramanain et al., (2016) were used to predict the effect of H_2 addition on SCW radiolysis (24 MPa, 148 kg·m^3 for vapour) after 1 s at 400 °C (Fig. 4.24). The LLRM predicted that H_2 addition could reduce (but not suppress) the radiolytic production of O_2; at 1 s the DO concentration is below 10^{11} mol·dm^{-3} when the initial concentration of added H_2 was $>10^{-4}$ mol·dm^{-3} (~10 mL·kg^{-1}).

However, adding H_2 is less effective at reducing the production of H_2O_2, the concentration of H_2O_2 at 1 s remaining at $\sim 10^{-7}$ mol·dm^{-3} even when the concentration of added H_2 was $>10^{-2}$ mol·dm^{-3} (on the order of 1000 mL·kg^{-1}); this hydrogen concentration is far higher than those used in existing WCRs. These results are similar to those of Yeh and Wang (2013). The VLRM predicts that addition of any reasonable amount of H_2 will not suppress the production of either O_2 or H_2O_2 (Fig. 4.24, a similar result is seen at longer times where the steady-state concentrations of O_2 and H_2O_2 are even higher). The only discernible effect of H_2 addition was a reduction in the concentration of •O with addition of H_2 at a concentration $>10^{-4}$ M due to the reaction:

$$\cdot O + H_2 \leftrightarrow \cdot OH + H\cdot \tag{4.8}$$

Both models predict that the addition of H_2 may not be effective in reducing the radiolytic production of H_2O_2. Fig. 4.24 also shows data for the oxygen concentration measured at the Beloyarsk NPP with added hydrogen and with added ammonia. The good agreement between the measured oxygen concentrations and those predicted by the VLRM model suggests the underlying soundness of the approach.

4.5 Chemistry control in a supercritical water-cooled reactor

While deployment of SCWRs is not expected until the 2020–30 timeframe, a set of water chemistry specifications with a solid technical basis is required for on-going general corrosion and EAC testing of candidate in-core materials as these phenomena are strongly influenced by the environment. Returning to Fig. 4.6, the discussions on corrosion product transport to the core, activity transport from the core and water radiolysis presented in this chapter allow the following three conclusions, each addressing one of the three processes identified in the figure, to be made:

1. It is probable that the in-core deposition of impurities and corrosion products can be minimized in an SCWR by adopting the best practices for feedtrain materials selection and chemistry control used in SCFPPs and BWRs. The data suggest that an oxygenated feedwater chemistry will result in fewer problems than the use of ammonia, hydrazine or LiOH. Long-term tests using realistic concentrations of expected corrosion products and impurities are needed, first in out-reactor loops and eventually using in-reactor test facilities, to assess the effects of transport and in-core deposition of these species.
2. Water radiolysis remains the key outstanding issue preventing definition of the SCWR in-core environment. Data clearly show that the high concentrations of oxidants produced by unsuppressed water radiolysis can lead to the release of chromium from alloy surfaces. Operating experience at Beloyarsk NPP with nuclear steam reheat suggests that hydrogen addition could suppress the net radiolysis production of oxidizing species. Recent modelling suggests that at lower SCW densities the production of oxygen and hydrogen peroxide by water radiolysis cannot be 'suppressed' by the addition of hydrogen at the core inlet but their concentrations can be reduced. Therefore it is recommended that hydrogen be added but no

Table 4.3 Proposed chemistry specifications for a supercritical water-cooled reactor

Parameter	Value		Unit
	Guzonas and Cook (2015)	Yurmanov et al. (2010)	
Chloride	0.25	<2	$\mu g \cdot kg^{-1}$
Sulphate	2.0	–	$\mu g \cdot kg^{-1}$
Iron	0.1	<5	$\mu g \cdot kg^{-1}$
Silica	300	<2	$\mu g \cdot kg^{-1}$
Sodium	–	<2	$\mu g \cdot kg^{-1}$
Initial feedwater dissolved oxygen	200		$\mu g \cdot kg^{-1}$
Final feedwater dissolved oxygen	50	<20	$\mu g \cdot kg^{-1}$
Hydrogen	To be determined	To be determined	–
pH	7.0	6.8–7.1	–
Conductivity	8	<0.1	$\mu S \cdot m^{-1}$
Oil	–	<100	$\mu g \cdot kg^{-1}$

specification for the concentration can be made at this time. Addition of hydrogen implies a 'dual' water chemistry in which the oxygenated feedwater chemistry is replaced by hydrogen water chemistry at the core inlet (Kysela et al., 2009). Materials testing should therefore be performed over a range of DO and hydrogen concentrations that bound the expected concentration of these species. The possibility that an explosive mixture of hydrogen and oxygen could exist in the SCW exiting the core will need to be considered during the design stage of any SCWR.

3. Simple calculations making reasonable assumptions show that the concentration of ^{60}Co in the SCWR coolant at the core outlet may not be significantly different than that found in current generation WCRs. While the solubilities of metal oxides in low-density SCW are low, they are still much higher than the solubilities of the same oxides in the lower pressure steam found in BWRs. Better data on the solubilities of relevant species under SCWR conditions are needed. The release and transport of particulate species must also be considered. It must be emphasized that, unlike BWRs where the phase change results in little activity transport out of the core in the steam, the solubilities of most metals in SCW, while low, is nonzero; the single-phase coolant can also easily transport particulates. Shielding of the main steam lines and the high pressure turbines will be necessary.

Kysela et al. (2009), Yurmanov et al. (2009), Guzonas (2010) and Guzonas and Cook (2015) have discussed possible water chemistry control strategies for an SCWR. Table 4.3 lists preliminary SCWR chemistry specifications proposed by Guzonas and Cook (2015) and by Yurmanov et al. (2010). The two sets of specifications are similar and form a good basis for materials testing, bearing in mind the comments in item 2 above concerning water radiolysis.

4.6 Molecular dynamics simulations

This final section serves as a transition to the last two chapters of the book by considering the structure of SCW at surfaces. At subcritical temperatures, it is now well established that an electrical double layer forms at metal surfaces because the electrically charged surface orders the water molecules and ions at the surface, creating a local imbalance in cation and anion density that gradually decays away with distance from the surface. Corrosion reactions consist of two half-reactions. The oxidation reaction, $M = M^{2+} + 2e^-$, produces a metal ion that can undergo hydrolysis reactions in solution; the electrons produced by the oxidation reaction are transported in the metal, while the local charge imbalance in solution resulting from the production or consumption of ions at the cathodic and anodic sites is balanced by ion transport through the solution.

At $T \gg T_c$, the SCW density is low and no electrical double layer can exist. Both experimental and modelling studies have shown that the local structure around ions in SCW is different than that of bulk SCW (Tucker, 1999). Taking into account this local structure and the heterogeneous, 'clustery' nature of SCW (Fig. 4.2), Guzonas and Cook (2012) proposed a possible surface structure at temperatures above 500 °C where the SCW density is low and the oxidation reaction is $M + O_2 \rightarrow MO_2$ (direct chemical oxidation, see Chapter 5). The near-surface region is predicted to consist of small clusters of 'liquidlike' water separated by regions of 'gaslike' water. Guzonas and Cook (2012) proposed a transition region at near-critical temperatures where discrete anodic and cathodic sites still exist, but are closer together, as the stability of the charge-conducting ions in solution decreases and ion pairs become the more favoured solution species. As a result, the charge balance (or imbalance) in solution must be more localized. The large density fluctuations that occur in the near-critical region in the bulk water are also expected to exist at the surface, and a continuous, stable double layer may not exist over the entire surface. The transition from a double-layer structure may occur gradually or suddenly as a surface phase transition[24]. It is expected that the random nature of the clusters in the bulk fluid would be organized differently at a metal or oxide surface, and the structure of SCW directly adjacent to the surface will be quite different than that of the bulk fluid, being more organized with a somewhat higher density. It was suggested that a continuous, liquidlike film may persist at the surface at temperatures well above T_c.

MD simulations are a valuable tool for obtaining detailed information of the reaction dynamics of water and solvated species in SCW (Kallikragas et al., 2015b). Water models are available that can accurately reproduce physical and thermodynamic properties, including self-diffusion coefficients, over a wide range of temperatures and pressures, making it possible to simulate environments of interest for SCWR

[24] As suggested by Prof. I. Svishchev.

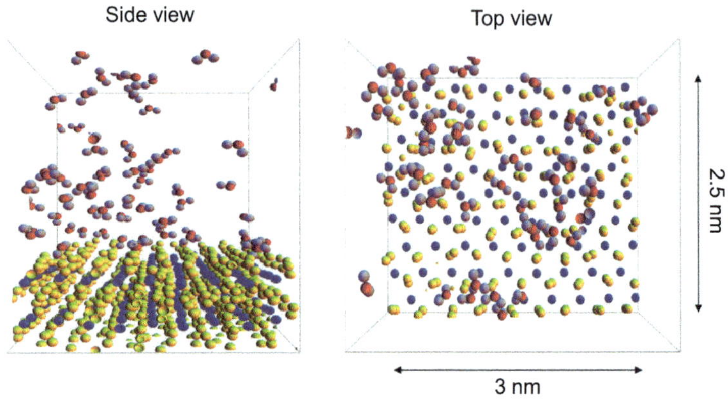

Figure 4.25 Atomic densities distribution at the supercritical water–iron (II) hydroxide interface at 715K. Surface iron atoms are shown in dark blue, surface oxygen atoms in yellow, water oxygen atoms in red and water hydrogen atoms in violet.
Adapted from Svishchev, I.M., Kallikragas, D.T., Plugatyr, A.Y., 2013b. Molecular dynamics simulations of supercritical water at the iron hydroxide surface. J. Supercrit. Fluids 78, 7–11.

development and use the insights gained to guide experimental programs on corrosion and EAC. The behaviour of SCW confined between electrostatically neutral but hydrophilic surfaces of iron (II) hydroxide was determined through MD simulations (Svishchev et al., 2013b) conducted at 715, 814 and 913K at water densities representative of those expected in an SCWR core. The surface water layer was found to establish itself quickly, reaching ~90% of its final density within the first 250 ps of simulation. The atomic density profile of the iron hydroxide–SCW system shows that the majority of the adsorbed water molecules are positioned within ~1.5 Å from the edge of the iron hydroxide surface with their dipoles oriented away from the surface. As predicted by Guzonas and Cook (2012), images of water configurations (Fig. 4.25) show localized clustering on the ionic surface at supercritical conditions, a nonuniform adsorption layer being formed with a higher (than bulk) density. The ratio of the density of water at the surface to the density in the bulk was found to increase in going from subcritical to supercritical water, and above T_c, this ratio was 2–3.5 depending on temperature and bulk water density (Fig. 4.26). This enhancement in surface water concentration may affect the corrosion mechanism (Chapter 5) in the near-critical region, especially the pressure (density) dependence, as the bulk SCW density will be lower than the surface water density, and will therefore not accurately reflect the surface chemistry conditions. As water radiolysis is also affected by the density, enhanced (or reduced) production of various radiolysis products in the near-surface region (and in pores and cracks) should be considered. The number of water molecules per exposed surface OH group decreased with temperature, from 5.71 at 567K to 0.4 at 913K (water density of 79 $kg \cdot m^{-3}$).

Kalligragas et al. reported the use of MD simulations to determine diffusion coefficients, hydration numbers and density profiles of both dilute chloride ion

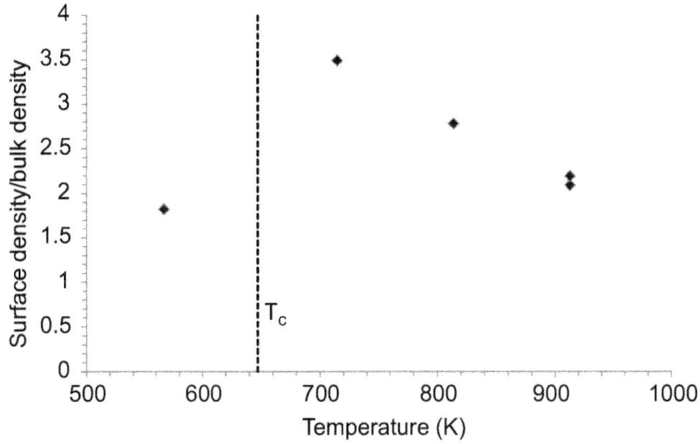

Figure 4.26 Relative densities of the surface water layer with respect to the bulk water densities at 567, 715, 814, and 913K.
Adapted from Kallikragas, D.T., Guzonas, D.A., Svishchev, I.M., 2015b. Properties of aqueous systems relevant to the SCWR via molecular dynamics simulations. AECL Nucl. Rev. 4, 9–21.

(Kalligragas et al., 2015a) and of dilute oxygen molecules (Kalligragas et al., 2016) confined between two closely spaced $Fe(OH)_2$ surfaces in high-temperature subcritical water as well as in SCW. Small spacings (1, 2 and 8 nm) were chosen to mimic the environment expected at crack tips; as noted by Staehle (2006, 2010), such an environment requires a molecular, not a continuum, analysis. The 1 nm spacing had a confining effect on the structure of the water that disappeared at the 8 nm spacing. Hydration increased with spacing size but diminished as the density decreased and the temperature increased. The number of H-bonded waters around the chloride ion remained relatively constant for all three surface separations at each state point. The size of the gap had little effect on the diffusion of the chloride ion but affected that of the water. For oxygen at $T > T_c$, it was found that the probability of finding a water molecule was at a minimum where the probability of finding an oxygen atom of O_2 was at a maximum, i.e., the O_2 was located outside of the water clusters.

For water, the largest diffusion coefficient was seen in the 8 nm systems and for chloride in the 1 nm systems. In both cases the diffusion coefficients in the confined systems were less than those in the simulations of the bulk. Clustering of the water in the 1 nm spacing increases in the middle of the gap, this ordering being lost at larger gap sizes, with the water approaching a bulklike behaviour in the larger 8 nm systems. Molecular configurations show that the chloride ion diffuses to locations near the surface and carries with it a small number of waters of hydration. The diffusion coefficient of O_2 generally increases as the spacing between the surfaces increased (the opposite behaviour to chloride ion). A key result of this work was that at supercritical temperatures O_2 diffused through nanometre-sized cracks more readily than in the bulk solution.

MD was also used to study the nucleation dynamics of NaCl nanoparticles in SCW (Svishchev and Guzonas, 2011). Particles of approximately 15–30 ions appeared critical for the SCW states examined. Nucleation rates appear to be on the order of 10^{28} $cm^{-3}\cdot s^{-1}$ indicating very fast nucleation dynamics. Critical salt nuclei in SCW were found to have an amorphous structure. The simulation results show that HCl and NaOH can partition between coexisting supercritical fluid and NaCl clusters, consistent with the presence of acidic condensates and alkaline deposits under hydrothermal conditions.

References

Adschiri, T., Sue, K., Arai, K., Watanabe, Y., 2001. Estimation of metal oxide solubility and understanding corrosion in supercritical water. In: *Corrosion 2001*, Paper 01354.

Akiya, N., Savage, P.E., 2002. Roles of water in chemical reactions in high-temperature water. Chem. Rev. 102, 2725–2750.

Alcorn, C., Brodovitch, J.-C., Ghandi, K., Kennedy, A., Percival, P.W., Smith, M., 2011. Kinetics of the reaction between H• and superheated water probed with muonium. In: 5th Int. Sym. SCWR (ISSCWR-5), Vancouver, British Columbia, Canada, March 13–16, 2011. Paper P130.

Alcorn, C.D., Brodovitch, J.-C., Percival, P.W., Smith, M., Ghandi, K., 2014. Kinetics of the reaction between H• and superheated water probed with muonium. Chem. Phys. 435, 29–39. https://doi.org/10.1016/j.chemphys.2014.02.016.

Aleksandrova, V.N., Veselkin, A.P., Levich, A.A., Lyutov, M.A., Sklyarov, V.P., Khandamirov, Yu.E., Shchapov, G.A., 1968. Investigation of the radioactivity of long-lived isotopes in the coolant of the I.V. Kurchatov nuclear power station at Beloyarsk. At. Energiya 24, 222.

Alexandrov, A.A., Grigoriev, B.A., 1999. The Tables of Thermophysical Properties of Water and Steam. MEI Press (in Russian).

Anisimov, M.A., Sengers, J.V., Levelt Sengers, J.M.H., 2004. Near-critical behavior of aqueous systems. In: Palmer, D.A., Fernàndez-Prini, R., Harvey, A.H. (Eds.), Aqueous Systems at Elevated Temperatures and Pressures. Elsevier Academic Press (Chapter 2).

Arkhipov, O.P., Verkhovskaya, A.O., Kabakchi, S.A., Ermakov, A.N., 2007. Development and verification of a mathematical model of the radiolysis of water vapor. At. Energy 103 (5), 870–874. https://doi.org/10.1007/s10512-007-0138-4.

Bakai, A.S., Guzonas, D.A., Boriskin, V.N., Dovbnya, A.N., Dyuldya, S.V., 2013. Supercritical water convection loop for SCWR materials corrosion tests under electron irradiation: first results and lessons learned. In: 6th International Symposium on Supercritical Water-cooled Reactors (ISSCWR-6), March 03–07, 2013, Shenzhen, Guangdong, China.

Bartels, D.M., Gosztola, D., Jonah, C.D., 2001. Spur decay kinetics of the solvated electron in heavy water radiolysis. J. Phys. Chem. A 105 (34), 8069–8072. https://doi.org/10.1021/jp012153u.

Bartels, D.M., Jonah, C., Edwards, E., Janik, D., Haygarth, K., Sims, H., Marin, T., Takahashi, K., Janik, I., Kanjara, K., 2010. Hydrogen water chemistry: can it work in a supercritical water reactor?. In: Proceedings of the 8th Int'l Radiolysis, Electrochemistry & Materials Performance Workshop, Quebec City, Canada, October 8, 2010.

Bartels, D.M., 2009. Comment on the possible role of the reaction $H\bullet + H_2O \rightarrow H_2 + \bullet OH$ in the radiolysis of water at high temperatures. Radiat. Phys. Chem. 78 (3), 191–194.

Bergeron, A., Hamilton, H., 2013. Fabrication and characterization of Canadian SCWR SIMFUEL. In: 6th International Symposium on Supercritical Water-cooled Reactors (ISSCWR6), March 03–07, 2013, Shenzhen, Guangdong, China.

Bevilacqua, F., Brown, G.M., 1963a. Chloride Deposition from Steam onto Superheater Fuel Clad Materials. General Nuclear Engineering Corporation Report GNEC 295.
Bevilacqua, F., Brown, G.M., 1963b. Chloride corrosion effects on BONUS superheater materials. In: Terminal Report Under Job No. 66–14, Chloride Corrosion Program. USAEC Report GNEC 257.
Bohnsack, G., 1987. The Solubility of Magnetite in Water and in Aqueous Solutions of Acid and Alkali. Hemisphere Publishing Corporation, Washington.
Bondarenko, G.V., Gorbaty, Yu.E., 1973. ν3 infrared spectra of HDO at high pressures and temperatures. Dokl. Phys. Chem. 210, 369.
Bonin, J., Janik, I., Janik, D., Bartels, D.M., 2007. Reaction of the hydroxyl radical with phenol in water up to supercritical conditions. J. Phys. Chem. A 111 (10), 1869–1878.
Brazhkin, V.V., Fomin, Yu.D., Lyapin, A.G., Ryzhov, V.N., Tsiok, E.N., 2011. Widom line for the liquid-gas transition in Lennard-Jones system. J. Phys. Chem. B 115, 14112–14115.
Brosseau, F., Guzonas, D.A., Tremaine, P., 2010. The solubility of magnetite and nickel ferrite under supercritical water reactor coolant conditions. In: 2nd Canada-China Joint Workshop on Supercritical Water-cooled Reactors (CCSC-2010), Toronto, Ontario, Canada, April 25–28, 2010.
Brosseau, F.J.R., 2010. Complexation and Hydrolysis of Aqueous Iron and Nickel With Chloride and Ammonia Under Hydrothermal Conditions (MSc. thesis). Faculty of Graduate Studies, University of Guelph.
Burns, W.G., Sims, H.E., 1981. Effect of radiation type in water radiolysis. J. Chem. Soc. Faraday Trans. 77, 2803–2813.
Burrill, K.A., 2000. Water chemistries and corrosion product transport in supercritical water in reactor heat transport systems. In: Proceedings of the 8th BNES Conference on Water Chemistry of Nuclear Reactor Systems, Bournemouth, UK, 2000, p. 357.
Bursik, A., Dooley, R.B., Gunn, D., Larkin, B.A., Oliker, I., Pocock, F.J., Ryan, D., Webb, L.C., 1994. Cycle Chemistry Guide for Fossil Plants: Oxygenated Treatment. Electric Power Research Institute Report TR-102285.
Burton, M., 1969. Chem. Eng. News 47, 86.
Busboom, H.J., Boyle, R.F., Harling, G., Hazel, V.E., 1966. Post-irradiation Examination of EVESR Mark III Superheat Fuel (0.008 Inch Cladding Failure). General Electric Report GEAP5135.
Carvajal-Ortiz, R.A., Choudhry, K.I., Svishchev, I.M., Guzonas, D.A., 2012a. Chemical species influences under flow conditions relevant to PWR and SCWR. In: Nuclear Plant Chemistry Conference (NPC 2012), Paris, France, September 23–28, 2012.
Carvajal-Ortiz, R.A., Plugatyr, A., Svishchev, I.M., 2012b. On the pH control at supercritical water-cooled reactor operating conditions. Nucl. Eng. Des. 248, 340–342.
Chialvo, A.A., Gruszkiewicz, M.S., Cole, D.R., 2010. Ion-pair association in ultrasupercritical aqueous environments: successful interplay among conductance experiments, theory, and molecular simulations. J. Chem. Eng. Data 55, 828–1836.
Choudhry, K.I., Carvajal-Ortiz, R.A., Svishchev, I.M., 2013. Thermochemical hydrogen production under SCWR conditions. In: 6th International Symposium on Supercritical Water-cooled Reactors (ISSCWR-6), March 03–07, 2013, Shenzhen, Guangdong, China. Paper ISSCWR6-13079.
Choudhry, K.I., Guzonas, D.A., Kallikragas, D.T., Svishchev, I.M., 2016. On-line monitoring of oxide formation and dissolution on Alloy 800H in supercritical water. Corros. Sci. 111, 574–582.
Christensen, H., 1981. Effects of water radiolysis on corrosion in nuclear reactors. Radiat. Phys.-Chem. 18, 147–158.

Chudnovskaya, I.I., Myakas, V.I., Buchis, S.V., Shtern, Z. Yu, 1988. The state of internal tube deposits when hydrazine water chemistry treatment is practised. Therm. Eng. 35, 96–99.

Cialone, H.J., Wright, I.G., Wood, R.A., Jackson, C.M., 1986. Circumferential Cracking of Supercritical Boiler Water-Wall Tubes. Electric Power Research Institute Report CS-4969.

Cohen, P., 1985. Water Coolant Technology of Power Reactors. American Nuclear Society.

Cook, W.G., Olive, R.P., 2012a. Corrosion product deposition on two possible fuel geometries in the Canadian-SCWR concept. In: 3rd China-Canada Joint Workshop on Supercritical-water-cooled Reactors (CCSC-2012), Paper 12064, Xi'an, China, April 18–20, 2012.

Cook, W.G., Olive, R.P., 2012b. Pourbaix diagrams for chromium, aluminum and titanium extended to high-subcritical and low-supercritical conditions. Corros. Sci. 58, 291.

Cook, W.G., Olive, R.P., 2013. Corrosion product transport and deposition in the Canadian supercritical water-cooled reactor. In: Proceedings of the 16th International Conference on the Properties of Water and Steam, September 1–5, 2013. University of Greenwich, London, UK.

Cowan, R., Ruhle, W., Hettiarachchi, S., 2011. Introduction to Boiling Water Reactor Chemistry, vol. 1. Advanced Nuclear Technology International, Sweden.

Cowan, R., Ruhle, W., Hettiarachchi, S., 2012. Introduction to Boiling Water Reactor Chemistry, vol. 2. Advanced Nuclear Technology International, Sweden.

Daigo, Y., Watanabe, Y., Sue, K., 2007a. Effect of chromium ion from autoclave material on corrosion behavior of nickel-based alloys in supercritical water. Corrosion 63, 277.

Daigo, Y., Watanabe, Y., Sue, K., 2007b. Corrosion mitigation in supercritical water with chromium ion. Corrosion 63, 1085.

Deeva, Z.V., Saichuk, L.E., Panova, G.N., Moreva, T.V., Belyaeva, M.V., 1986. The results of operation and investigation of neutral/hydrazine water treatment in supercritical power-generating units. Therm. Eng. 33 (5), 265–267.

Deng, J., Deng, F., 2012. The setting of sipping test devices for irradiated fuel in nuclear power plant. J. Energy Power Eng. 6, 1935.

Dickinson, N.L., Keilbaugh, W.A., Pocock, F.J., 1958. Some physico-chemical phenomena in supercritical water. In: ASME Paper No. 58-A-267 Presented at the ASME Annual Meeting of Committee on Boiler Feedwater Studies and Power Division, New York.

Dollezhal, N.A., Krasin, A.K., Aleshchenkov, P.I., Galanin, A.N., Grigoryants, A.N., Emelyanov, I.Ya., Kugushev, N.M., Minashin, M.E., Mityaev, U.I., Florinsky, B.V., Sharpov, N.N., 1958. Uranium-graphite reactor with superheated high pressure steam. In: Proceeding 2nd Int. Conf. on the Peaceful Uses of Atomic Energy, 8, p. 398. UN, Geneva.

Dollezhal, N.A., Aleshchenkov, P.I., Evdokimov, Yu.V., Emel'yanov, I.Ya., Ivanov, B.G., Kochetkov, L.A., Minashin, M.E., Mityaev, Yu.I., Nevskii, V.P., Shasharin, G.A., Sharapov, V.N., Orlov, K.K., 1969. Operating experience with the Beloyarsk nuclear power station. At. Energiya 27, 379.

Dooley, R.B., Ball, M., Bursik, A., Rziha, M., Svoboda, R., 2004. Water chemistry in commercial water-steam cycles. In: Palmer, D.A., Fernàndez-Prini, R., Harvey, A.H. (Eds.), Aqueous Systems at Elevated Temperatures and Pressures. Elsevier Academic Press (Chapter 17).

Edwards, M., Rousseau, S., Guzonas, D., 2014. Corrosion wall loss and oxide film thickness on SCWR fuel cladding. In: 2014 Canada-China Conference on Advanced Reactor Development (CCCARD-2014) Niagara Falls, Ontario Canada, April 27–30, 2014. Paper CCCARD2014-49.

Elliott, A.J., Bartels, D.M., 2009. The Reaction Set, Rate Constants and G-Values for the Simulation of the Radiolysis of Light Water over the Range 20 to 350 °C Based on Information Available in 2008. Atomic Energy of Canada Ltd. Report AECL-153-127160-001.

EPRI, 2004. Radiation Field Control Manual. Electric Power Research Institute Report 1003390, Palo Alto CA.

Franck, E.U., Roth, K., 1967. Infrared absorption of deuterium hydroxide in water at high pressures and temperatures. Discuss. Faraday Soc. 43, 108–114.

Fujiwara, K., Watanabe, K., Domae, M., Katsumura, Y., 2007. Stability of chromium oxide film formed by metal organic chemical vapor deposition in high temperature water up to supercritical region. In: Corrosion 2007, Nashville, Tennessee, March 11–15, 2007.

Gabrielli, F., Schwevers, H., 2008. Design factors and water chemistry practices - supercritical power cycles. In: Proceedings of the 15th International Conference on the Properties of Water and Steam, September 7–11, 2008, Berlin, Germany.

Galkin, A.A., Lunin, V.V., 2005. Subcritical and supercritical water: a universal medium for chemical reactions. Russ. Chem. Rev. 74, 21–35.

Galobardes, J.F., Van Hare, D.R., Rogers, L.B., 1981. Solubility of sodium chloride in dry steam. J. Chem. Eng. Data 26, 363.

Ghandi, K., Addison-Jones, B., Brodovitch, J.-C., McKenzie, I., Percival, P.W., 2002. Near-diffusion-controlled reactions of muonium in sub- and supercritical water. Phys. Chem. Chem. Phys. 4, 586–595.

Ghandi, K., Addison-Jones, B., Brodovitch, J.-C., Kecman, S., McKenzie, I., Percival, P.W., 2003. Muonium kinetics in sub- and supercritical water. Phys. B 326 (1–4), 55–60.

Ghandi, K., McFadden, R.M.L., Cormier, P.J., Satija, P., Smith, M., 2012. Radical kinetics in sub- and supercritical carbon dioxide: thermodynamic rate tuning. Phys. Chem. Chem. Phys. 14, 8502–8505.

Gorbaty, Yu.E., Kalinichev, A.G., 1995. Hydrogen bonding in supercritical water. 1. Experimental results. J. Phys. Chem. 99, 5336–5340.

Goulet, T., Patau, J.P., Jay-Gerin, J.P., 1990. Influence of the parent cation on the thermalization of subexcitation electrons in solid water. J. Phys. Chem. 94 (18), 7312–7316.

Gruzdev, N.I., Shchapov, G.A., Tipikin, S.A., Boguslavskii, V.B., 1970. Investigating the water conditions in the second unit at Beloyarsk. Therm. Eng. 17 (12), 22–25 (Translated from Russian).

Guzonas, D.A., Cook, W.G., 2012. Cycle chemistry and its effect on materials in a supercritical water-cooled reactor: a synthesis of current understanding. Corros. Sci. 65, 48.

Guzonas, S., Cook, W., 2015. Water chemistry specifications for the Canadian supercritical water-cooled reactor concept. In: 7th International Symposium on Supercritical Water-cooled Reactors (ISSCWR-7), March 15–18, 2015, Helsinki, Finland. Paper ISSCWR7-2089.

Guzonas, D.A., Qiu, L., 2004. A predictive model for radionuclide deposition around the CANDU heat transport system. In: Proceedings of International Conference Water Chemistry of Nuclear Reactor Systems, San Francisco, October 11–14, 2004.

Guzonas, D.A., Qiu, L., 2013. Activity transport in a supercritical water-cooled reactor. In: 6th International Symposium on Supercritical Water-cooled Reactors, March 03–07, 2013, Shenzhen, Guangdong, China.

Guzonas, D.A., Wills, J., Do, T., Michel, J., 2007. Corrosion of candidate materials for use in a supercritical water CANDU® reactor. In: Canadian Nuclear Society – 13th International Conference on Environmental Degradation of Materials in Nuclear Power Systems 2007, pp. 1250–1261.

Guzonas, D.A., Wills, J., Dole, H., Michel, J., Jang, S., Haycock, M., Chutumstid, M., 2010. Steel corrosion in supercritical water: an assessment of the key parameters. In: 2nd Canada-China Joint Workshop on Supercritical Water-cooled Reactors (CCSC-2010), Toronto, Ontario, April 25–28, 2010.

Guzonas, D., Tremaine, P., Brosseau, F., Meesungnoen, J., Jay-Gerin, J.-P., 2012. Key water chemistry issues in a supercritical-water-cooled pressure-tube reactor. Nucl. Technol. 179, 205.

Guzonas, D., Qiu, L., Livingstone, S., Rousseau, S., 2016. Fission product release under supercritical water-cooled reactor conditions. J. Nucl. Eng. Rad. Sci. 2, 021010–021011, 021010–021016.

Guzonas, D.A., 2010. Chemistry control strategies for a supercritical water-cooled reactor. In: Nuclear Plant Chemistry Conference 2010, Quebec City, Canada, October 3–7, 2010.

Guzonas, D., 2014. Extreme water chemistry-how Gen IV water chemistry research improves Gen III water-cooled reactors. In: 19th Pacific Basin Nuclear Conference (PBNC 2014), Vancouver, B.C., Canada, August 24–28, 2014.

Halvarsson, M., Tang, J.E., Asteman, H., Svensson, J.-E., Johansson, L.-G., 2006. Microstructural investigation of the breakdown of the protective oxide scale on a 304 steel in the presence of oxygen and water vapour at 600 °C. Corros. Sci. 48, 2014.

Han, Z., Muroya, Y., 2009. Development of a new method to study elution properties of stainless materials in subcritical and supercritical water. In: 4th International Symposium on Supercritical Water-cooled Reactors, Heidelberg, Germany, March 8–11, 2009.

Haygarth, K., Bartels, D.M., 2010. Neutron and β/γ radiolysis of water up to supercritical conditions. 2. SF_6 as a scavenger for hydrated electron. J. Phys. Chem. A 114, 7479–7484.

Hazel, V.E., Boyle, R.F., Busboom, H.J., Murdock, T.B., Skarpelos, J.M., Spalaris, C.N., 1965. Fuel Irradiations in the ESADE-VBWR Nuclear Superheat Loop. General Electric Report GEAP-4775.

Hazelton, R.F., 1987. Characteristics of Fuel Crud and its Impact on Storage, Handling and Shipment of Spent Fuel. Pacific Northwest Laboratory. PNL-6273.

Henshaw, J., McGurk, J.C., Sims, H., Tuson, A., Dickinson, S., Deshon, J., 2006. A model of chemistry and thermal hydraulics in PWR fuel crud deposits. J. Nucl. Mater. 353, 1–11.

Ho, P.C., Palmer, D.A., Wood, R.H., 2000. Conductivity measurements of dilute aqueous LiOH, NaOH and KOH solutions to high temperatures and pressures using a flow-through cell. J. Phys. Chem. B 104, 12084–12089.

Hodes, M., Marrone, P.A., Hong, G.T., Smith, K.A., Tester, J.W., 2004. Salt precipitation and scale control in supercritical water oxidation – Part A: Fundamentals and research. J. Supercrit. Fluids 29, 265–288.

Hoffmann, M.M., Conradi, M.S., 1997. Are there hydrogen bonds in supercritical water? J. Am. Chem. Soc. 119, 3811–3817.

Holcomb, G.R., 2008. Calculation of reactive-evaporation rates of chromia. Oxid. Met. 69, 163.

Holcomb, G.R., 2009. Steam oxidation and chromia evaporation in ultrasupercritical steam boilers and turbines. J. Electrochem. Soc. 156, 292.

Holster, W.T., Schneer, C.J., 1961. Hydrothermal magnetite. Geol. Soc. Am. Bull. 72, 369–386.

IAEA, 1994. Decontamination of Water Cooled Reactors. International Atomic Energy Agency Technical Report 365, Vienna.

IAEA, 2002. Fuel failure in water reactors: causes and mitigation. In: Proceedings of a Technical Meeting Held in Bratislava, Slovakia, IAEA-TECDOC-1345, June 2002.

Imre, A.R., Deiters, U.K., Kraska, T., Tiselj, I., 2012. The pseudocritical regions for supercritical water. Nucl. Eng. Des. 252, 179–183.

Ishigure, K., Kawaguchi, M., Oshoma, K., 1980. The effect of radiation on the corrosion product release from metals in high temperature water. In: Proceedings of the 4th BNES Conference on Water Chemistry of Nuclear Reactor Systems, Bournemouth, UK. Paper 43.

Janik, D., Janik, I., Bartels, D.M., 2007. Neutron and beta/gamma radiolysis of water up to supercritical conditions. 1. Beta/gamma yields for H(2), H(.) atom, and hydrated electron. J. Phys. Chem. A 111, 7777–7786.

Joseph, J.M., Seon Choi, B., Yakabuskie, P., Wren, J.C., 2008. A combined experimental and model analysis on the effect of pH and O_2(aq) on γ-radiolytically produced H_2 and H_2O_2. Radiat. Phys. Chem. 77, 1009–1020.

Kalinichev, A.G., Bass, J.D., 1997. Hydrogen bonding in supercritical water. 2. Computer simulations. J. Phys. Chem. A 101, 9720–9727.

Kallikragas, D.T., Plugatyr, A.Yu., Guzonas, D.A., Svishchev, I.M., 2015a. Effect of confinement on the hydration and diffusion of chloride at high temperatures. J. Supercrit. Fluids 97, 22–30.

Kallikragas, D.T., Guzonas, D.A., Svishchev, I.M., 2015b. Properties of aqueous systems relevant to the SCWR via molecular dynamics simulations. AECL Nucl. Rev. 4, 9–21.

Kallikragas, D.T., Choudhry, K.I., Svishchev, I.M., 2016. Diffusion and hydration of oxygen confined in $Fe(OH)_2$ nano-cracks at high temperatures. J. Supercrit. Fluids 120, 345–354.

Karakama, K., Rogak, S.N., Alfantazi, A., 2012. Characterization of the deposition and transport of magnetite particles in supercritical water. J. Supercrit. Fluids 71, 11–18.

Karasawa, H., Fuse, M., Kiuchi, K., Katsumura, Y., 2004. Radiolysis of supercritical water. In: 5th Int. Workshop on LWR Coolant Water Radiolysis and Electrochemistry, San Francisco, USA, October 15, 2004.

Kontorovich, L.Kh., Rogal'skaya, I.A., 1987. Effects of water-chemical conditions on growth of deposits and metal corrosion in supercritical pressure boilers. Energomashinostoenie 6 (1987), 24–26 in Russian (English translation in 1987. Sov. Energy Technol. 6, 48–52).

Kontorovich, L.K., Vasilenko, G.V., Sutotskii, G.P., Rogal'skaya, I.A., Evtushenko, V.M., 1982. The effect of water chemistry conditions on the resistance to corrosion of steel in the condensate-feed circuit of supercritical power generating units. Therm. Eng. 29 (6), 337–339.

Kritsky, V.G., 1999. Water Chemistry and Corrosion of Nuclear Power Plant Structural Materials. American Nuclear Society.

Kysela, J., Růžičková, M., Petr, J., 2009. Water chemistry specification for supercritical water cooled reactors – possible options. In: 4th International Symposium on Supercritical Water-cooled Reactors, Heidelberg, Germany, March 8–11, 2009. Paper No. 68.

Larsen, O.H., Blum, R., Daucik, K., 1996. Chemical and mechanical control of corrosion product transport. In: Proceedings from Power Plant Chemical Technology, Kolding, Denmark, September 4–6, 1996, pp. 11.1–11.17.

Leblanc, R., Ghandi, K., Hackman, B., Liu, G., 2014. Extrapolation of rate constants of reactions producing H_2 and O_2 in radiolysis of water at high temperatures. In: 19th Pacific Basin Nuclear Conference (PBNC 2014), Vancouver, British Columbia, Canada, August 24–28, 2014.

LeSurf, J.E., Johnson, R.E., Allison, G.M., 1963. Radiolysis in a Steam Loop. Atomic Energy of Canada Limited Report AECL-1800.

Leusbrock, I., Metz, S.J., Rexwinkel, G., Versteeg, G.F., 2008. Quantitative approaches for the description of solubilites of inorganic compounds in near-critical and supercritical water. J. Supercrit. Fluids 47, 117.

Lewis, B.J., MacDonald, R.D., Ivanoff, N.V., Iglesias, F.C., 1993. Fuel performance and fission product release studies for defected fuel elements. Nucl. Technol. 103, 220.

Likhanskii, V.V., Evdokimov, I.A., Sorokin, A.A., Kanukova, V.D., 2009. Applications of the RTOP-CA code for failed fuel diagnosis and predictions of activity level in WWER primary circuit. In: Proceedings of Top Fuel 2009, Paris, France, 2009, p. 2054.

Lin, M., Katsumura, T., He, H., Muroya, Y., Han, Z., Miyazaki, T., Kudo, H., 2005. Pulse radiolysis of 4,4′-bipyridyl aqueous solutions at elevated temperatures: spectral changes and reaction kinetics up to 400 °C. J. Phys. Chem. A 109, 2847–2854.

Lin, M., Muroya, Y., Baldacchino, G., Katsumura, Y., 2010. Radiolysis of supercritical water. In: Wishart, J.F., Rao, B.S.M. (Eds.), Recent Trends in Radiation Chemistry. World Scientific Publishing Co. PTE. Ltd.

Liu, G., Du, T., Toth, L., Beniger, J., Ghandi, K., 2016. Prediction of rate constants of important reactions in water radiation chemistry in sub- and supercritical water: equilibrium reactions. CNL Nucl. Rev. 5, 345–361.

Locke, D.H., 1972. The behaviour of defective reactor fuel,. Nucl. Eng. Des. 21, 318.

Majer, V., Sedlbauer, J., Wood, R.H., 2004. Calculation of standard thermodynamic properties of aqueous electrolytes and non-electrolytes. In: Palmer, D.A., Fernàndez-Prini, R., Harvey, A.H. (Eds.), Chapter 12 in "Aqueous Systems at Elevated Temperatures and Pressures". Elsevier Academic Press (Chapter 4).

Marchaterre, J.F., Petrick, M., 1960. Review of the Status of Supercritical Water Reactor Technology. Argonne National Laboratory Report ANL-6202.

Margulova, T.K., 1978. An investigation of 'neutral' water treatment in supercritical power generating units. Therm. Eng. 25 (10), 39–45.

Marin, T.W., Jonah, C.D., Bartels, D.M., 2005. Reaction of hydrogen atoms with hydroxide ions in high-temperature and high-pressure water. J. Phys. Chem. A 109 (9), 1843–1848.

Martynova, O.I., 1973. Solubility of inorganic compounds in subcritical and supercritical water. In: High Temperature, High Pressure Electrochemistry in Aqueous Solutions": International Corrosion Conference Series, NACE-4, University of Surry, England, January 7–12, 1973, pp. 131–138.

Mayanovic, R.A., Anderson, A.J., Dharmagunawardhane, H.A.N., Pascarelli, S., Aquilanti, G., 2012. Monitoring synchrotron x-ray-induced radiolysis effects on metal (Fe, W) ions in high-temperature aqueous fluids. J. Synchrotron Rad. 19, 797.

Meesungnoen, J., Sanguanmith, S., Jay-Gerin, J.-P., 2013. Density dependence of the yield of hydrated electrons in the low-LET radiolysis of supercritical water at 400 °C: influence of the geminate recombination of subexcitation-energy electrons prior to thermalization. Phys. Chem. Chem. Phys. 15, 16450.

Metatla, N., Jay-Gerin, J.-P., Soldera, A., 2011. Molecular dynamics simulation of sub-critical and supercritical water at different densities. In: 5th International Symposium on Supercritical Water-cooled Reactors (ISSCWR-5), Vancouver, British Columbia, Canada, March 13–16, 2011.

Metatla, N., Lafond, F., Jay-Gerin, J.-P., Soldera, A., 2016. Heterogeneous character of supercritical water at 400 °C and different densities unveiled by simulation. RSC Adv 6, 30484–30487.

Moreau, S., Fenart, M., Renault, J.P., 2014. Radiolysis of water in the vicinity of passive surfaces. Corros. Sci. 83, 255–260.

Mravca, A.E., Simpson, D.E., 1961. Nuclear Superheat Meeting Number 5. September 13–15, 1961, San Jose, California. United States Atomic Energy Commission Div. of Technical Information Report TID-7630.

Muroya, Y., Lin, M., de Waele, V., Hatano, Y., Katsumura, Y., Mostafavi, M., 2010. First observation of picosecond kinetics of hydrated electrons in supercritical water. J. Phys. Chem. Lett. 1 (1), 331–335.

Muroya, Y., Sanguanmith, S., Meesungnoen, J., Lin, M., Yan, Y., Katsumura, Y., Jay-Gerin, J.-P., 2012. Time-dependent yield of the hydrated electron in subcritical and supercritical water studied by ultrafast pulse radiolysis and Monte-Carlo simulation. Phys. Chem. Chem. Phys. 14 (41), 14325–14333.

Murray, J.L., 1965. EVESR-nuclear Superheat Fuel Development Project Thirteenth Quarterly Report. General Electric Report GEAP-4941.
Oliker, I., Armor, A.F., 1992. Supercritical Power Plants in the USSR. Electric Power Research Institute Report TR-100364.
Olive, R.P., 2012. Pourbaix Diagrams, Solubility Predictions and Corrosion-Product Deposition Modelling for the Supercritical Water-cooled Reactor. PhD Thesis. University of New Brunswick, Department of Chemical Engineering.
Palmer, D.A., Simonson, J.M., Jensen, J.P., 2004. Partitioning of electrolytes to steam and their solubilities in steam. In: Palmer, D.A., Fernàndez-Prini, R., Harvey, A.H. (Eds.), Aqueous Systems at Elevated Temperatures and Pressures. Elsevier Academic Press (Chapter 12).
Pencer, J., Edwards, M.K., Guzonas, D., Edwards, G.W.R., Hyland, B., 2011. Impact of corrosion product deposition on CANDU-SCWR lattice physics. In: The 5th Int. Sym. SCWR (ISSCWR-5) Vancouver, British Columbia, Canada, March 13−16, 2011.
Percival, P.W., Brodovitch, J.-C., Ghandi, K., McCollum, B.M., McKenzie, I., 2007. H atom kinetics in superheated water studied by muon spin spectroscopy. Rad. Phys. Chem. 76, 1231−1235.
Plugatyr, A., Carvajal-Ortiz, R.A., Svishchev, I.M., 2011a. Ion-pair association constant for LiOH in supercritical water. J. Chem. Eng. Data 56, 3637−3642.
Plugatyr, A., Hayward, T.M., Svishchev, I.M., 2011b. Thermal decomposition of hydrazine in sub and supercritical water at 25 MPa,. J. Supercrit. Fluids 55, 1014−1018.
Procaccia, I., Gitterman, M., 1981. Slowing down of chemical reactions near thermodynamic critical points. Phys. Rev. Lett. 46 (17), 1163.
Qiu, L., Guzonas, D.A., 2010. An overview of corrosion products solubilities in subcritical and supercritical water. In: 2nd Canada-China Joint Workshop on Supercritical Water-cooled Reactors (CCSC-2010), Toronto, Ontario, Canada, April 25−28, 2010.
Qiu, L., 2014. Solubilities of fission products under supercritical water-cooled reactor conditions. In: Canada-China Conference on Advanced Reactor Development (CCCARD-2014), Niagara Falls, Ontario Canada, April 27−30, 2014.
Red'kin, A.F., Savelyeva, N.I., Sergeyeva, E.I., Omel'yanenko, B.I., Ivanov, I.P., Khodakovsky, I.L., 1989. Investigation of uraninite UO_2(c) solubility under hydrothermal conditions. Sci. Géologiques Bull. 42, 329.
Red'kin, A.F., Savel'yeva, N.I., Sergeyeva, E.I., Omel'yaneko, B.I., Invanov, I.P., Khodakovsky, I.L., 1990. Experiment 89, Informative Volume, p. 79. Moscow, Nauka.
Roduner, E., 2005. Hydrophobic solvation, quantum nature, and diffusion of atomic hydrogen in liquid water. Rad. Phys. Chem. 72 (2−3), 201−206.
Roof, R.R., 1966. EVESR-nuclear Superheat Fuel Development Project Seventeenth Quarterly Report. General Electric Report GEAP-5269.
Ru, X., Staehle, R.W., 2013. PART I− review: historical experience providing bases for predicting corrosion and stress corrosion in emerging supercritical water nuclear technology. Corrosion 69 (3), 211−229.
Saito, N., Tsuchiya, Y., Yamamoto, S., Akai, Y., Yotsuyanaji, T., Domae, M., Katsumura, Y., 2006. Chemical thermodynamic considerations on corrosion products in supercritical water-cooled reactor coolant. Nucl. Technol. 155, 105−112.
Sanguanmith, S., Muroya, Y., Meesungnoen, J., Lin, M., Katsumura, Y., Mirsaleh Kohan, L., Guzonas, D.A., Stuart, C.R., Jay-Gerin, J.-P., 2011. Low-linear energy transfer radiolysis of liquid water at elevated temperatures up to 350 °C: Monte-Carlo simulations. Chem. Phys. Lett. 508 (4−6), 224−230.

Sanguanmith, S., Meesungnoen, J., Guzonas, D.A., Stuart, C.R., Jay-Gerin, J.-P., 2013. Effects of water density on the spur lifetime and the "escape" e^{-}_{aq} yield in γ-irradiated supercritical water at 400 °C: a re-examination. In: 6th International Symposium on Supercritical Water Cooled Reactors (ISSCWR-6), Shenzhen, Guangdong, China, March 03–07, 2013. Paper ISSCWR6-13093.

Sanguanmith, S., 2012. Low-linear Energy Transfer Radiolysis of Liquid Water at Elevated Temperatures up to 350 °C: Monte-Carlo Simulations (M.Sc. thesis). Faculty of Medicine and Health Sciences, Université de Sherbrooke, Sherbrooke, Québec, Canada.

Shields, K., Cutler, F.M., Sadler, M.A., 2006. Condensate Polishing State of Knowledge Assessment: Technology Needs for Fossil Plants. Electric Power research Institute Report 1012208.

Simeoni, G.G., Bryk, T., Gorelli, F.A., Krisch, M., Ruocco, G., Santoro, M., Scopigno, T., 2010. The widom line as the crossover between liquid-like and gas-like behavior in supercritical fluids. Nat. Phys. 6, 503–507.

Solberg, H.L., Brown, T.R., 1968. The effect of high-temperature steam corrosion upon the heat transfer through superheater tube alloys. In: Lien, G.E. (Ed.), Behavior of Superheater Alloys in High Temperature, High Pressure Steam. The American Society of Mechanical Engineers, New York.

Soper, A.K., Bruno, F., Ricci, M.A., 1997. Site-site pair correlation functions of water from 25 to 400 °C: revised analysis of new and old diffraction data. J. Chem. Phys. 106, 247–254.

Spalaris, C.N., Boyle, R.F., Evans, T.F., Esch, E.L., 1961. Design, Fabrication and Irradiation of Superheat Fuel Element SH-4B in VBWR. General Electric Report GEAP-3796.

Spinks, J.W.T., Woods, R.J., 1990. An Introduction to Radiation Chemistry, third ed. Wiley, New York.

Staehle, R., 2006. The small dimensions of stress corrosion cracking – tight cracks. In: Proc. Int. Conf. on Water Chemistry of Nuclear Plant Systems, October 23–26, 2006, Jeju Island, Korea.

Staehle, R., 2010. Fundamental approaches to predicting stress corrosion: quantitative micro-nano approach to predicting stress corrosion cracking in water cooled nuclear plants. In: Proceedings of the Nuclear Plant Chemistry Conference (NPC2010), Quebec City, Canada.

Stellwag, B., Landner, A., Weiss, S., Huttner, F., 2011. Water chemistry control practices and data of the European BWR fleet. Powerpl. Chem. 13, 167.

Styrikovich, M.A., 1969. Steam solutions. Vestn. Akad. Nauk. SSSR 39, 1969.

Subramanian, V., Joseph, J.M., Subramanian, H., Noël, J.J., Guzonas, D.A., Wren, J.C., 2016. Steady-state radiolysis of supercritical water: model predictions and validation. J. Nucl. Eng. Rad. Sci. 2, 021021, 021021–021026.

Svishchev, I.M., Guzonas, D.A., 2011. Supercritical water and particle nucleation: implications for water chemistry control in a GEN IV supercritical water cooled nuclear reactor. J. Supercrit. Fluids 60, 121–126.

Svishchev, I.M., Carvajal-Ortiz, R.A., Choudhry, K.I., Guzonas, D.A., 2013a. Corrosion behavior of stainless steel 316 in sub- and supercritical aqueous environments: effect of LiOH additions. Corros. Sci. 72, 20–25.

Svishchev, I.M., Kallikragas, D.T., Plugatyr, A.Y., 2013b. Molecular dynamics simulations of supercritical water at the iron hydroxide surface. J. Supercrit. Fluids 78, 7–11.

Sweeton, F.H., Baes Jr., C.F., 1970. Solubility of magnetite and hydrolysis of ferrous ion in aqueous solutions at elevated temperatures. J. Chem. Thermodyn. 2, 479–500.

Swiatla-Wojcik, D., Buxton, G.V., 2005. On the possible role of the reaction $H\bullet + H_2O \rightarrow H_2 + \bullet OH$ in the radiolysis of water at high temperatures. Radiat. Phys. Chem. 74 (3–4), 210–219.

Swiatla-Wojcik, D., Buxton, G.V., 2010. Reply to comment on the possible role of the reaction H•+H_2O→H_2+•OH in the radiolysis of water at high temperatures. Radiat. Phys. Chem. 79, 52–56.

Tabor, D., 1969. Gases, Liquids and Solids: And Other States of Matter. Cambridge University Press.

Tremaine, P., Arcis, H., 2014. Solution calorimetry under hydrothermal conditions. Rev. Mineral. Geochem. 76, 219–263.

Tremaine, P.R., LeBlanc, J.C., 1980. The solubility of magnetite and the hydrolysis and oxidation of Fe^{2+} in water to 300 °C. J. Solut. Chem. 9, 415.

Tremaine, P., Zhang, K., Bénézeth, P., Xiao, C., 2004. Ionization equilibria of acids and bases under hydrothermal conditions. In: Palmer, D.A., Fernàndez-Prini, R., Harvey, A.H. (Eds.), Aqueous Systems at Elevated Temperatures and Pressures. Elsevier Academic Press (Chapter 13).

Tse, J.S., Klein, M.L., 1983. Are hydrogen atoms solvated by water molecules? J. Phys. Chem. 87 (25), 5055–5057.

Tucker, S.C., 1999. Solvent density inhomogeneities in supercritical fluids. Chem. Rev. 99, 391–418.

Vasilenko, G.V., 1976. The effect of the pH of the feedwater on corrosion of the low pressure heaters of supercritical power generating units. Therm. Eng. 23 (8), 51–52.

Vasilenko, G.V., 1978. The pattern of precipitation of iron compounds in supercritical steam generators under different water conditions. Therm. Eng. 25 (3), 26–30.

Veselkin, A.P., Lyutov, M.A., Khandamirov, Yu.E., 1968. Radioactive deposits on the surfaces of the technological equipment of the I.V. Kurchatov nuclear power station at Beloyarsk. At. Energiya 24, 219.

Veselkin, A.P., Beskrestnov, N.V., Sklyarov, V.P., Khandamirov, Yu.E., Yashnikov, A.I., 1971. Radiation safety in the design and operation of channel power reactors. At. Energiya 30, 144.

Walther, J.V., 1986. Mineral solubilities in supercritical H_2O solutions. Pure Appl. Chem. 58, 1585–1598.

Wesolowski, D.J., Palmer, D.A., Machesky, M.L., Anovitz, L.M., Hyde, K.E., Hayashi, K.-I., 2000. Solubility and surface charge of magnetite in hydrothermal solutions. In: Tremaine, P.R., Hill, P.G., Irish, D.E., Balakrishnan, P.V. (Eds.), Steam, Water and Hydrothermal Systems: Physics and Chemistry Meeting the Needs of Industry, Proc. of the 13th Int. Conf. on the Properties of Water and Steam. NRC Research Press.

Wesolowski, D.J., Ziemniak, S.E., Anovitz, L.M., Machesky, M.L., Bénézeth, P., Palmer, D.A., 2004. Solubility and surface adsorption characteristics of metal oxides. In: Palmer, D.A., Fernàndez-Prini, R., Harvey, A.H. (Eds.), Aqueous Systems at Elevated Temperatures and Pressures. Elsevier Academic Press (Chapter 14).

Woolsey, I.S., 1989. Review of Water Chemistry and Corrosion Issues in the Steam-Water Circuits of Fossil-Fired Supercritical Plants. Central Electricity Generating Board Report RD/L/3413/R88, PubID 280118.

Wren, J.C., Ball, J.M., 2001. LIRIC 3.2 an updated model for iodine behaviour in the presence of organic impurities. Radiat. Phys. Chem. 60, 577–596.

Wren, J.C., Glowa, G.A., 2000. A simplified kinetic model for the degradation of 2-butanone in aerated aqueous solutions under steady-state gamma-radiolysis. Radiat. Phys. Chem. 58, 341–356.

Yakabuskie, P.A., Joseph, J.M., Wren, J.C., 2010. The effect of interfacial mass transfer on steady-state water radiolysis. Radiat. Phys. Chem. 79, 777–785.

Yakabuskie, P.A., Joseph, J.M., Stuart, C.R., Wren, J.C., 2011a. Long-term γ-radiolysis kinetics of NO3(-) and NO2(-) solutions. J. Phys. Chem. A 115, 4270–4278.

Yakabuskie, P.A., Joseph, J.M., Keech, P., Botton, G.A., Guzonas, D.A., Wren, J.C., 2011b. Iron oxyhydroxide colloid formation by gamma-radiolysis. Phys. Chem. Chem. Phys. 13, 7198–7206.

Yeh, T.K., Wang, M.Y., 2013. Evaluation of the effectiveness of hydrogen water chemistry in a supercritical water reactor. In: Proceedings of the 16th International Conference on the Properties of Water and Steam, September 1–5, 2013. University of Greenwich, London, UK.

Yeh, T.K., Wang, M.Y., Liu, H.M., Lee, M., 2012. Modeling coolant chemistry in a supercritical water reactor. In: Topsafe 2012, Helsinki, Finland, April 22–26, 2012.

Yetisir, M., Gaudet, M., Rhodes, D., 2013. Development and integration of Canadian SCWR concept with counter-flow fuel assembly. In: ISSCWR-6, March 3–7, 2013, Shenzhen, Guangdong, China.

Yurmanov, V.A., Belous, V.N., Vasina, V.N., Yurmanov, E.V., 2009. Chemistry and corrosion, issues in supercritical water reactors. In: IAEA Int. Conf. on Opportunities and Challenges for Water-cooled Reactors in the 21st Century, October 27–30, 2009, Vienna, Austria, IAEA-CN-164-5S12.

Yurmanov, V.A., Belous, V.N., Vasina, V.N., Yurmanov, E.V., 2010. Chemistry and corrosion, issues in supercritical water reactors. In: Proceedings of the Nuclear Plant Chemistry Conference 2010 (NPC 2010), October 3–8, 2010, Quebec City, Canada. Paper 11.02.

Zenkevitch, Yu.V., Sekretar, V.E., 1976. The formation of iron oxide deposits on the tubes of supercritical boilers. Therm. Eng. 23, 60.

Zhou, S., Turnbull, A., 2002. Steam Turbine Operating Conditions, Chemistry of Condensates, and Environment Assisted Cracking – A Critical Review. National Physical Laboratory Report MATC (A) 95.

Ziemniak, S.E., Jones, M.E., Combs, K.E.S., 1995. Magnetite solubility and phase stability in alkaline media at elevated temperatures. J. Solut. Chem. 24, 837–877.

Zimmerman, G.H., Arcis, H., Tremaine, P., 2014. Ion-pair formation in aqueous strontium chloride and strontium hydroxide solutions under hydrothermal conditions by AC conductivity measurements. Phys. Chem. Chem. Phys. 16.

Corrosion

5

5.1 Introduction

Of the various degradation modes of supercritical water-cooled reactor (SCWR) materials addressed in this book, it seems safe to say that general corrosion in SCW has been by the far the most studied and reviewed (Longton, 1966; Montgomery and Karlsson, 1995; Fry et al., 2002; Kritzer, 2004; Was and Teysseyre, 2005; Was et al., 2007; Sun et al., 2009c; Wright and Dooley, 2010; Allen et al., 2012; Ru and Staehle, 2013a,b,c; Sarrade et al., 2017; Guzonas et al., 2017). The recent reviews of corrosion of materials under SCWR conditions, spanning a little more than a decade, provide an overview of the significant progress made in understanding the phenomenon as it relates to the SCWR. In 2005, the situation was concisely summarized by Was and Teysseyre (2005): 'For the range of candidate materials proposed for the SCWR, few studies of corrosion in supercritical water have been reported in the open literature and the existing database is limited'. In 2017, it is more appropriate to state that '...conclusions can be based on the results of a plethora of corrosion resistance measurements'.

While the amount of R&D effort related to SCWR corrosion has increased over the last decade, there were many relevant studies prior to this recent renaissance. Boyd and Pray (1957) reported the weight gain of 12 alloys after exposures of up to 130 days (3168 h) to SCW at 427, 538 and 738 °C, 34.5 MPa. Spalaris (1963) noted that testing in the Superheat Advanced Demonstration Experiment loop in the Vallecitos boiling water reactor (between 1959 and 1962) had shown that even 304 SS had a sufficiently low general corrosion rate for use in a nuclear superheater with a cladding surface temperature of 677 °C and a peak cladding temperature of 732 °C in 15–20 $mg \cdot kg^{-1}$ dissolved oxygen (DO) and a stoichiometric amount of H_2 for a 3–4 year service life. Longton (1966) is an excellent review of this early work (up to ~1966) summarizing the results of six studies covering temperatures from 427 to 765 °C, pressures from 6.9 to 34.5 MPa and test durations from 1000 to 10,100 h. Some tests examined the effects of oxygen additions and of heat transfer. Data for 65 alloys were reported. Longton's conclusions were succinct and many have stood the test of time. He noted that most data show an initially high corrosion rate, eventually settling at a rate that was essentially linear with time. He concluded that pressure had only a minor effect, and that there was only a twofold difference between isothermal and heat transfer tests. He discussed at length the differences between weight gain and weight loss measurements, and cautioned that the former '...is capable of yielding reliable information only when it is certain that no oxide becomes detached from the surface and is lost to the system'. He noted that oxide losses of up to 50% had been reported. Of some

Materials and Water Chemistry for Supercritical Water-cooled Reactors
https://doi.org/10.1016/B978-0-08-102049-4.00005-2

significance is his conclusion that alloys existed at that time whose penetrations after 3 years were only 13 μm at 600 °C and 41 μm at 700 °C.

Studies of corrosion in support of SCWR development have examined many material classes, including ferritic–martensitic (F/M) steels, austenitic steels, Ni-based alloys, Zr-based alloys and Ti-based alloys. When data obtained in support of supercritical fossil power plants (SCFPPs) and in programs to develop supercritical water oxidation (SCWO) systems[1] are considered, some corrosion data relevant to the development of an SCWR exist for ~100 alloys (see, for example, Gu et al., 2010). While there are few alloys for which the data sets are sufficiently complete to allow confident prediction of end-of-life corrosion penetration, large data sets do exist for several of the top candidate alloys. For example, Guzonas et al. (2016a) carefully evaluated 14 studies of Alloy 800H performed at temperatures between 500 and 800 °C for exposure times of 500–14,000 h to develop an empirical model of corrosion as a function of time and temperature. An important caveat is that test conditions often differ significantly between studies; in the aforementioned Alloy 800H data, the pressure ranged from 0.1 to 34.5 MPa and the surface finish ranged from as-received to polished. As discussed in Chapter 2 (Section 2.3.2.1), interlaboratory variations can occur even under nominally the same test conditions due to differences in flow rate, test specimen mounting, etc. Therefore it is important to understand the effects of the various test parameters on the measured corrosion rates.

While ceramic components are used in some SCWR concepts, the degradation of ceramics in SCW is not considered in this book, except in the context of coatings. Sun et al. (2009c) provides a useful summary.

In the text and figures that follow, the alloy designation used is the same as that used in the original reference[2]. When possible, figures have been redrawn to use $mg \cdot dm^{-1}$ or $mg \cdot dm^{-1} \cdot d^{-1}$ (mdd) as the units of weight change to facilitate comparisons, and pressures have been converted to MPa.

5.1.1 Performance criteria

The ultimate goal of engineering studies of general corrosion is to define the total metal loss (metal penetration) at the end of in-service life, enabling designers to add a corrosion allowance to the component thickness required on the basis of other (typically mechanical) material properties. For example, internal pressurization of the SCWR fuel cladding imposes requirements on yield strength and creep strength that define the minimum wall thickness; from these considerations the maximum allowable corrosion penetration for the High-Performance Light-Water Reactor fuel cladding has been defined as 140 μm after 20,000 h (Schulenberg, 2013). Zhang et al. (2012) discussed the strength requirements for a pressurized fuel cladding and proposed that for a cladding thickness of 0.5–0.6 mm, cladding penetration (including general corrosion and stress corrosion cracking, SCC) should not exceed 25–30 μm after 3–4 operating

[1] Kritzer (2004) lists ~80 references related to corrosion in SCWO systems.

[2] Some specific alloys have multiple designations, e.g., HCM12A and T122 have the same composition.

cycles. In the Canadian SCWR concept, the fuel cladding is collapsible and must have a minimum thickness of ~0.4 mm to prevent longitudinal ridging (Xu et al., 2014). However, for neutron economy the cladding thickness must be no more than ~0.6 mm. As a result, the maximum allowable metal loss (including average oxide penetration along grain boundaries, as well as fretting and other forms of wear) is 0.2 mm at the end of in-service life (~3 years) (Guzonas et al., 2016a).

5.2 Alloy composition

Alloy composition is perhaps the key material parameter with respect to corrosion resistance. The corrosion resistance of alloys in SCW has been found to be a strong function of alloy class, as illustrated in Fig. 5.1, which shows data from corrosion tests of alloys from five classes (Ti-based, Zr-based, F/M steels, austenitic steels and Ni-based) performed in SCW at 500 °C, 25 MPa for 1000 h by different research groups; in most tests the DO concentration was $<25\ \mu g \cdot kg^{-1}$ at the test section inlet. Of the five classes of materials in the figure, oxidation occurs most rapidly for F/M steels and nickel-based alloys have the lowest rate. Within each alloy class, some alloys exhibit better oxidation resistance, e.g., for the F/M steels in Fig. 5.1, the corrosion rate decreases in the order T91 > T92 > T122.

Higher chromium content has long been known to improve corrosion resistance of steels; as shown in Fig. 5.1, the austenitic steels have a much lower corrosion rate than the F/M steels. Fig. 5.2 shows the dependence of the parabolic corrosion rate constant K_p for various steels at 650, 700 and 750 °C as a function of Cr content. The dependence on Cr concentration is clear for both the F/M steels

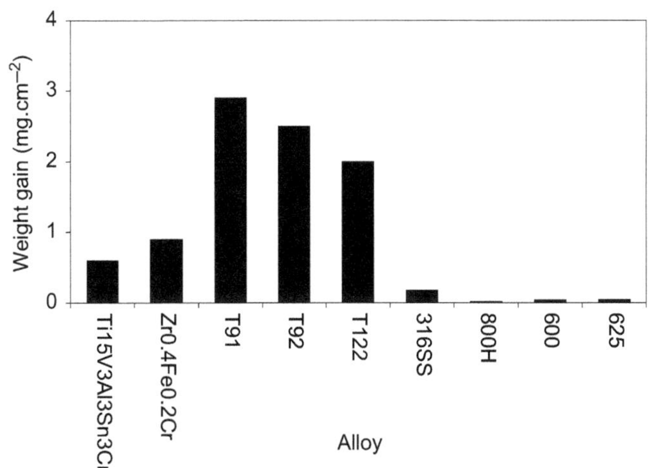

Figure 5.1 Weight change for selected Ti-based alloys, Zr-based alloys, F/M steels, austenitic steels and Ni-based alloys in supercritical water at 500 °C, 25 MPa after 1000 h. The dissolved oxygen concentration was $<25\ \mu g \cdot kg^{-1}$ with the exception of the Ti-based alloy ($8\ mg \cdot kg^{-1}$) and 316 SS ($8\ mg \cdot kg^{-1}$).

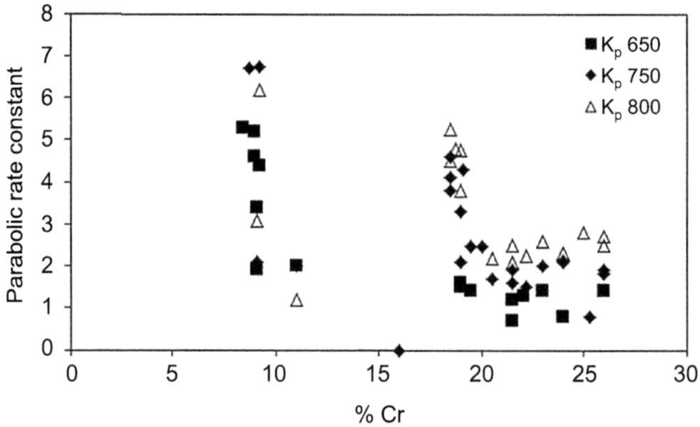

Figure 5.2 Dependence of the parabolic corrosion rate constant K_p on the Cr content of various F/M and austenitic stainless steels at 650, 700 and 750 °C.
Adapted from Sarver, J., 2009. The oxidation behavior of candidate materials for advanced energy systems in steam at temperatures between 650 °C and 800 °C. In: 14th Int. Conf. on Environmental Degradation of Materials in Nuclear Power Systems, Virginia Beach, VA, August 23–27, 2009.

(large decrease in rate at 9 wt% Cr) and austenitic steels (large decrease in rate at 19 wt% Cr). A higher Cr content is required to form a protective Cr oxide on austenitic steels compared with ferritic steels due to the lower diffusivity of Cr in the fcc austenitic lattice. This protective film reduces outward Fe diffusion and prevents formation of a less-protective $(Fe,Cr)_3O_4$ outer layer.

Wright and Dooley (2010) provide an excellent summary of the data on the effect of Cr on corrosion for ferritic and F/M steels, drawing on the extensive database that exists for these alloys due to their widespread use in SCFPP boilers. The data clearly show that a Cr concentration of at least 9% is needed to see a reduction of the oxidation rate in SCW. Jang et al. (2005a) measured the oxidation of four ferritic steels with Cr content ranging from 9 to 20 wt% (ground to a 600 grit finish) at 500 and 627 °C in air-saturated 25 MPa SCW and found a 5-fold reduction in weight change between 9 and 12 wt% Cr and an additional 14-fold decrease in weight gain in increasing the Cr content to 20 wt%. Zurek et al. (2008) suggested that F/M steels with Cr contents in the range 10%–12% should be distinguished from ferritic steels with Cr > 12%, the former exhibiting oxidation behaviour that varied depending on factors such as test duration, temperature, surface treatment and the concentrations of minor alloying elements. This variability was attributed to the nature of the oxide formed during the initial period of oxide formation for these alloys, which contained only the necessary and sufficient amount of Cr required to form a continuous protective oxide; small changes in initial conditions could impair or improve the formation of a continuous oxide layer, which would have a long-term impact on the corrosion resistance. There have been fewer studies of higher Cr F/M steels (Cr > 12 wt%) in support of SCWR development. Artymowicz et al. (2010) and Cook et al. (2010) reported preliminary data on the oxidation of ferritic steels with 14 and 25 wt% Cr exposed to SCW

(500 °C, 25 MPa) for 500 h in the as-cast, as-rolled and rolled plus homogenized conditions. The 14 wt% Cr materials all showed weight gains, following power law kinetics with the exponent being about 0.9. The 25 wt% Cr materials showed a slight weight loss with oxidation being restricted to small nodules on the surface. Cho et al. (2004) showed that a Cr content greater than 13 wt% reduced the corrosion rate of oxide dispersion strengthened (ODS) steels. Brar et al. (2015) found that the high chromium ferritic steel E-brite[3] performed as well as austenitic steels and nickel-based alloys when exposed to SCW at 550 °C, 25 MPa, low DO.

Fig. 5.2 suggests a plateau in the corrosion resistance for austenitic steels above 20 wt% Cr, implying a maximum Cr content above which further additions confer no additional benefit. Otsuka et al. (1989) measured the oxidation of a number of austenitic steels (13–25 wt% Cr, ~15 wt% Ni, ~0.5 wt% Si, ~1.5 wt% Mn, average grain size 110 μm) in superheated steam (SHS) (T = 500–900 °C, 0.1 MPa) and reported that a Cr content > 21 wt% was necessary for formation of a uniform external Cr_2O_3 film on the surface of coarse-grained austenitic steels for T < 750 °C. When the grain size was reduced to 20 μm, the protective film was formed up to 900 °C. As with F/M steels, a Cr content just above the minimum required to form a continuous protective Cr-rich layer may be sensitive to small changes in initial conditions that could affect protective oxide formation, which could impair long-term corrosion resistance. Mahboubi (2014) found that Alloy 33 (33.4 wt% Cr) was better able to form a Cr_2O_3 oxide than Alloy 800HT (20.6 wt% Cr).

While austenitic steels have been found to have better corrosion resistance in SCW than F/M steels, there has been little systematic study of the effects of nickel on corrosion resistance. Brar et al. (2015) measured the weight change as a function of time for a group of commercially available iron- and nickel-based alloys with relatively similar Cr content and nickel contents ranging from 0.5 wt% to 72 wt% in SCW (550 °C, 25 MPa) for a period of up to 2500 h. There was little difference in weight change for all alloys except 304 SS, which exhibited a much higher weight change (Fig. 5.3). Peraldi and Pint (2004) studied the oxidation of a number of ferritic and austenitic alloys in a mixture of air plus 10% water vapour and found that a higher Ni content delayed (but did not prevent) the onset of accelerated corrosion and/or reduced spalling.

Silicon is known to improve corrosion resistance of steels; Wright and Dooley (2010) present a concise summary of data for F/M steels. Li (2012) studied the effects of additions of 1.5 wt% Si, Al, Mn, Ti or V on the oxidation of model 9 wt% Cr and 12 wt% Cr alloys at 500 °C, 25 MPa and 5 or 200 ppb DO. Si additions were found to confer the largest benefit, attributed to the formation of Si-rich particles ahead of the oxidation front. These particles served as nucleation sites that facilitated the rapid formation of Fe-Cr mixed oxides. The additions of Mn, Ti, Al, or V had little effect on weight gain although these ternary alloys had thinner oxides than the binary oxides.

Liu (2013) presented data on the oxidation of model F-M steels (~12 wt% Cr) with Si content from 0.6 wt% to 2.2 wt% and Mn content from 0.6 wt% to 1.8 wt%.

[3] 72 wt% Fe, 26 wt% Cr, 0.5 wt% Ni, 0.4 wt% Si, 0.7–1.5 wt% Mo, 0.4 wt% Mn.

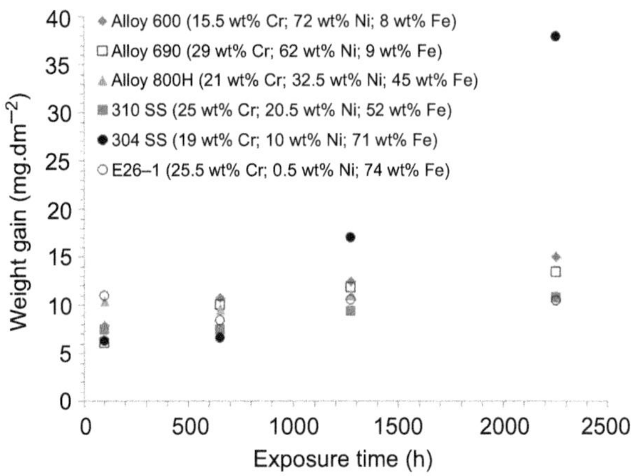

Figure 5.3 Weight gain of six alloys with varying Ni content as a function of exposure time at 550 °C, 25 MPa in a static autoclave. Alloy compositions are shown in the legend. Adapted from Brar, C., Davis, W., Mann, E., Ridley, C., Wirachowsky, B., Rangacharyulu, C., Guzonas, D., Leung, L., 2015. GEN IV SCWR cladding analysis project: nickel content in SCWR cladding material. In: Proceedings 5th Int. Youth Conference on Energy, Pisa, Italy, May 27–30, 2015, pp. 103–110.

Samples were ground to a 1200 grit finish and exposed to SCW at 500 °C, 25 MPa. Increasing the Si content increased the oxidation resistance of F-M steel and promoted the formation of a more uniform oxide layer (Fig. 5.4). However, increasing the Mn content decreased the corrosion resistance and resulted in the formation of a

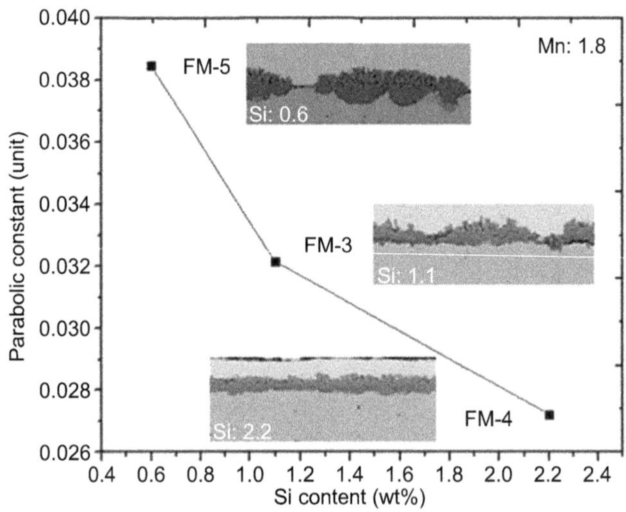

Figure 5.4 Plot of parabolic constant as a function of Si content for the oxidation of model F-M steels (~12 wt% Cr with 1.8 wt% Mn) (Dong et al., 2014). SEM cross-sections of the three samples tested are shown as insets.

nonuniform corrosion film. Mn is a strong spinel-forming element that can promote the formation of $MnCr_2O_4$ spinel[4] in Cr-containing alloys. Gilewicz-Wolter et al. (2006) measured the diffusion rates of ^{51}Cr, ^{54}Mn and ^{59}Fe in $MnCr_2O_4$ and $FeCr_2O_4$ and found that metal diffusion was higher in $MnCr_2O_4$.

The volatilization rate of a Mn-enriched oxide has been shown to be less than that of a pure Cr_2O_3 oxide (Holcomb and Alman, 2006). Mahboubi et al. (2014) reported that the oxide formed on 310S SS after 100 h exposure to 8 $mg \cdot kg^{-1}$ oxygenated SCW at 550 °C, 25 MPa contained more Fe and Mn at the oxide/SCW interface and less Cr. They proposed that a protective oxide layer containing Mn separated the Cr-rich oxide from the SCW and reduced its dissolution rate, this benefit disappearing after 250 h exposure. The diffusion of Mn in Cr_2O_3 is much faster than that of Fe, Cr and Ni. The metal release data for Alloy 800H reported by Choudhry et al. (2016) support this hypothesis; the oxide that forms at short exposure times appears to contain sufficient Mn to prevent Cr release (Fig. 5.45), but as the exposure time increases the Mn dissolves into the coolant, being replaced by Fe. The Mn concentration in the bulk alloy is insufficient to maintain the required Mn concentration in the oxide and composition was assumed to change to $Cr_xFe_yMn_wO_z$ with $x > y > w$. The data suggested that the protection afforded by Mn may be greater at higher temperatures. Abe and Yoshida (1985) found that the formation of a relatively thick $MnCr_2O_4$ layer above the Cr_2O_3 layer on Alloy 800 exposed to SHS (800 °C, 4 MPa) prevented the formation of nodular oxides, whereas Ni-based alloys with lower Mn content exhibited nodular corrosion under the same conditions.

Kish and Jiao (unpublished data) characterized the behaviour of UNS S21400, an austenitic Fe–Cr–Mn stainless steel (16.7 wt% Cr, 12.4 wt% Mn, 0.7 wt% Ni, 0.5 wt% Si, 0.3 wt% N, 0.06 wt% C) ground to an 800 grit finish and exposed to SCW at 550 °C, 25 MPa for 542 h in a static autoclave. The initial DO concentration was 8 ppm. The replacement of Ni by Mn (also an austenite former) was considered attractive for SCWR use as nickel-based alloys are strongly affected by radiation damage (Chapter 3). The authors suggested that the transformation of the austenite phase to a ferrite phase in the Mn-depleted zone could reduce the risk of SCC of the cladding, since ferritic steels are known to be less susceptible to environmentally assisted cracking (EAC) (Chapter 6). However, the alloy showed excessive oxide spallation and the general corrosion rate was unacceptably high. It was suggested that the Mn content of the alloy could be optimized to balance general corrosion with surface ferrite formation.

ODS has been used to improve the mechanical performance of alloys (in particular F/M steels) for use at high temperatures, resulting in high creep strength and good radiation resistance. These alloys have also been found to have somewhat improved corrosion resistance over the non-ODS version of the same alloy (Chen et al., 2007b; Motta et al., 2007b). Cho et al. (2004) found that 9–12 wt% Cr ODS steels

[4] Spinel refers to a class of compounds of general formulation $A^{2+}B^{3+}{}_2O^{2-}{}_4$ that crystallize in the cubic (isometric) crystal system. The oxygen anions are arranged in a cubic close-packed lattice and the cations A and B occupy some or all of the octahedral and tetrahedral sites in the lattice. Magnetite and nickel ferrite are examples of iron oxides having the (inverse) spinel structure.

showed similar corrosion behaviour to the equivalent non-ODS alloys. Zhang et al. (2012) reported results of tests of 12 wt% Cr, 14 wt% Cr and 18 wt% Cr ODS steels exposed to SCW at 550 °C, 25 MPa and also found that the corrosion resistance of these alloys was no better than that of non-ODS steels. Cho et al. (2004), Zhang et al. (2012) and Gong et al. (2015) all reported that the addition of aluminum to the ODS alloys improved their corrosion resistance.

The limited work on Ti- and Zr-based alloys is summarized below. The remainder of the chapter will then focus on the various Fe- and Ni-based alloys that have been the focus of the majority of the SCWR corrosion studies.

5.2.1 Zr-based alloys

Zirconium-based alloys have been considered for in-core SCWR applications (fuel cladding in all designs, pressure tubes in a pressure tube design) because of their good neutron economy, although current commercial alloys lack the high-temperature strength for load bearing application at SCW temperatures. As discussed in Chapter 3, the Canadian SCWR concept retains the use of Zr alloy pressure tubes by using an insulated fuel channel concept. It has been known since the 1970s that Zr alloys exhibit very high corrosion rates in SCW (Cox, 1973). Initial studies on Zircaloy-2 and Zircaloy-4 (Cox, 1973; Jeong et al., 2005; Motta et al., 2007a; Peng et al., 2007; Li et al., 2007; Khatamian, 2013) found that these alloys had extremely high oxidation rates at $T > 450$ °C making them unacceptable for use at SCWR core temperatures. Further studies on a number of model Zr alloy classes (Zr–Nb, Zr–Fe–Cr and Zr–Cu–Mo systems) in both static autoclaves and flow loops at 25.2 MPa, <400 $\mu g \cdot kg^{-1}$ DO, as well as in a static autoclave at 10.3 MPa, showed that optimized alloy compositions existed whose oxidation rates were lower than those of F/M steels, although still approximately three times higher than those of austenitic steels. The corrosion kinetics of most of the alloys studied followed a roughly cubic rate law. Khatamian (2013) found that these alloys could also exhibit breakaway corrosion. Oxidation results for Zr0.4Fe0.2Cr are shown in Fig. 5.1.

5.2.2 Ti-based alloys

Ti-based alloys had been found to be good candidate materials for use in SCWO systems and therefore were examined for possible use in an SCWR. Data on the corrosion of several Ti-based alloys in SCW have been reported (Kasahara, 2003; Kaneda et al., 2005); the alloys examined include Ti–15Mo–5Zr–3Al, Ti–3Al–2.5 V, Ti–6Al–4 V and Ti–15V–3Al–3Sn–3Cr. These alloys were exposed in static autoclaves with 8000 $\mu g \cdot kg^{-1}$ DO for 500 h at temperatures of 290, 380 and 550 °C. While weight gains of samples exposed at 290 and 380 °C were similar, the weight gain of the samples exposed at 550 °C was significantly higher. At 550 °C, Ti–15V–3Al–3Sn–3Cr and Ti–15Mo–5Zr–3Al showed a factor of about three times lower weight gain than the Ti–3Al–2.5 V and Ti–6Al–4V alloys. Kaneda et al. (2005) reported that the weight changes of Ti–15V–3Al–3Sn–3Cr and Ti–15Mo–5Zr–3Al were about the same as those for 304 and 316 SS exposed under the same test conditions. There has been little subsequent work on Ti-based alloys for SCWR development.

5.3 Effects of key parameters

For a particular alloy, the relative contributions of various physical and chemical environmental parameters to the general corrosion rate over the temperature range 350–600 °C were ranked as follows (in order of decreasing importance) (Guzonas et al., 2010)[5]:

$$\text{Temperature} \approx \text{Surface Finish} \approx \text{Grain Size} > \text{Water Chemistry} > \text{SCW Density}$$

Irradiation will affect the corrosion rate of in-core materials by changing the water chemistry (water radiolysis) and by interaction of radiation with the alloy surface region; the former effect is expected to be larger. Until recently, there have been few systematic studies of the effects of flow rate, heat transfer and thermal ageing under SCWR conditions, and these factors were not included in the ranking. The effects of each of these parameters are explored in the next sections.

5.3.1 Temperature

Generally, at $T \gg T_c$, temperature has been found to have the greatest influence on oxidation/corrosion of materials. With an increase of exposure temperature, most measures of corrosion (weight loss or gain, oxide thickness) increase significantly. Because of the large change in temperature across an SCWR core, corrosion resistance for materials to be used in the core must be evaluated for three cases: high-temperature subcritical water ($\sim$260 °C $< T < T_c$)[6], near-critical and low temperature SCW ($T_c < T <$ 450–500 °C) and high-temperature SCW (450–500 °C $< T <$ peak cladding temperature). As corrosion is a surface phenomenon the coolant temperature and density at the fuel cladding surface are of interest, not the bulk coolant values. The core can be divided into three regions defined by the bulk fluid temperature and density:

1. A region (1) just downstream of the fuel string inlet at temperatures below T_c,
2. A region (2) with relatively constant, but supercritical, temperature but rapidly decreasing density, and
3. A region (3) of rapidly increasing temperature and low, relatively constant density.

Regions 1 and 2 correspond to the ‘evaporator’ region (Fig. 2.4) and region 3 to the ‘superheater’ region.

[5] SCWR in-core water chemistry conditions are expected to be neutral to slightly alkaline, oxidizing (1–20 ppm DO, some H_2O_2) with low concentrations of aggressive impurities (e.g., Cl^-, SO_4^-); see Chapter 4.

[6] The lower temperature will be defined by the core inlet temperature of each SCWR concept.

Corrosion studies performed in support of the development of SCWO systems suggested a maximum in the corrosion rate at temperatures around the critical point (Kritzer, 2004), where the water density increases significantly (Chapter 4). However, these studies were typically carried out at near-critical temperatures where the properties of SCW (e.g., density, dielectric constant) vary significantly over a small temperature range, using chemically aggressive conditions such as very acidic pH and high concentrations of oxidants and aggressive anions such as chloride.

In Regions 1 and 2, where the SCW density is high and the fluid is 'liquidlike', 'traditional' electrochemical processes that occur in subcritical high-temperature water are expected. Lillard (2012) summarizes the basic principles of corrosion in high-temperature subcritical water. In subcritical water, weight losses are commonly observed, especially under conditions where the solubilities of the surface oxides are high and chemistry conditions are such that the surface is no longer passive. In this region, the rate determining step can be a charge transfer reaction (activation polarization) or mass transport of reagents to the surface (concentration polarization). The situation is made more complex in that an electrochemical corrosion reaction involves two coupled half-cells (anodic and cathodic) that can be spatially separated on the surface and requires the use of mixed potential theory. Under conditions where such active corrosion takes place, alloys will have relatively short in-service lives. The ability to utilize metal alloys in high-temperature water depends on the formation of a passive film that reduces the rate of oxidation to an acceptable value (Macdonald, 1999).

As with radiolysis reaction rate constants (Chapter 4), in the near-critical region rates of corrosion reactions show a non-Arrhenius behaviour. Zhou et al. (2002) used electrochemical noise analysis to measure the corrosion rate of 304 SS in water containing 0.1 $mol \cdot kg^{-1}$ NaCl and 0.01 $mol \cdot kg^{-1}$ HCl, 25 MPa, over the temperature range 150–390 °C, at flow rates from 0.375 to 1.00 $mL \cdot min^{-1}$. (The authors reported data for two other flow rates but the dependence on T behaved in a similar manner and these data were omitted from the figure for clarity.) They noted (Fig. 5.5) a maximum in the rate at 350 °C at all flow rates measured (only the highest and lowest flow rates are shown in the figure) consistent with later modelling work (e.g., Kriksonuv and Macdonald (1995a), discussed in Section 5.6). Guan and Macdonald (2009) later reported electrochemical noise data measured in deaerated 0.01 m HCl, deaerated 0.01 m NaOH and deaerated pure water over the temperature range 25–500 °C at 25 MPa and found that the maxima observed for water and NaOH were much weaker than that observed for HCl. No explanation could be given at the time.

Figs. 5.6 and 5.7 show weight change data measured over a much larger range of temperatures, showing that as one moves away from the near-critical region, one observes the expected Arrhenius behaviour, the weight change increasing rapidly with increasing temperature. In Fig. 5.6, a change from a weight loss to weight gain is observed for T91 at temperatures above T_c. Fig. 5.7 shows data for the corrosion of 316 SS compiled from different authors; a distinct change in rate near T_c is observed, easily seen in the Arrhenius plot. Above T_c, the data from the different sources are in good agreement. At $T \gg T_c$ a similar behaviour (exponentially increasing rate with increasing T) is observed for all alloys studied in SCW to date. Additional examples

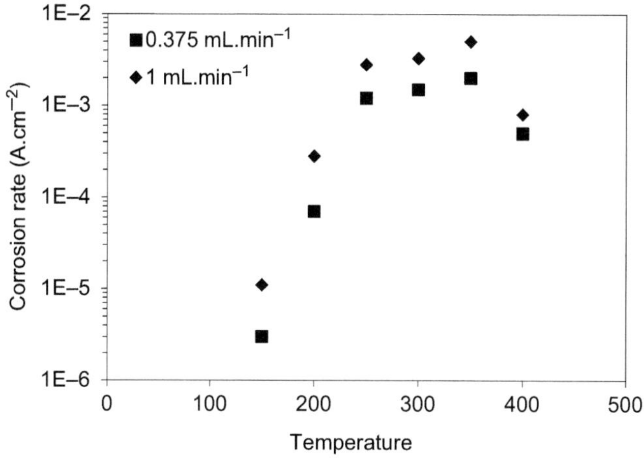

Figure 5.5 Plot of corrosion rate of 304 SS from electrochemical noise analysis as a function of test temperature in water containing 0.1 mol·kg^{-1} NaCl and 0.01 mol·kg^{-1} HCl at 25 MPa at two linear flow rates.
Adapted from Zhou, Z.Y., Lvov. S.N., Wei, X.J., Benning, L.G., Macdonald, D.D., 2002. Quantitative evaluation of general corrosion of Type 304 stainless steel in subcritical and supercritical aqueous solutions via electrochemical noise analysis. Corros. Sci. 44, 841–860.

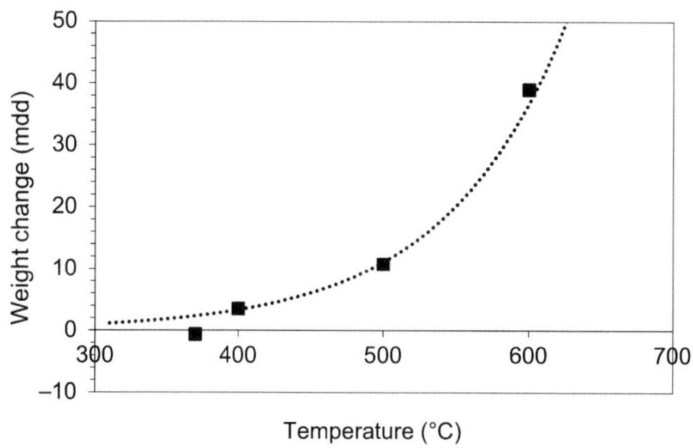

Figure 5.6 Corrosion rate of T91 in deaerated water (dissolved oxygen < 10 μg·kg^{-1}), 25 MPa, pH 6.5. Coupons ground to 600 grit finish. The *dotted line* is an exponential fit to the data above T_c.
Data from Hwang, S.S., Lee, B.H., Kim, J.-G., Jang, J., 2005. Corrosion behavior of F/M steels and high Ni alloys in supercritical water. In: Proceedings of Global 2005, Tsukuba, Japan, October 9–13, 2005, Paper 043.

of the temperature dependencies of the oxidation rate of alloys can be found in Allen et al. (2012) as well as in the other reviews listed at the start of the chapter.

Potter and Mann (1963) suggested that a change in the nature of the oxidation process occur at T_c. Glebov et al. (1976) suggested the coexistence of electrochemical

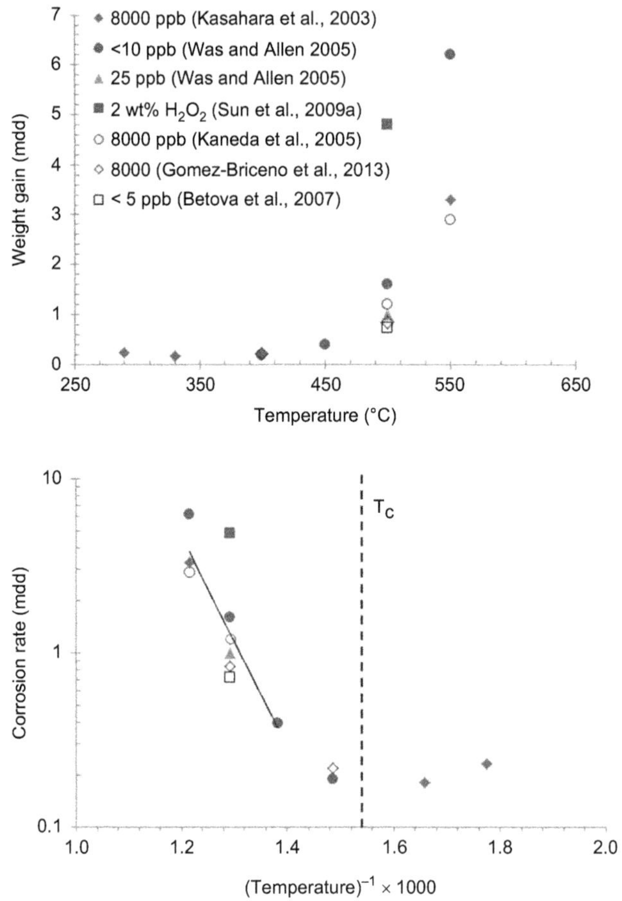

Figure 5.7 (Top) Weight gain rate versus temperature for 316 and 316L exposed to supercritical water containing various concentrations of dissolved oxygen (DO). Source references are listed in the legend. (Bottom) Corrosion rate versus 1/T for the same data. The *vertical dashed line* denotes T_c; the line through the data points is a least squares fit to the data above 450 °C.

oxidation (EO) and chemical oxidation (CO) mechanisms in SCW as a means of explaining the observed effect of pH on corrosion above T_c (and attributed similar ideas to N.D. Tomashov as early as 1959). Robertson (1991) examined data on the temperature dependence of 304 SS in subcritical water and SHS and proposed that the observed local maximum in the parabolic rate constant could be the result of the existence of two parallel corrosion mechanisms with different activation energies. Kriksunov and Macdonald (1995a) and Guan and Macdonald (2009) modelled the near-critical corrosion behaviour and attributed the maximum to a change in corrosion mechanism from an EO at higher water densities (and higher dielectric constants) to a CO below some threshold water density. The CO mechanism is based on molecular interactions of the metal with oxygen and/or water and the transport of cation and anion defects in the oxide. Yi et al. (2006) concluded that corrosion in SCW was similar to that in gaseous conditions, where oxide formation occurs without metal dissolution. Betova et al. (2007) concluded that at temperatures up to 500 °C, the oxidation of stainless steels proceeds in a manner analogous to that in high-temperature subcritical water, whereas at higher temperatures the oxidation kinetics seems to be closer to that in water vapour. In the near-critical region, a **local** maximum in corrosion rate at temperatures around T_c is predicted (Section 5.6). Measurements of the corrosion rate under SCWR conditions are generally carried out at much higher temperatures and such a maximum is seldom observed. As the SCWR coolant will pass through the critical point in the SCWR core, additional studies of corrosion near the critical point would be helpful.

5.3.2 Surface finish and grain size

It has long been recognized that surface finish (e.g., grinding, polishing) has a strong effect on corrosion rate (e.g., Ruther et al., 1966; Berge, 1997). Numerous studies have shown that while severe cold work reduces general corrosion in SCW, preparations that give a strain-free surface result in high general corrosion. Maekawa et al. (1968) measured the weight change of 304 SS exposed to subcritical water and SHS. Surfaces were either abraded, pickled or electropolished before exposure. Fig. 5.8 plots $\log(WG_{pickled}/WG_{abraded})$ as a function of temperature. Above T_c, the abraded surface shows improved corrosion resistance compared with the pickled surface, the effect increasing as the temperature increases. At 600 °C, the abraded surface

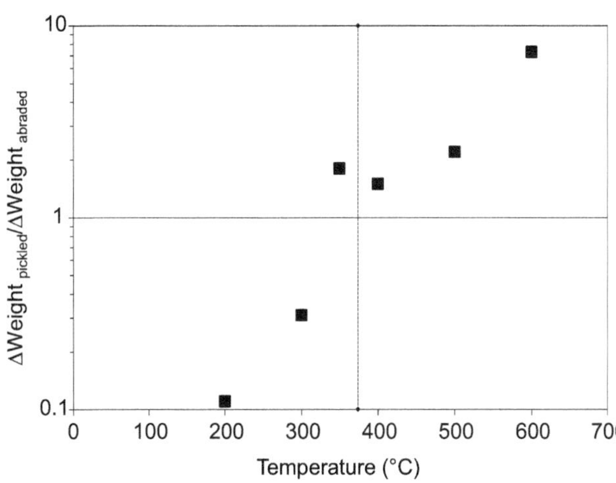

Figure 5.8 Logarithm of the ratio of the weight gain of pickled specimens ($WG_{pickled}$) to the weight gain of abraded specimens ($WG_{abraded}$) for 304 SS exposed to subcritical water and superheated steam (SHS) as a function of temperature.
Data from Maekawa, T., Kagawa, M., Nakajima, N., 1968. Corrosion behaviors of stainless steel in high-temperature water and superheated steam. Trans. Jpn. Inst. Metals 9 (2), 130–136.

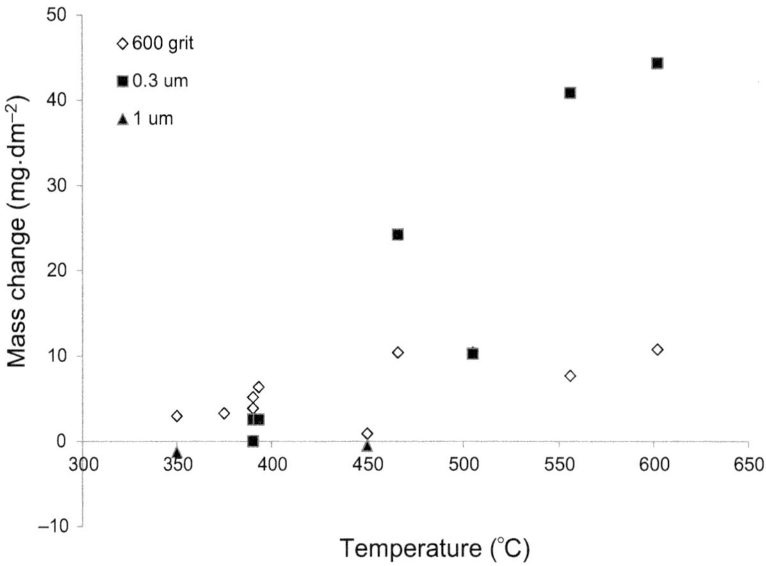

Figure 5.9 Mass change as a function of temperature for 304 SS with three surface finishes exposed to supercritical water in a static autoclave at 25 MPa.
Adapted from Guzonas, D., Wills, J., Dole, H., Michel, J., Jang, S., Haycock, M., Chutumstid, M., 2010. Steel corrosion in supercritical water: an assessment of the key parameters. In: 2nd Canada-China Joint Workshop on Supercritical Water-cooled Reactors (CCSC-2010), Toronto, Ontario, Canada, April 25–28, 2010.

is about 10 times more corrosion resistant than the pickled surface. A similar increase in corrosion resistance for abraded versus polished surfaces was reported for 304 SS in SCW at 25 MPa (Guzonas et al., 2010); as seen in Fig. 5.9, the weight gains for the abraded and polished samples are similar in the near-critical region, the difference increasing with increasing temperature. At 600 °C, the abraded sample shows a roughly fourfold reduction in weight gain compared with the polished sample. Below T_c, the **opposite** effect is observed in Fig. 5.8, cold work decreasing corrosion resistance. In the near-critical region, there is a plateau in the ratio (slightly greater than one). Zemniak et al. (2008) found a reduction in the parabolic rate constant for the corrosion of 304 SS in subcritical water for electropolished samples compared with machined specimens. Robertson (1991), Zemniak et al. (2008) and Guzonas and Cook (2012) have proposed that the different (opposite) behaviour above and below T_C is related to a change in the corrosion mechanism. The benefits of electropolishing for reducing corrosion rates and radioactivity uptake on surfaces in subcritical water are well known and have been exploited by the nuclear industry (Asay, 1990; Hudson and Ocken, 1996; Sawochka and Leonard, 2006; Hsu et al., 1994).

Leistikow (1972) measured the weight change of a number of iron- and nickel-based alloys in SHS (600 °C, 0.1 MPa, $6 \pm 2\ mg \cdot kg^{-1}$ DO, flow rate $2–4\ cm \cdot s^{-1}$) as a function of exposure time up to 5000 h with pickled or ground surfaces. The pickled surfaces all exhibited weight gain, whereas the ground surfaces showed weight losses, except for 304 SS, which showed essentially no weight change after ~2000 h. The fraction of metal lost to the coolant by dissolution was significantly higher for the ground samples and showed a monotonically increasing dependence on the nickel content of the alloy. Analysis of the steam condensate showed that the majority of the metal loss was chromium, likely as Cr(VI) species. When the oxygen content of the steam was reduced to 1 mg·kg^{-1}, ground Alloy 800 samples showed a weight gain rather than a weight loss.

Fig. 5.10 compares the effects of grinding, sandblasting and polishing on the weight gain of Super 304 (19.04 wt% Cr, 8.93 wt% Ni) exposed at 700 °C for 50 h at 0.1 and 26 MPa (Yuan et al., 2013). Both grinding and sandblasting introduce cold work into an alloy surface. Interestingly, the pressure was observed to have a much larger effect on the polished samples than on the cold worked sample.

Penttilä and Toivonen (2013) reported that the weight gains of mill-annealed 316L and 316Ti tube specimens were over one order of magnitude higher than those of plane-milled 316L. Huang and Guzonas (2014) found that a Ni20wt%Cr5wt%Al model alloy ground to a 600 grit finish showed negligible weight change during a 6300 h exposure to SCW at 500 °C, 25 MPa. Samples polished with diamond paste showed a very high weight gain over the first 2000 h of exposure, followed by decreases in the weight gain at longer exposure times, attributed to spalling of the thick oxide during autoclave shutdown for sample removal. A similar reduction in weight gain was found for a model Fe–20Cr–6Al–Y alloy (Huang et al., 2015a).

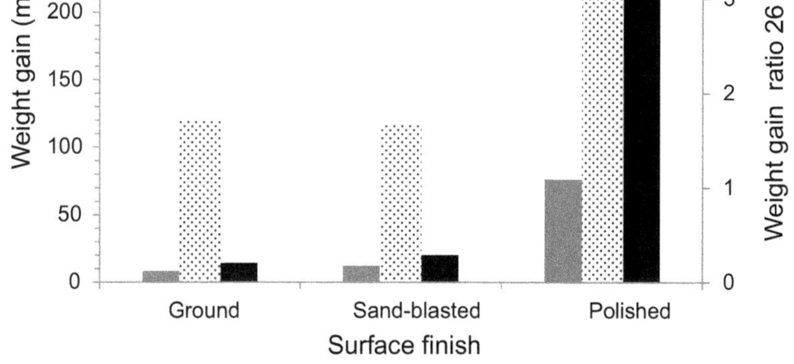

Figure 5.10 Comparison of the effects of grinding, sandblasting and polishing on the weight gain of Super 304 exposed to supercritical water at 700 °C for 50 h at 0.1 and 26 MPa (Yuan et al., 2013). The ratios of the weight gains at 0.1 and 26 MPa are shown by the crosshatched bars.

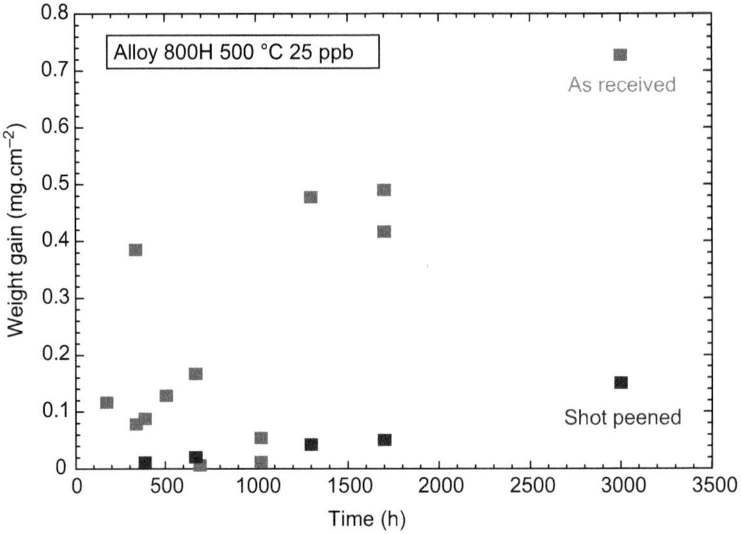

Figure 5.11 Comparison of weight gain for Alloy 800H in large- and small-grained metallurgical conditions exposed to supercritical water at 500 °C for times up to 3000 h (Allen et al., 2012).

Fig. 5.11 illustrates the improvement in oxidation resistance of Alloy 800H through grain refinement (Tan et al., 2008a; Allen et al., 2012). Shot peening introduced a surface deformation layer 70 μm deep. After shot peening, the surface grain size of the Alloy 800H was around 15 nm, and during exposure to SCW the grain size grew to approximately 180 nm. Shot peening and the subsequent grain size reduction in the near-surface region reduced the oxidation rate by approximately 60% and also improved oxide adhesion, significantly reducing spalling (note the erratic weight changes of the as-received material in Fig. 5.11). The microstructure of the shot-peened material was found to be simpler than that of the as-received samples and was dominated by Cr-rich oxides.

While early studies suggested no benefit of cold work on the corrosion resistance of F/M steels (Eberle and Kitterman, 1968), recent work has shown this not to be the case. Ren et al. (2008) reported that shot peening of T91, HT9, NF616 and two model Fe–Cr alloys (Fe–15%Cr and Fe–18% Cr) led to a reduction of corrosion in SCW (500 °C, <25 $\mu g \cdot kg^{-1}$ DO, up to 3000 h exposure) of 20%–40%. Artymowisc et al. (2014) studied the effects of various surface finishes on the corrosion of a model Fe14Cr1.5Si alloy in SCW (500 °C, 25 MPa, 5 $\mu g \cdot kg^{-1}$ DO, 250 h). The surface was found to be covered with a mixture of two types of oxide; a thin, compact oxide (200 nm thick Cr_2O_3 on top of a 50 nm thick Si-rich layer adjacent to the metal-oxide interface), and a thick oxide consisting of a Fe_3O_4 layer growing outward from the surface on top of an (Fe, Cr)$_3$O$_4$ layer growing inward from the surface, typical of alloys whose Cr content is too low to form a protective Cr_2O_3 layer. The fraction of the surface covered by the thicker oxide was found to increase as the extent of surface cold work decreased (Fig. 5.12).

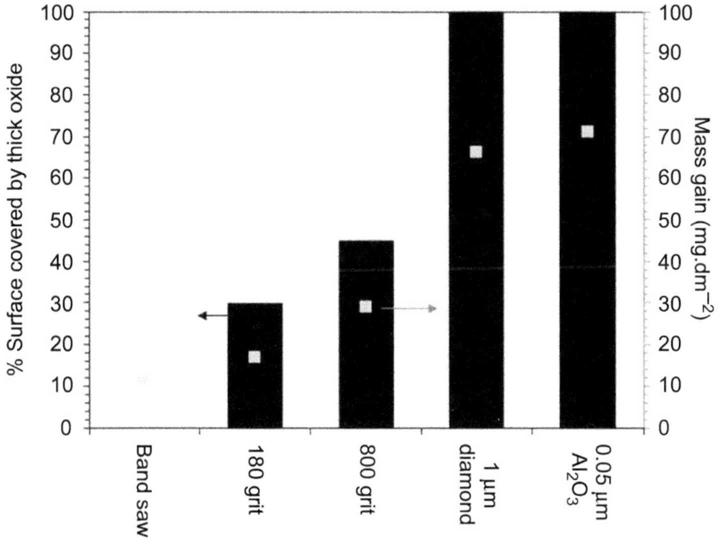

Figure 5.12 Effects of various surface finishes on the corrosion of a model Fe14Cr1·5Si alloy in supercritical water (500 °C, 25 MPa, 5 ppb DO, 250 h)
Adapted from Artymowicz, D., Andrei, C., Cook, W.G., Newman, R.C., 2014. Influence of the cold work on the supercritical water oxidation of Fe14Cr1.5Si model alloy. In: Canada-China Conference on Advanced Reactor Development (CCCARD-2014), Niagara Falls, Canada, April 27–30, 2014.

Dong et al. (2015) found that ultrasonic impact peening of model 12 wt% Cr F/M steels with 1.7–1.8 wt% Mn and 0.6 or 2.2 wt% Si resulted in a more uniform and compact oxide and a twofold reduction in weight gain after 1000 h exposure to SCW (500 °C, 25 MPa, initial DO 8 $mg \cdot kg^{-1}$). Fig. 5.13 shows cross-sectional microhardness data for the as-received and peened samples. The hardness at the surface of the peened samples had increased significantly and decreased with depth until it reached the value for the as-received material at a depth between 150 and 200 μm. The oxide on the as-received sample was found to form preferentially on the lath-shaped martensite phase. Li (2012) noted that the differences in hardness between the martensite and ferrite grains would affect the depth of the deformed layer.

Fig. 5.14 shows data for the minimum and maximum oxide thickness on four austenitic steels after SCWR exposure for 2000 h. The steels had either a ground (1200 grit) or machined finish. The benefits of both higher Cr content and surface cold work on reducing corrosion are apparent, with the benefit of surface cold work possibly diminishing at higher Cr content. The beneficial effect of reducing the grain size (347H vs. 347HFG) can also be seen in the figure.

Fig. 5.15 shows photographs and SEM cross-section images of 316L tube samples with machined and polished surfaces after exposure to SCW (650 °C, 25 MPa, 3000 h). The machined surface has a thin, uniform and compact surface oxide layer, while the polished surface oxide is thick with significant internal oxidation. The maximum oxide thickness on the machined surface was on the order of 1 μm, while the maximum oxide thickness on the polished sample was on the order of 100 μm.

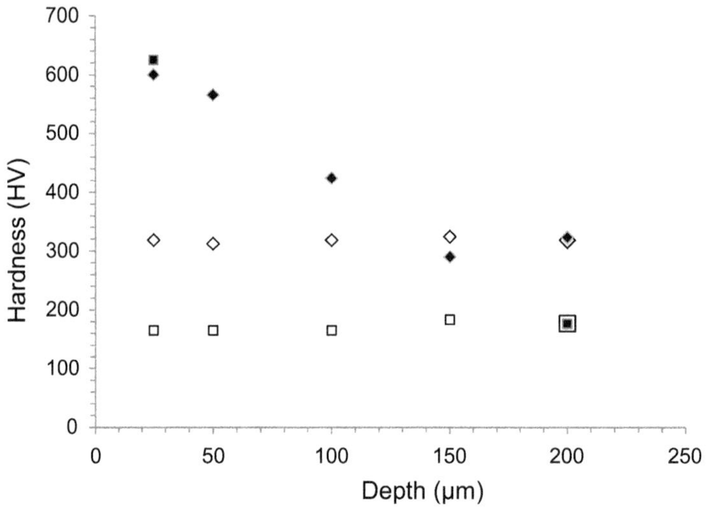

Figure 5.13 Cross-sectional microhardness data for as-received (*open symbols*) and ultrasonic impact peened (*filled symbols*) samples of a model 12 wt% Cr F/M alloy containing 2.2 wt% Si and 1.7 wt% Mn.
Adapted from Dong, Z., Liu, Z., Li, M., Lou, J-L., Chen, W., Zheng, W., Guzonas, D., 2015. Effect of ultrasonic impact peening on the corrosion of ferritic-martensitic steels in supercritcal water. J. Nucl. Mater. 457, 266–272.

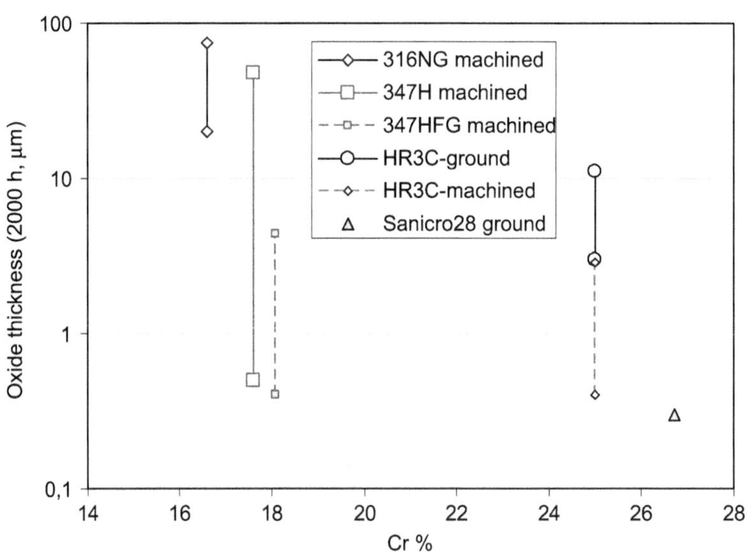

Figure 5.14 Minimum and maximum oxide thickness obtained from SEM cross-sections of four austenitic steels as a function of Cr content after 2000 h exposure to supercritical water at 650 °C, 25 MPa with 150 $\mu g \cdot kg^{-1}$ dissolved oxygen. The steels had either a ground (1200 grit) or machined finish.

Figure 5.15 Photographs and corresponding SEM cross-section images after 3000 h exposure at 650 °C, 25 MPa for 316L SS tube samples with (a) machined and (b) #1200 emery paper finish.

The reduction in corrosion rate for cold worked surfaces above T_c is attributed to increased near-surface defect density, which enhances Cr diffusion to the surface and facilitates formation of a uniform protective oxide film. Sarver and Tanzosh (2013) reported that no benefit from shot peening of austenitic stainless steels was observed at 750 and 800 °C in SHS. Microhardness testing showed that the near-surface cold work was rapidly reduced at temperatures above 750 °C. Li et al. (2014) used FIB techniques to examine the surfaces of 316L samples that had been (1) machined or (2) ground to a 600 grit finish. The subsurface region (1–2 μm thick) of the machined sample had a heavily deformed microstructure with shear bands. Annealing the sample at 650 °C for 100 h resulted in recrystallization of the surface layer to a thin, fine-grained layer. Further annealing at 850 °C resulted in a coarse-grained microstructure. After exposure of the machined sample to SCW (650 °C, 25 MPa, 1000 h) the deformed subsurface layer had been replaced by a recrystallized layer similar to that found on the sample annealed at 650 °C. On the ground sample a thicker, iron-rich oxide layer was formed, with internal oxidation extending deep into the substrate, formation of voids at grain boundaries, and severely oxidized grain interiors. While the data in Fig. 5.8 show increased benefit of cold work as the temperature increases above T_C because of faster diffusion rates and more diffusion pathways, the data of Sarver and Tanzosh (2013) and Li et al. (2014) suggest that as the temperature increases a competition between the rates of protective oxide formation and microstructural changes (recrystallization) may determine the benefit of surface cold work on corrosion resistance.

Since grain boundaries are fast diffusion pathways, reducing the bulk grain size will also allow elements such as Cr that form protective layers to quickly diffuse to the surface. The concentrations of protective oxide–forming elements in the bulk metal typically determine if smaller grains are beneficial for mitigating oxidation. Montgomery and Karlsson (1995) summarize data for 347 SS showing that a decrease in grain size gives increased corrosion resistance at 650 and 700 °C. Fig. 5.14 compares the oxide thicknesses on 347H and 347HFG (fine grained) showing an approximately 10-fold decrease in the maximum oxide thickness for the fine-grained alloy. Tsuchiya et al. (2007) compared the corrosion of large grain size (~30 μm) and fine-grain (~3 μm) 304L and 310L SS in SCW at 550 °C, 25 MPa, 8 $mg \cdot kg^{-1}$ DO. The oxide thickness on the fine-grained 304L sample was about 1/10 the thickness of the large-grained material. The metal loss into the test solution (obtained from the descaled weight loss data) showed a large fraction of the oxidized metal was released into solution from the fine-grained material (~50% for both 304L and 310S). This is expected given the enhanced rates of metal transport to the surface. Ren et al. (2010) found that Equal Channel Angular Processing (ECAP)[7] to produce smaller grains on T91 reduced corrosion in SCW (500 °C, 25 MPa, <25 $\mu g \cdot kg^{-1}$ DO) by ~30% after 3000 h, about the same amount as shot peening. Before the grain refinement treatments, the grain size was 5 μm, and following treatment the ECAP material had an average grain size of 1.25 μm and the shot-peened material had a grain size of ~1.3 μm.

The ability of grain boundary engineering (GBE) to reduce oxide spallation was tested on Alloy 800H by applying one cycle of thermomechanical processing (cold-rolling with a thickness reduction of ~7%) followed by annealing (1050 °C for 90 min) to improve the spallation resistance of the protective oxide layer (Tan et al., 2006a). Oxide spallation has been a serious concern of SCFPPs as it can cause flow blockage and damage to turbines and has been a topic of much investigation (Wright and Tortorelli, 2007). In an SCWR, oxide spallation can also result in flow blockage, particularly in fuel assemblies, can damage turbines and is a mechanism for activity transport. In addition, the now bare surfaces can undergo increased corrosion. In the work of Tan et al. (2006a), samples were polished to a 1 μm finish. Weight change measurement and SEM evaluations after SCW corrosion tests (500 °C, 25 MPa, 25 $\mu g \cdot kg^{-1}$ DO) showed that extensive oxide spallation occurred in the as-received samples but not in the GBE-treated samples (Fig. 5.16). Electron backscatter diffraction (EBSD) examination of cross-sections identified austenite, magnetite, Fe–Cr

[7] A technique that uses severe plastic deformation to create a fine-grained microstructure throughout the bulk of an alloy.

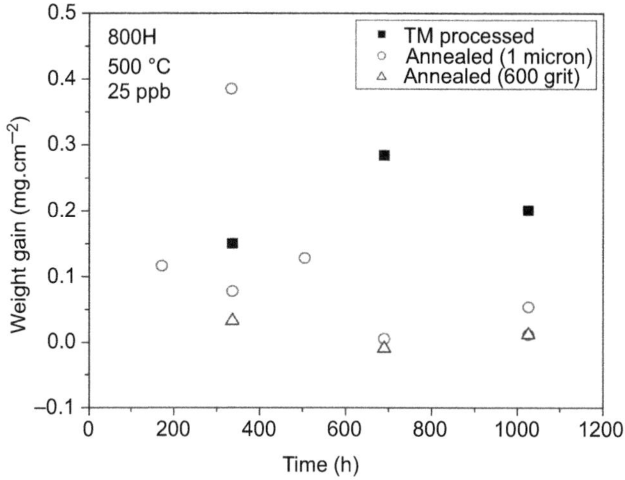

Figure 5.16 Weight gain versus time for Alloy 800H exposed at 500 °C. TM refers to thermomechanically processed condition. Annealed samples were polished to 1 μm and 600 grit finish prior to exposure. TM samples were polished to a 1 μm finish prior to exposure. Figure from Allen T.R., Chen Y., Ren X., Sridharan K., Tan L., Was G.S., West E., Guzonas, D.A., 2012. Material performance in supercritical water. In: Konings, R.J.M. (Ed.), Comprehensive Nuclear Materials, vol. 5. Elsevier, Amsterdam, pp. 279–326.

spinel, and hematite in both types of samples (Tan and Allen, 2005). The GBE-treated sample showed a higher fraction of hematite (with a smaller grain size) mixed with a smaller amount of magnetite in the outer layer, and an inner layer of a spinel phase mixed with phases identified as austenite. The strain distribution (obtained from the local average misorientation between each EBSD data point measurement and its neighbours excluding any higher angle boundaries (>5 degrees)) showed strain accumulation close to the interface between the inner and the outer layers for the as-received sample, whereas the strain distribution for the GBE-treated sample was relatively uniform. The relative strain intensity as a function of the location across the oxide scale showed a significant reduction in strain change between the inner and outer oxides after GBE treatment. It was suggested that the lower extent of oxide spallation on the GBE-treated samples was due to the relatively isotropic oxide texture and greater hematite fraction. GBE was also shown to improve the corrosion resistance of Alloy 617 (Tan et al., 2008b). Nezakat et al. (2014) applied thermomechanical processing (multistep cross-rolling at room temperature followed by annealing in 1050 °C for 5 min per mm of thickness) to 316L and then assessed the corrosion resistance. Treated samples were exposed to SCW (600 °C, 25 MPa, 150 $\mu g \cdot kg^{-1}$ DO) for 100, 300 and 1000 h. Texture was found to have a significant effect on oxidation resistance, samples with {103} texture and {110} texture showing the worst and the best oxidation resistance, respectively. It was also found that thermomechanical processing

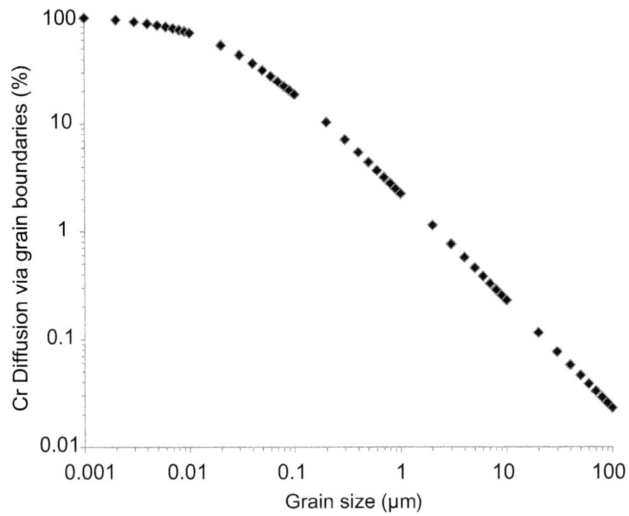

Figure 5.17 Calculated percentage of the total (lattice plus grain boundary) Cr diffusion along grain boundaries for F/M steels.
Adapted from Ren, X., Sridharan, K., Allen, T.R., 2010. Effect of grain refinement on corrosion of ferritic–martensitic steels in supercritical water environment. Mater. Corros. 61, 748–755.

improved adhesion between oxide scale and austenite due to reduction of the fraction of high-energy grain boundaries.

Ren et al. (2010) modelled the effect of grain size on Cr diffusion using an effective diffusion coefficient that considered the diffusion coefficients through the lattice and grain boundary, the grain shape, the grain size and the grain boundary width (estimated to be 0.5 nm). They used values for the lattice and grain boundary diffusion coefficients for 12CrMoV steel (close in composition to the materials used in their study). By assuming negligible transfer of Cr between a grain and its boundary, they calculated the flux of Cr along the lattice and the grain boundary (Fig. 5.17). For the grain sizes in their tests, the percentage of the total (lattice plus grain boundary) Cr diffusion along grain boundaries was found to increase from 3% in the as-received condition to 31% in the grain-refined condition. Using a similar approach, Kim et al. (2015) concluded that a grain size of <8 μm was required to ensure formation of a continuous Cr_2O_3 layer on austenitic steels, in good agreement with experimental data.

5.3.3 *Water chemistry*

Water chemistry factors include pH, additive and impurity concentrations, and the concentrations of oxidants including radiolysis products. Water chemistry effects are expected to be alloy-class specific, as they are determined by the interaction of water, additive or impurity with specific elements in the alloy. The effects of impurities on general corrosion under SCWR conditions have not been studied.

5.3.3.1 pH

In the near-critical region, the effect of pH can be understood with reference to Pourbaix diagrams[8]. Pourbaix diagrams for a number of relevant systems have been reposed by various authors (Kriksunov and Macdonald, 1997; Saito et al., 2006; Cook and Olive, 2012a,b). Diagrams for the iron-water (Kriksunov and Macdonald, 1997; Saito et al., 2006; Cook and Olive, 2012a), chromium-water (Saito et al., 2006; Cook and Olive, 2012b) and nickel-water (Saito et al., 2006; Cook and Olive, 2012c) systems have been reported; diagrams developed by Cook and Olive (2012a,b,c) are shown in Fig. 5.18. One shortcoming of Pourbaix diagrams for high temperature use is that they are often based on extrapolations of thermodynamic data obtained at lower temperatures. Cook and Olive (2012a) note that at 400 °C and 25 MPa, the SCW density (167 kg m^{-3}) is outside of the range over which the revised HKF model has been validated. Chapter 4 discusses additional limitations of the thermodynamic data. Kritzer et al. (1999) discussed additional limitations of Pourbaix diagrams, which include (1) their inability to predict oxide morphology (e.g., porous oxides are less protective); (2) they do not account for kinetic factors, e.g., a thermodynamically unstable oxide with a slow rate of conversion to the stable phase and (3) their inability to easily account for local pH effects (e.g., pits, crevices), which must be modelled separately and whose chemistry parameters may be hard to measure. A further difficulty with Pourbaix diagrams in SCW is the lack of a standardized pH scale, as discussed by Kriksunov and Macdonald (1995b), who proposed a practical (and, as the authors note, somewhat arbitrary) pH scale based on available high-temperature data for the dissociation constants of HCl and NaOH and the activity of water.

Mitton et al. (2000) compared experimental results obtained by Kritzer et al. (1998) on the effect of pH on metal release in support of SCWO development with thermodynamic predictions. At 350 °C, 24 MPa metal release decreased as the pH increased, and the relative fractions of the various released elements changed, being most similar to the relative proportion in the alloy at neutral pH and the results being in reasonable agreement with thermodynamic predictions. Guzonas et al. (2010) reported little difference in corrosion at 450 °C in several proposed SCWR feedwater chemistries (oxygenated water, deoxygenated water and water plus ammonia and hydrazine at $pH_{25} = 8.5–9.5$) except for the combined mode (300 $\mu g \cdot kg^{-1}$ DO + ammonium hydroxide to give a pH_{25} of 8.0–8.5) (Fig. 5.19). They found higher metal release and oxide thickness for deoxygenated water chemistries that contained ammonia, the increase in oxide thickness correlating well with ammonia concentration. Glebov et al. (1976) reported a reduction in corrosion rate of four steels (12Cr1MoV, EI756, Cr18Ni12Ti and 12Cr2Mo1) at pH 9.5–10.5 compared with pH 6.5–7.5 at 28 MPa and temperatures between 380 and 520 °C. For Cr18Ni12Ti the reduction was a factor of 2–3. Fig. 5.20 shows the measured reduction in steady-state Fe and

[8] A Pourbaix diagram is a plot of the predominance of species (as determined by thermodynamics) as a function of pH and the electrochemical potential. The electrochemical potential is related to the Gibbs free energy by $\Delta G = -nFE$.

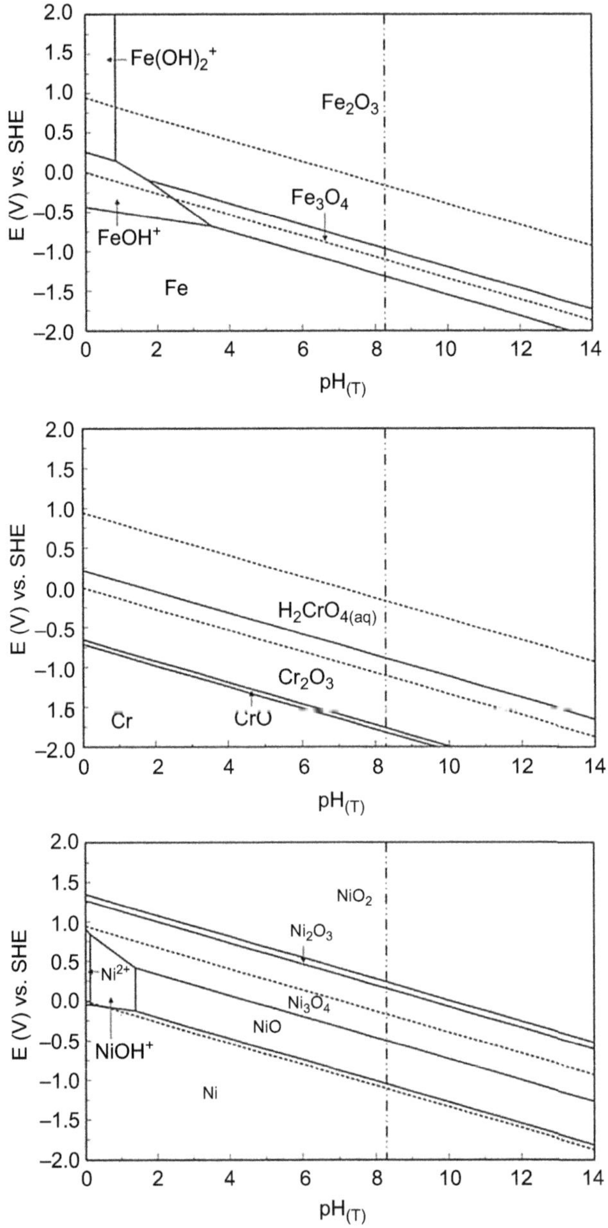

Figure 5.18 Pourbaix diagrams for (top) iron, (middle) chromium and (bottom) nickel at 400 °C, 25 MPa and 10^{-8} mol kg^{-1}.
Figure courtesy Prof. W. Cook, University of New Brunswick.

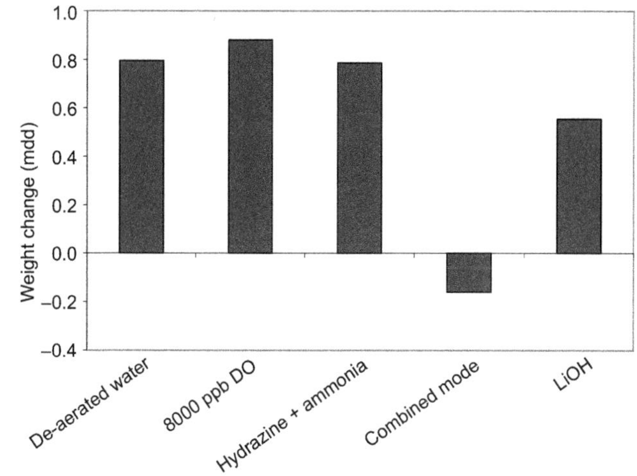

Figure 5.19 Weight gain for 304 SS in supercritical water (450 °C, 25 MPa) for several candidate supercritical water-cooled reactor water chemistries.
Adapted from Guzonas, D., Wills, J., Dole, H., Michel, J., Jang, S., Haycock, M., Chutumstid, M., 2010. Steel corrosion in supercritical water: an assessment of the key parameters. In: 2nd Canada-China Joint Workshop on Supercritical Water-cooled Reactors (CCSC-2010), Toronto, Ontario, Canada, April 25–28, 2010.

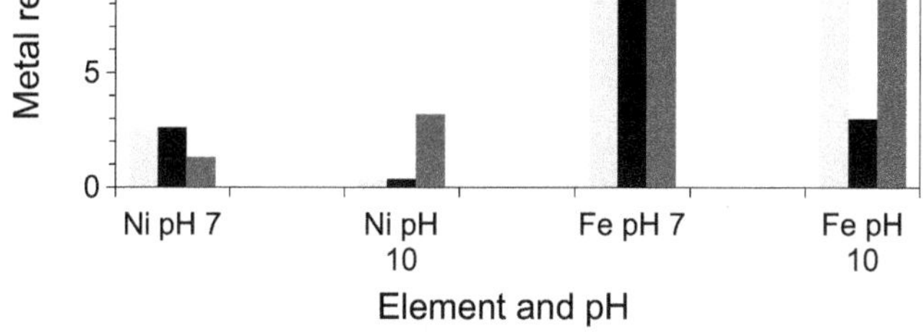

Figure 5.20 Steady-state Fe and Ni release to the test solution from 316 SS exposed to supercritical water containing LiOH at pH 7 and pH 10 at 25 MPa.
Adapted from Svishchev, I.M., Carvajal-Ortiz, R.A., Choudhry, K.I., Guzonas, D.A., 2013. Corrosion behavior of stainless steel 316 in sub- and supercritical aqueous environments: effect of LiOH additions. Corros. Sci. 72, 20–25.

Ni release from 316 SS to the test solution at pH 10 versus pH 7 at 350 and 400 °C, but similar Fe release at 600 °C for both pH values and higher Ni release (Svishchev et al., 2013).

In general, the effects of pH on corrosion in SCW in the near-critical region are consistent with thermodynamic predictions as typically presented in Pourbaix diagrams. There is very little dissociation of ionic species at temperatures above 550 °C at 25 MPa, and pH has little effect on corrosion rate above this temperature (Chapter 4, Section 4.2). At these higher temperatures, Pourbaix diagrams have little meaning and Ellingham diagrams[9] are more relevant. Qiu and Guzonas (2012) discuss the two approaches and illustrate their convergence in the near-critical region. The pH of the feedwater entering the 'evaporator' region at the core inlet is likely to have some effect on corrosion and corrosion product deposition, the effect diminishing as the coolant is heated above 400 °C.

5.3.3.2 *Dissolved oxygen*

As noted in Chapter 4, in-core conditions are expected to be oxidizing (DO concentrations on the order of 1–20 $mg \cdot kg^{-1}$) due to water radiolysis, and therefore the effect of DO concentration on corrosion has been the topic of many studies. In general, testing under SCWR conditions has found that at DO concentrations between 0 and 8 $mg \cdot kg^{-1}$, increasing the DO concentration increases the corrosion rate, but the effect is not large. The weight gain of HCM12A at 500 °C, 25 MPa (Fig. 5.21) shows a decrease with increasing DO concentration between 10 and 300 $\mu g \cdot kg^{-1}$ and an increase at 2000 $\mu g \cdot kg^{-1}$ (Allen et al., 2012). This minimum in oxidation at low DO concentrations is well known and is exploited in the oxygenated and combined water chemistry strategies (Chapter 4) in which small amounts of oxygen are added to the feedwater to reduce the corrosion rate. Fig. 5.22 compares data for the ferritic steels T91 and T92 exposed to SCW at 500 °C as a function of exposure time from tests carried out under similar conditions but a number of DO concentrations. There is relatively good agreement between the data sets, suggesting little effect of DO at these relatively low concentrations within the spread of the data and a possible benefit at DO concentrations of 8000 $\mu g \cdot kg^{-1}$. A similar behaviour is observed for austenitic steels. Fig. 5.7 shows that the weight gain of 316 and 316L at 500 and 550 °C exhibits only a weak dependence on the DO concentration at concentrations up to 8 $mg \cdot kg^{-1}$. Fig. 5.23 shows that a 40-fold increase in DO concentration results in only a roughly 2-fold increase in weight loss of 304 SS after 500 h exposure to SCW at temperatures from 400 to 550 °C at 25 MPa.

In a study of the oxidation of HCM12A (500 °C, 25 MPa, 1 μm surface finish) exposed to 25 $\mu g \cdot kg^{-1}$ and 2 $mg \cdot kg^{-1}$ DO for up to 576 h (Tan et al., 2006b; Tan et al., 2007), it was found that the DO concentration had little effect on oxide composition, resulted in a slight increase in oxide thickness and weight gain, but had a

[9] An Ellingham diagram is a plot of the free energy of oxide stability as a function of temperature for specified concentrations of reactant (e.g., $p(O_2)$). The Gibbs energy change on the vertical axis corresponds to the electrode potential in a Pourbaix diagram ($\Delta G = -nFE$).

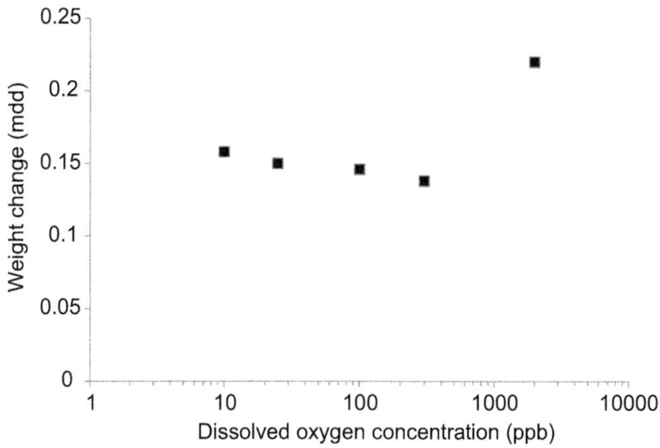

Figure 5.21 Weight gain as a function of temperature for HCM12A exposed to various dissolved oxygen concentrations in supercritical water at 500 °C, 25 MPa for 1026 h. As noted in Allen et al. (2012), data were compiled from two different sources.
Adapted from Allen T.R., Chen Y., Ren X., Sridharan K., Tan L., Was G.S., West E., Guzonas, D.A., 2012. Material performance in supercritical water. In: Konings, R.J.M. (Ed.), Comprehensive Nuclear Materials, vol. 5. Elsevier, Amsterdam, pp. 279–326.

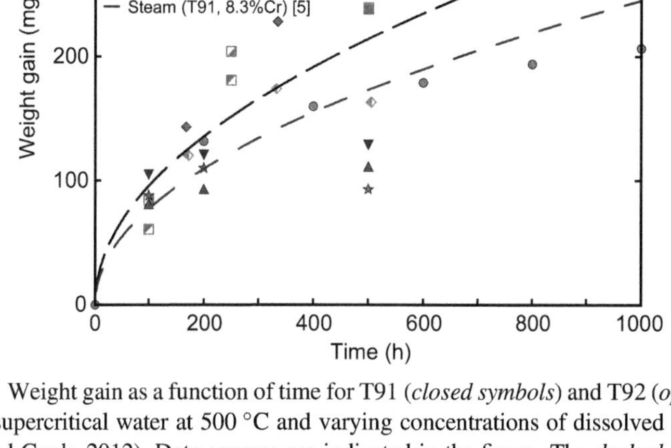

Figure 5.22 Weight gain as a function of time for T91 (*closed symbols*) and T92 (*open symbols*) exposed to supercritical water at 500 °C and varying concentrations of dissolved oxygen (Guzonas and Cook, 2012). Data sources are indicated in the figure. The *dashed lines* are derived from activation energies taken from the indicated sources.

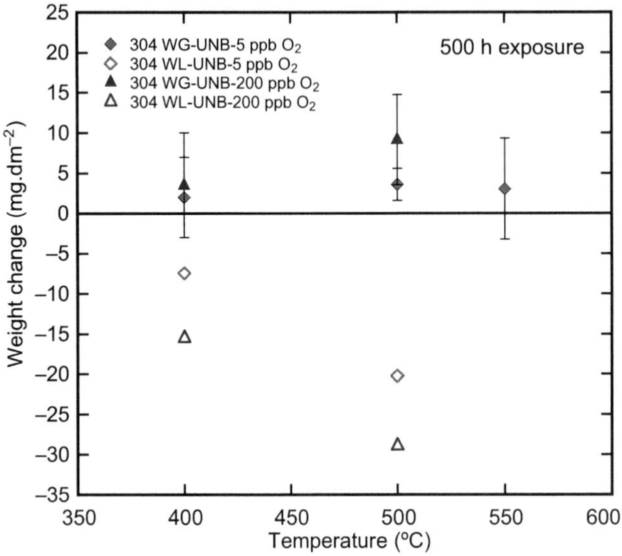

Figure 5.23 Effect of dissolved oxygen concentration on corrosion in supercritical water. From Guzonas, D.A., Cook, W.G., 2012. Cycle chemistry and its effect on materials in a supercritical water-cooled reactor: a synthesis of current understanding. Corros. Sci. 65, 48–66.

significant effect on oxide morphology, the relative differences in the oxide thickness decreasing with increasing exposure time. Coupons exposed at the higher DO concentration showed Cr_2O_3 crystals oriented preferentially perpendicular to the surface and an outer layer of hematite. Ren et al. (2007) found that the oxides formed on Alloy 625 and Alloy 718 in SCW at 500 and 600 °C in 2 $mg \cdot kg^{-1}$ DO were denser than the oxides formed under deaerated conditions, and the extent of pitting was reduced.

At DO concentrations greater than 2 $mg \cdot kg^{-1}$, different behaviour is observed. Choudhry et al. (2015) studied the effect of high DO concentrations (20 $mg \cdot kg^{-1}$) on 316 SS and Alloy 800H. Fig. 5.24 shows the data for Alloy 800H (annealed at 1177 °C followed by water quenching, grain size ASTM 4) exposed to SCW at a 650 °C, 25 MPa, flow rate 0.1 $mL \cdot min^{-1}$ (corresponding to residence times of about 140 s). Experiments were performed with deoxygenated feedwater (DO < 10 $\mu g \cdot L^{-1}$) and high D (20 and 40 $mg \cdot L^{-1}$). The higher DO concentrations resulted in a much finer and more compact surface oxide (Fig. 5.24). While EDS line scans of the cross-sectioned oxide layers showed little compositional differences, XRD analysis of the oxide on the sample exposed in 20 $mg \cdot kg^{-1}$ SCW showed that the outer oxide layer was composed of both magnetite and hematite.

Choudhry et al. (2015) also noted that as the temperature increased, the effect of added DO changed. At temperatures up to 700 °C, metal release from Alloy 800H in 20 $mg \cdot kg^{-1}$ DO was lower than that in deaerated water, but at 750 °C 20 $mg \cdot kg^{-1}$ DO resulted in higher metal release. Liu (2013) studied the oxidation of model Fe12Cr alloys with additions of various concentrations of Si (0.6–2.2 wt%) and Mn (0.6–1.8 wt%) in SCW (500 °C, 25 MPa) at 8 $mg \cdot kg^{-1}$ and 5 vol% H_2O_2-water

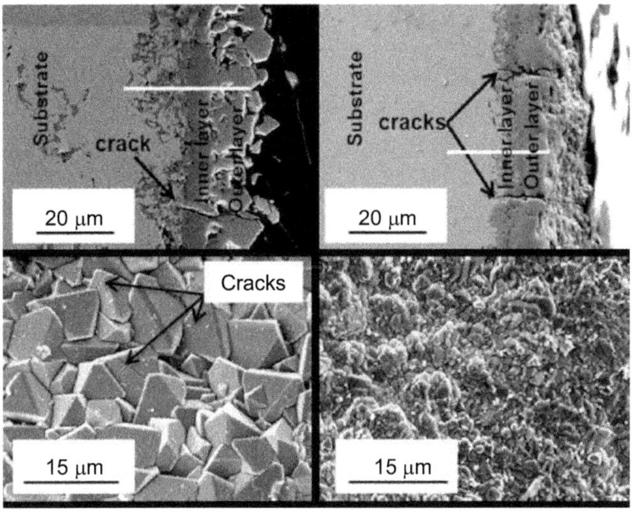

Figure 5.24 Cross-sectional (top) and surface morphologies (bottom) of the oxide layers formed on Alloy 800H in supercritical water at 650 °C, 25 MPa with 20 $mg \cdot kg^{-1}$ dissolved oxygen (DO) (right) and with deoxygenated feedwater (DO < 10 $\mu g \cdot kg^{-1}$) (left) (Choudhry et al., 2016).
Adapted from an original figure reported in Choudhry, K.I., Mahboubi, S., Botton, G.A., Kish, J.R., Svishchev, I.M., 2015. Corrosion of engineering materials in a supercritical water cooled reactor: characterization of oxide scales on Alloy 800H and stainless steel 316. Corros. Sci. 100, 222–230.

solution. They found the oxidation kinetics changed from parabolic at 8 $mg \cdot kg^{-1}$ DO to an exponent of 0.7–0.8 in SCW containing a high concentration of DO, and the weight gain increase by a factor of ~2 after 1000 h. While the oxides were similar in overall structure (two-layer oxides), the oxide formed in SCW containing 5 wt% H_2O_2 was found to be more uniform and continuous than that formed at 8 $mg \cdot kg^{-1}$ DO, and the outer layer formed at high DO concentration was hematite rather than magnetite as found at low DO concentration. Leistikow (1972) found that while Alloy 800 exposed to SHS (620 °C, 0.1 MPa, flow rate 2–4 $cm \cdot s^{-1}$) at 6 $mg \cdot kg^{-1}$ DO lost weight (32% metal loss to the steam phase by dissolution of Cr), at 1 $mg \cdot kg^{-1}$ DO the weight increased with time and there was only 8% dissolution of Cr. Watanabe et al. (2002) and Watanabe and Daigo (2005) reported that in SCW containing 0.01 $mol \cdot kg^{-1}$ H_2SO_4 at 400 °C, 30 MPa, the weight loss of a number of Fe- and Ni-based alloys decreased with increasing Cr content of the alloy, but that this relationship was not true at much higher oxygen concentrations. The corrosion resistance of 316 SS was found to improve at 800 $mg \cdot kg^{-1}$ oxygen, while the corrosion rate of most Ni-based alloys increased at the higher DO concentrations. The improved corrosion resistance of the Fe-based alloys was attributed to formation of protective Fe_2O_3 oxides on the surface, while the reduced corrosion resistance of the Ni-based alloys was attributed to the formation of soluble Cr(VI) species, consistent with the Pourbaix diagrams for iron and chromium (Fig. 5.17).

As discussed in Chapter 4, Cr(VI) species are soluble (volatile) in SCW and lead to the dissolution of protective Cr-rich oxides. Fujiwara et al. (2007) measured the solubility of Cr_2O_3 films deposited onto Pt substrates by chemical vapour deposition at 350 and 450 °C and various DO concentrations. They concluded that Cr_2O_3 was stable at 450 °C at DO concentrations <400 $\mu g \cdot kg^{-1}$. Dong et al. (2012) studied the stability of Cr_2O_3 ceramics in SCW containing various amounts of DO, including measurements of the concentration of Cr dissolved in the test solution. XPS analysis of one sample exposed to SCW at 600 °C for 25 h at an estimated DO content of 168 $g \cdot kg^{-1}$ showed only Cr^{6+} species. They also noted that some of the samples exposed to high DO had molybdenum oxide crystals precipitated on the surface. The effect of oxygen was related to the grain size of the Cr_2O_3 samples, being lowest for a sample with a grain size of 10–20 μm and highest for a sample with a grain size of 0.2 μm.

In a discussion of gas-phase oxidation, Hauffe (1976) notes that if adsorption of oxygen on the surface is the rate-limiting step then a large dependence on oxygen partial pressure (equivalent to DO concentration) is expected, whereas if solid-state diffusion of oxygen is rate-limiting then a much weaker dependence is expected. However, even at low pressures it is well established that oxidation of metals in the presence of water is different than oxidation in dry air. The mechanisms of oxidation in high-density SCW will be far more complex than oxidation in low-density SCW (SHS) because the bulk density of water molecules is two to three orders of magnitude higher, and recent MD work has shown that the density of water molecules at an $Fe(OH)_2$ surface (a model for the outermost oxide surface in contact with water) is higher than in the bulk (Chapter 4). The number of adsorbed waters per exposed OH group (surface coverage) was found to decrease with temperature. While the fraction of the surface covered by water decreases due to the much lower density (Fig. 4.26), even at the lowest density the alloy surface will be covered mostly by adsorbed water. Simulations also show that water is found at the surface in nanopores, and, significantly, at supercritical temperatures O_2 diffused through nanometre-sized cracks more readily than in the bulk solution.

Saunders et al. (2008) noted that all aspects of oxide growth are altered in the presence of water vapour compared with similar processes in dry conditions. The first stages of the reaction (adsorption and dissociation) are controlled by the nature of oxides, with data suggesting that dissociation of gas molecules is favoured by surface defects. Water adsorbed on the oxide surface dissociates as:

$$H_2O \leftrightarrow O^{2-} + 2H^+ \tag{5.1}$$

or

$$H_2O \leftrightarrow H^+ + OH^- \tag{5.2}$$

At high temperatures, O_2 is formed by the thermal decomposition of water so that even a nominally deaerated system will contain $\mu g \cdot kg^{-1}$ concentrations of oxygen at temperatures higher than about 600 °C. In aerated systems the added O_2 provides additional oxidant. More significantly, in an SCWR core, water radiolysis can produce high concentrations of O_2 and H_2O_2 in addition to transient radical oxidizing and reducing species. The overall reactions of oxygen with iron are:

$$2Fe + 2H_2O + O_2 \leftrightarrow 2Fe(OH)_2 \tag{5.3}$$

$$4Fe(OH)_2 + 2H_2O + O_2 \leftrightarrow 4Fe(OH)_3 \tag{5.4}$$

Similar reactions can be written for the other alloy constituents. The metal hydroxides formed will dissolve (evaporate) into the SCW coolant at low concentrations (Chapter 4) or undergo dehydration reactions at the surface to form metal oxides.

Both O^{2-} and OH^- can diffuse through the oxide (Galerie et al., 2001), and analyses of surface oxides suggest that O^{2-}, OH^-, O_2 and H_2O can all penetrate into the oxide layers either by solid-state diffusion (O^{2-} and OH^-) or via cracks and pores in the oxides (O_2 and H_2O). The hydrogen produced by the reactions in Eqs. (5.1) and (5.2) can form H_2 and pass into solution or it can diffuse through the oxide and metal and eventually be released into the environment. The presence of O^{2-}, OH^- and adsorbed H_2O has been reported in oxide layers by XPS by Rodriguez et al. (2014) and Rodriguez and Chidambaram (2015). Yamamura et al. (2005) reported the presence of CrOOH in the oxides formed on 304 SS in SCW (450 °C, 50 MPa, neutral pH_{25}) using XPS, CrOOH accounting for ~90% of the Cr in the oxide. Was et al. (2004) reported the presence of iron and chromium hydroxides in the oxide layers formed on 304 and 316L exposed to SCW. The amounts of iron present as oxide and hydroxide were similar, and they suggested that most of the Cr present in the oxide was in the form of a hydroxide. Galerie et al. (2001) note that OH^- can easily fill sites in oxides usually filled by O^{2-} ions, transporting through the oxide in an analogous manner.

The work of Zhu et al. (2016) provides some insight into the role of DO on corrosion. They exposed P92, polished to a 1 μm finish, to 2 $mg \cdot kg^{-1}$ isotopically labelled oxygen (>95% purity $^{18}O_2$) in unlabelled water ($H_2^{16}O$) at 600 °C, 25 MPa for 100 h, followed by SIMS depth profiling to monitor the content of ^{18}O and ^{16}O in the oxide. At this temperature, the amount of O_2 in equilibrium with water is 3 $\mu g \cdot kg^{-1}$. The concentration of water was about six orders of magnitude higher than the concentration of oxygen ignoring any surface effects. The ratio $^{16}O/^{18}O$ was ~500 and was relatively constant throughout the oxide with a slight increase in the ratio near the metal-oxide interface and in the transition zone. The authors suggested that the main role of oxygen was to increase the oxygen potential in the oxide, increasing the diffusion of cations. Jacob et al. (2002) used a similar isotopically labelled system to study the role of water vapour in the oxidation of powder metallurgical (PM) chromium at 800 and 1000 °C. Coupons were exposed to a mixture of Ar plus 20% O_2 with either very low water content ($P_{H_2O} < 13$ Pa) or a high water content ($p_{H_2O} = 1$ kPa). In the 'wet' gas stream, water was found to be the main source of the oxygen in the oxide, making up roughly 2/3 of the total, even though it accounted for only 1/20 (partial pressure basis) of the total gas stream. They found that the defect structure of the oxide was not changed in any way that would significantly influence the diffusion mechanisms of either Cr or oxygen in the oxide. They noted that SIMS data showed a small but unquantified amount of OH^- in the oxide.

There have been few studies of corrosion in irradiated SCW, where as noted water radiolysis can result in high concentrations of oxidants. Karasawa et al. (2004) reported the results of studies of γ-ray irradiation of several Fe-based alloys (304, 316 and 310 SS), Ni-based alloys (Alloy 600, Alloy 625, X-750, and Hastelloy C276) and Ti and a Ti-based alloy (containing 21.2% V, 3.7% Al and 0.08% Fe). Three water chemistry conditions were studied; 2000 ppb O_2, 50 ppb H_2 and degassed. They found that the results were complex and depended on the alloy composition. Irradiation altered the oxide formed on the surface as determined by Raman spectroscopy. For stainless steel, they found that the weight gain was best described by $WG = (kT)^{0.25}$, while for Ni- and Ti-based alloys they found the data could be described by $WG = (kT)^{0.5}$. They concluded that OH• radical or H_2O_2 oxidized the surface, and on 316 SS this promoted the formation of nickel ferrite. Fig. 5.25 shows

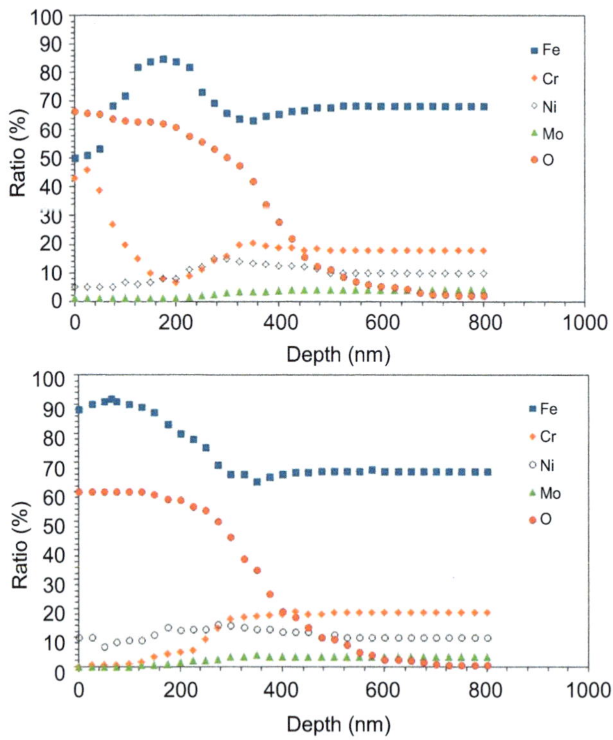

Figure 5.25 Effect of irradiation on corrosion of 316 SS in supercritical water at 400 °C, 24.7 MPa, 400 μg·kg^{-1} dissolved oxygen nominal. Top figure – unirradiated; bottom figure – irradiated (3 × 10^4 Gy·h^{-1}).
Adapted from Karasawa, H., Fuse, M., Kiuchi, K., Katsumura, Y., 2004. Radiolysis of supercritical water. In: 5th Int. Workshop on LWR Coolant Water Radiolysis and Electrochemistry, 2004 October 15, San Francisco, USA.

that in the presence of irradiation, the surface film formed on 316 SS in 400 $\mu g \cdot kg^{-1}$ DO was depleted in Cr compared to the unirradiated surface film. Bakai et al. (2015) showed increased release of Cr in irradiated SCW due to the radiolysis of water (Fig. 4.23).

5.3.3.3 Supercritical water pressure/density

Most corrosion testing in support of SCWR development has been performed at constant pressure, such that the SCW density changes as the temperature changes, especially in Regions 1 and 2 (the 'evaporator' region). While experimentally simple and directly applicable to the SCWR (which will operate at essentially constant in-core pressure), this type of testing complicates data interpretation because the SCW properties change significantly in the vicinity of the critical point (Chapter 4). Few systematic studies of the dependence of corrosion rate on pressure or density have been performed, in part because of the experimental difficulties and in part because the SCWR will operate at essentially constant pressure. Data presented by Montgomery and Karlsson (1995) and Wright and Dooley (2010) suggest only a small effect of pressure on weight change or oxide thickness. The pressure effect is of practical importance because testing at lower pressures is experimentally simpler than testing at 25 MPa, and the advantage of such testing for materials screening at very high temperatures ($T > 500\ ^\circ C$) is obvious. At these temperatures, oxide solubilities are low and dissolution, which can occur at higher water densities and is expected to have little effect on oxide layer thickness. As a significant body of corrosion data in SHS, including in-reactor testing, was obtained during nuclear steam superheat development programs, neglect of the pressure dependence would make these data more relevant for SCWR development.

Macdonald (2004) suggested three possible effects of pressure on reaction rates when active dissolution is taking place: an effect on the activation energy; an effect on the volumetric concentrations of reactants and an effect on the dissociation of acids, bases, etc. If the reaction rates are controlled by a passive film, then other considerations apply. When metal diffusion in an oxide is dominant, Wright and Dooley (2010) proposed that the parabolic rate constant should be proportional to P^x (i.e., $\log(K_p) = x \cdot \log(P_T)$), with $x = 0.11$ to 0.15 for magnetite. Holcomb (2014) proposed that increasing the pressure could increase the critical concentration of Cr required to form a protective chromium oxide film.

It seems reasonable, in light of the discussions in Chapter 4 and data already presented in this chapter to divide the regions of interest into (1) near-critical and (2) $T \gg T_c$. In the near-critical region, changing the pressure at constant T can change the values of properties such as density and dielectric constant from 'liquidlike' to 'gaslike'. At higher temperatures, changing the pressure has only a small effect on these other properties. It is therefore a priori expected that the pressure dependencies of corrosion in these two regions should be different.

Near-critical region

Watanabe et al. (2001) studied the pressure dependence of the corrosion of five alloys in SCW containing $0.01\ mol \cdot kg^{-1}$ H_2SO_4 and $0.025\ mol \cdot kg^{-1}$ DO at 400 °C

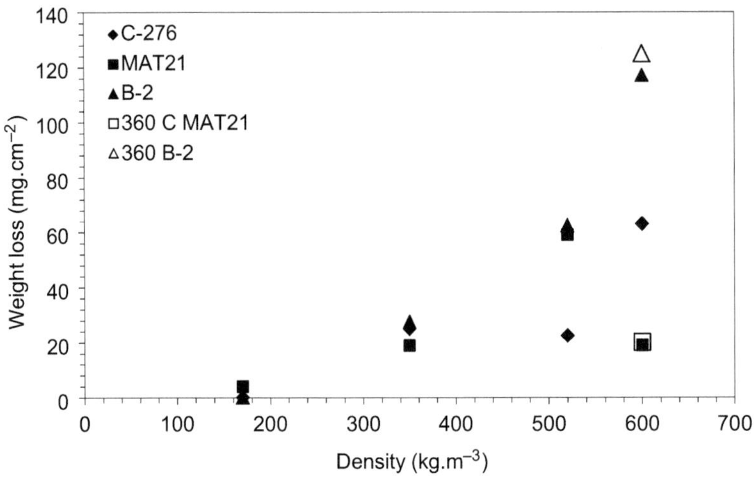

Figure 5.26 Dependence of weight loss on supercritical water density for five alloys in 0.01 mol·kg^{-1} H_2SO_4 and 0.025 mol·kg^{-1} dissolved oxygen at 400 °C. The *open symbols* were measured at 360 °C.
Adapted from Watanabe, Y., Kobayashi, T., Adschiri, T., 2001. Significance of Water Density on Corrosion Behavior of Alloys in Supercritical Water. NACE Corrosion 2001, Paper 01369.

(Fig. 5.26). In general, the corrosion (weight change after 100 h) increased with increasing density between ~150 and 600 kg·m^{-3}. The effect of pressure in this region is in agreement with predictions based on changes in water properties with density. Experiments were performed at a constant temperature of 390 °C and variable density to measure the density dependence of the corrosion of 304 stainless steel (Guzonas, 2013) and Alloy 625 (Li et al., 2009). As the SCW density increased, the weight gain (a measure of both metal loss and oxide formation) decreased (Fig. 5.27) but the weight loss (a measure of corrosion) remained almost constant (not shown, see original papers), reflecting the increase in oxide film solubility as the SCW density increases (Guzonas, 2013). As discussed in Chapter 4, a reasonable approximation for the dependence of solubility (S) on density is given by log(S) ≈ log(ρ) (see, for example Fig. 4.13), while the temperature dependence is relatively weak. Guzonas (2013) assumed that the mass (thickness) of the outer oxide layer depended inversely on S, resulting in a simple relationship between the logarithm of the weight change and SCW density:

$$\log(\mathrm{WC}) = -\mathrm{k}\cdot\log(\rho) + \mathrm{C} \quad (5.5)$$

where k and C are constants. This expression was found to fit the descaled weight loss data reasonably well.

High T region

Cowen et al. (1966) reported data for the weight gain of four alloys (Nimonic PE 16, 20/25/Nb, FV 555 and Alloy 800) at 700 °C exposed to SCW at pressures of 6.9, 24.1

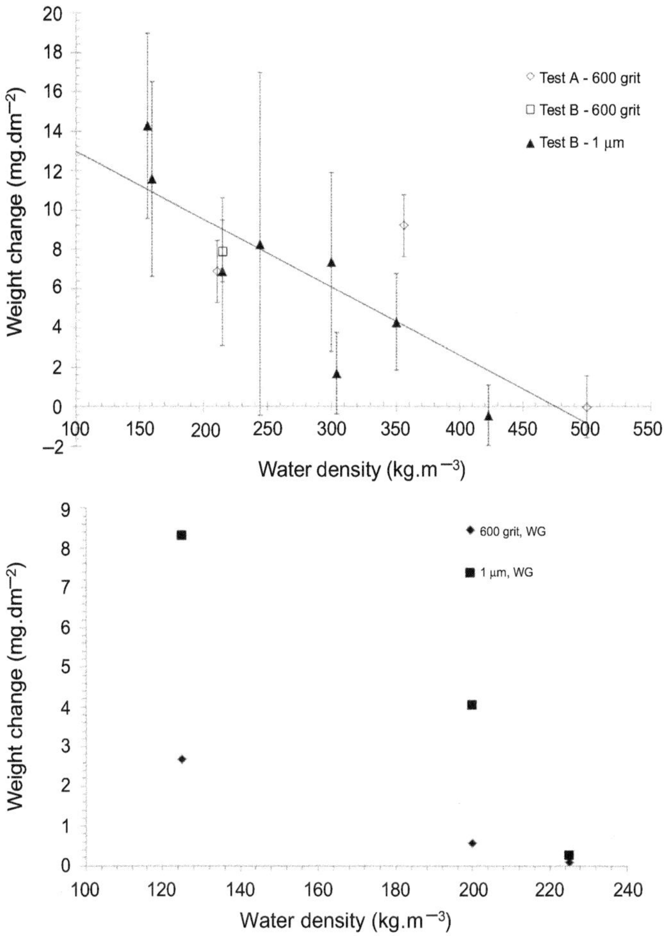

Figure 5.27 Average weight change for 304 SS (top) and Alloy 625 (bottom) exposed to supercritical water (SCW) versus SCW density, adapted from Guzonas et al. (2013) (top) and Li et al. (2009) (bottom). *Vertical error bars* in the upper figure represent the standard deviation of the average value of weight change.

and 34.5 MPa. Fig. 5.28 shows increasing the pressure increases the weight gain of Alloy 800, the difference increasing slowly with time. No systematic differences were observed for the other three alloys tested, although at the longest exposures the weight gains of PE 16 and 20/25/Nb at $P > P_c$ were higher than those at $P < P_c$ by about the same relative amount as Alloy 800. In all cases the effect was small; after 3500 h the weight gain of Alloy 800 at 34.5 MPa was only about 1.75 times that at 6.9 MPa. Fig. 5.29 shows the ratio of the weight gain at 24.1 MPa to the weight gain at 6.9 MPa at 600, 650 and 700 °C for three alloys (20/25/Nb, Alloy 600 and Alloy 800) exposed to SCW containing low DO, H_2 released by corrosion for an exposure time of 2000 h. Most samples were ground to 180 grit and some were cold-rolled

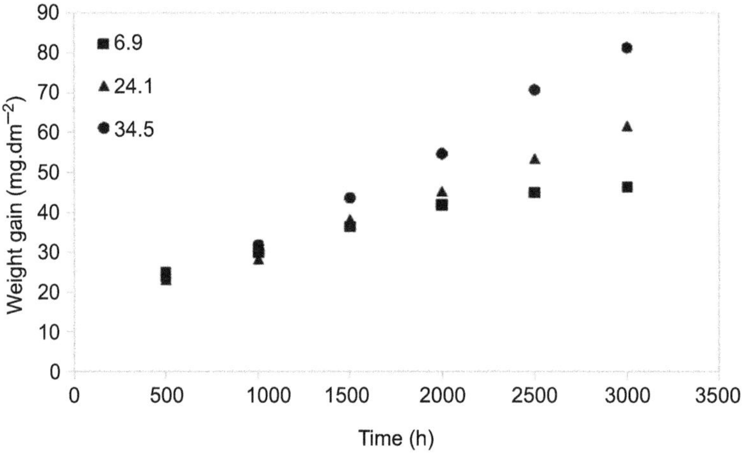

Figure 5.28 Weight gain versus time for Alloy 800 exposed to supercritical water at 700 °C at three pressures (in MPa).
Adapted from Cowen, H.C., Longton, P.B., Hand, K., 1966. Corrosion of Stainless Steels and Nickel-Base Alloys in Supercritical Steam. U.K. Atomic Energy Authority TRG Memorandum 3399 (C).

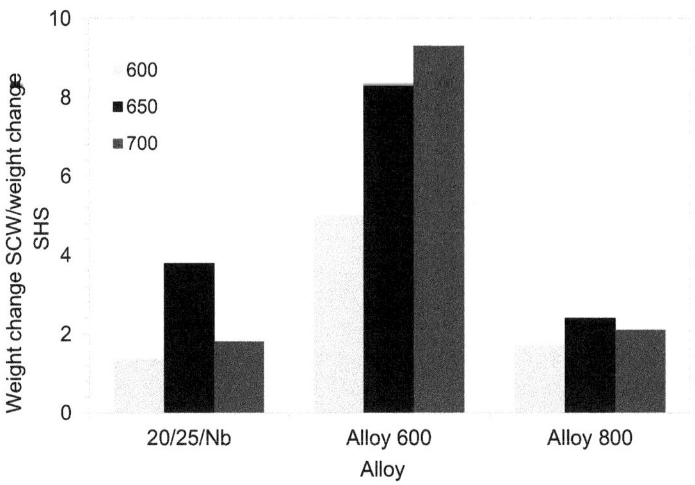

Figure 5.29 Ratio of weight gain at 24.1 MPa to weight gain at 6.9 MPa for three alloys at 700 °C.
Adapted from Cowen, H.C., Longton, P.B., Hand, K., 1966. Corrosion of Stainless Steels and Nickel-Base Alloys in Supercritical Steam. U.K. Atomic Energy Authority TRG Memorandum 3399 (C).

to introduce cold work. Alloy 600 showed the largest effect of pressure, the ratio increasing with increasing temperature; at 700 °C the weight gain at 24.1 MPa was 9.3 times that at 6.9 MPa. Alloy 800 and 20/25/Nb showed only a slight increase in ratio with increasing T.

Bischoff (2011) and Bischoff et al. (2012) reported results for the corrosion of T91, NF616, HCM12A, 9CrODS and 14CrODS in SCW at 500 and 600 °C and in 500 °C steam. A recirculating loop was used for the SCW tests and the water contained between 10 and 25 $\mu g \cdot kg^{-1}$ DO. The steam corrosion tests were performed in a static deaerated autoclave at a pressure of 10.8 MPa. Corrosion at 10.8 MPa was lower than at 25 MPa, and the weight gain and oxide layer thicknesses were about 1.5–2 times lower at the lower pressure. This was attributed to changes in the diffusion of iron, possibly as the result of differences in water density on the sample surface. Unfortunately, the weight loss was not reported. The rate law was close to parabolic in SCW but closer to cubic in SHS. The oxide formed in SCW was more porous, especially the inner layer, and the porosity in the outer layer was more spread-out in SCW and localized at the inner layer/outer layer interface in SHS. The diffusion layer appeared to contain more oxide precipitates in SHS. The SHS samples contained more Cr_2O_3 in the inner and diffusion layers. The Fe/Cr ratio in the SCW samples was not significantly different than the SHS samples, but the outer layer/inner layer thickness ratio was consistently higher in SCW than SHS (~10% for NF616 and HCM12A, ~30% for 14Cr ODS). Bischoff noted that if iron migration was the rate-limiting step then increased solubility in SCW would increase the corrosion rate. Jeong et al. (2005) performed a similar comparison of corrosion in SHS and SCW at 500 °C using a range of zirconium alloys. Again, a slightly higher rate was reported for SCW than SHS, and it was noted that the close agreement between the data obtained at the two pressures could enable the use of SHS as a surrogate for materials screening in lieu of using SCW.

Yi et al. (2001) studied the pressure dependence of the corrosion of four steels (HCM12A, NF616, HCM2S and STBA24) in a static autoclave using deaerated water. The coupons were ground with 1500 grit SiC. They concluded that the parabolic rate constant decreased by a factor of about 2 at 600 °C as the pressure increased, but that there was little effect at 700 °C. Fig. 5.30 compares the data of Yi et al. (2001) with those of Bischoff (2011) and Bischoff et al. (2012) for NF616; although the rate law in the work of Bischoff et al. was closer to cubic, it was considered useful for comparison purposes. The Bischoff data lie above the expected K_p value at 500 °C from the Yi et al. (2001) data set, as expected for surfaces that had been finely polished, which suggest that surface finish has a larger effect on K_p than pressure.

Gómez-Briceño et al. (2013) studied the effect of pressure on the oxidation of 316L SS, Alloy 600 and Alloy 625 at 400 and 500 °C for exposures up to 780 h in SCW containing 8 $mg \cdot kg^{-1}$ DO. The test samples were abraded to a 600 grit finish. At 400 °C the weight gain of 316L SS was essentially constant after 200 h at 30 MPa, whereas the weight gain at 25 MPa increased with time. At 500 °C, the weight gain at 30 MPa was roughly 50% of the value at 25 MPa, the oxide formed at 25 MPa being consistent with that reported by other authors for this material under similar test conditions (chromium-rich inner oxide, iron-rich outer layer and underlying diffusion zone), whereas at 30 MPa the oxide layer was much thinner and the diffusion zone much thicker. For Alloy 600, pressure had little effect at both temperatures within the estimated experimental uncertainty of the figures presented.

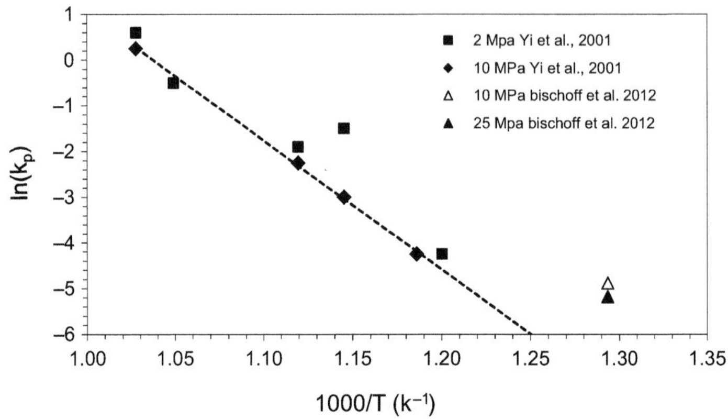

Figure 5.30 Plot of the logarithm of the parabolic rate constant K_p as a function of inverse temperature for the corrosion of NF616 in supercritical water. Data are from the work of Yi et al. (2001) and Bischoff et al. (2012). In the study of Yi et al. (2001) the samples were ground with 1500 grit SiC and exposed in deaerated water. In the work of Bischoff et al. (2012) the samples were polished to 0.05 μm to enable marker deposition. The *solid line* is the linear least squares fit to the 10 MPa data of Yi et al. (2001).

Holcomb et al. (2016) examined experimental data for the pressure dependence of the parabolic rate constants for the oxidation of 18-8 austenitic stainless steels and Ni-based alloys from a number of sources. At 700 °C the effect of pressure on the oxidation of Ni-based alloys was found to be better described by $\log(K_p) \propto P_T$, rather than $\log(K_p) \propto 1 \; \log P_T$ as predicted by models based on defect transport. For fine-grained austenitic steels, $\log(K_p) \propto \log P_T$ better fit the data, while for coarse-grained austenitic steels, no relationship between the parabolic rate constant and pressure could be found.

Choudhry et al. (2016) reported data for the steady-state concentrations of metal release from Alloy 800H at 25 MPa as a function of temperature (500, 550, 650, 700 and 750 °C). The steady-state concentrations of Cr and Ni increased between 500 and 550 °C and then decreased with further increases in temperature. As the corrosion rate was expected to increase with increasing temperature based on literature measurements, the decrease in Cr and Ni was attributed to the change in oxide solubility. They found a good correlation between log(metal concentration) and log(SCW density) (Fig. 5.31) indicating that most of the observed decrease in concentration can be attributed to the change in SCW density, although the magnitude of the effect was small. This conclusion is consistent with the calculated solubilities shown in Fig. 4.10, which shows that above 500 °C the solubilities of both NiO and Cr_2O_3 increase with increasing temperature. This shows that the effect of changes of SCW density, although small, may still play a role in the measured pressure dependence of weight gain at these high temperatures. It is expected that the descaled weight loss would still show an increase with increasing temperature.

While the data show that corrosion rates do not show a strong density (pressure) dependence at $T \gg T_c$, the morphologies of the resultant oxide layers do (e.g., the work of Bischoff (2011) discussed earlier). Sanchez (2014), Li et al. (2015) and

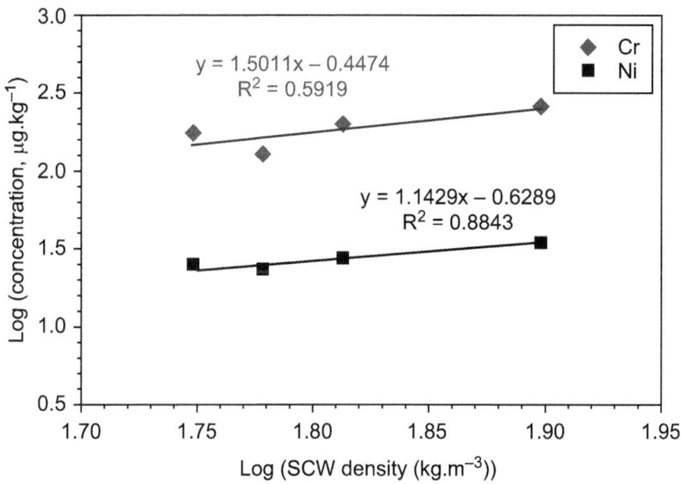

Figure 5.31 Ni and Cr release from Alloy 800H as a function of log supercritical water (SCW) density (Choudhry et al., 2016).

Huang et al. (2015b) reported the results of studies of four alloys (304 SS, 310 SS, A-286 and Alloy 625) exposed to SCW for 1000 h at 625 °C at 8 and 29 MPa (static autoclave) and at 0.1 MPa (flowing steam in a tube furnace). The coupons were ground to a 600 grit finish. The weight gain (oxide thickness) peaked at a density of $20\ kg \cdot m^{-3}$ (8 MPa) for the three iron-based alloys but showed a slight weight loss for Alloy 625 (Fig. 5.32). Sanchez (2014) reported little difference in weight gain of Alloy 625 and 310 SS at 8 and 29 MPa after 3000 h but a large increase in weight for A-286. Alloy 625 showed little difference in weight change at any of the densities studied, whereas the weight change for 304 SS and A-286 at 0.1 MPa was 10–100 times lower than at 8 and 29 MPa. Alloy 304 was found to be completely covered by oxide after 1000 h at 8 MPa but only partially covered at 29 MPa. The sample exposed at 0.1 MPa was only partially covered by a thin oxide. 310 SS formed a Cr-rich oxide layer with the spinel structure with an underlying recrystallized γ layer depleted in Cr (Fig. 5.33, Table 5.1). The spinel-structure oxide had essentially the same lattice parameters at each pressure; the inner oxides had roughly similar compositions (Table 5.1), while the outer oxide compositions showed a pressure (density) dependence.

In this same study, a detailed microstructural characterization of the surface region was performed for A-286 (Huang et al., 2015b). At 0.1 MPa, a continuous layer of Cr_2O_3 was formed, which was underlain by both a nearly continuous SiO_2 layer and Ti-rich layer. At 29 MPa, a nearly continuous Cr-rich layer was formed, covered by a layer of Fe_2O_3; in these regions the oxide was thinner and there was a 1–2 µm deep recrystallization layer below the oxide. In regions with thicker oxide, oxidation was found to have progressed deeply into the substrate, and a (Cr, Mo, $Fe)_2O_3$ phase was found in the internal oxidation region. At 8 MPa, the surface was completely covered by a ~15–20 µm outer Fe oxide and a 15–20 µm partially oxidized inner layer.

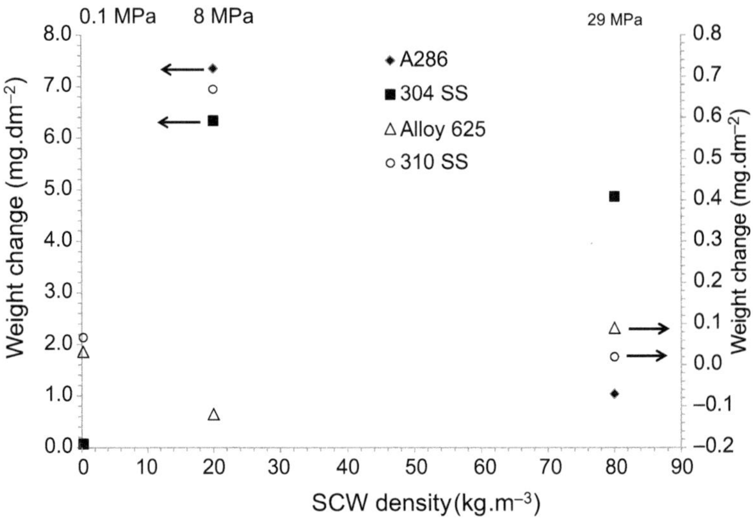

Figure 5.32 Density (pressure) dependence of the weight change of four alloys after 3000 h exposure at 625 °C. Note that the points for Alloy 625 and 310 SS overlap at a density of 80 $kg \cdot m^{-3}$.
Adapted from Li, W., Woo, O.T., Guzonas, D., Li, J., Huang, X., Sanchez, R., Bibby, C.D., 2015. Effect of pressure on the corrosion of materials in high temperature water. In: Characterization of Minerals, Metals, and Materials. John Wiley & Sons Inc., Hoboken, NJ (Chapter 12).

5.3.4 Flow rate

Ruther and Greenberg (1964) reported that at 650 °C, 42 $kg \cdot cm^{-1}$, 0.05 $mg \cdot kg^{-1}$ DO, changing the steam velocity from 30 $m \cdot s^{-1}$ to 91 $m \cdot s^{-1}$ had little effect on the metal loss for 304 SS and 406 SS for a 5-week exposure. Lukaszewicz et al. (2013) studied the effects of flow rate on the oxidation of T23, T92, TP347HFG and Super 304 at 650 and 700 °C at steam flow rates of 4, 16 and 40 $mm \cdot s^{-1}$ in deoxygenated, deionized water. The samples were placed in the test furnace in specific orientations to evaluate differences in direct and indirect flow. Samples were ground to a 600 grit finish. A large effect was observed for T23, the damage increasing significantly between the highest and lowest flow rates. For T92, the increase in steam flow rate promoted the formation of an outermost hematite layer, and the oxide formed had a higher number of gaps or voids. For T347HFG, the thickest, most adherent oxides were formed on the parts of the surface facing direct flow. With increasing exposure time the protective oxide was observed to break down and nodules formed. The behaviour of Super 304 was similar but the oxides were thinner and the nodules smaller. Fig. 5.34 shows the dependence of the weight change for Super 304 at the different flow rates studied. The authors attributed the observed flow rate effects to a higher pO_2 at the surface, the result of the removal of the H_2 produced by the thermal decomposition of water at the surface. The breakdown of the oxides on the austenitic alloys was the result of

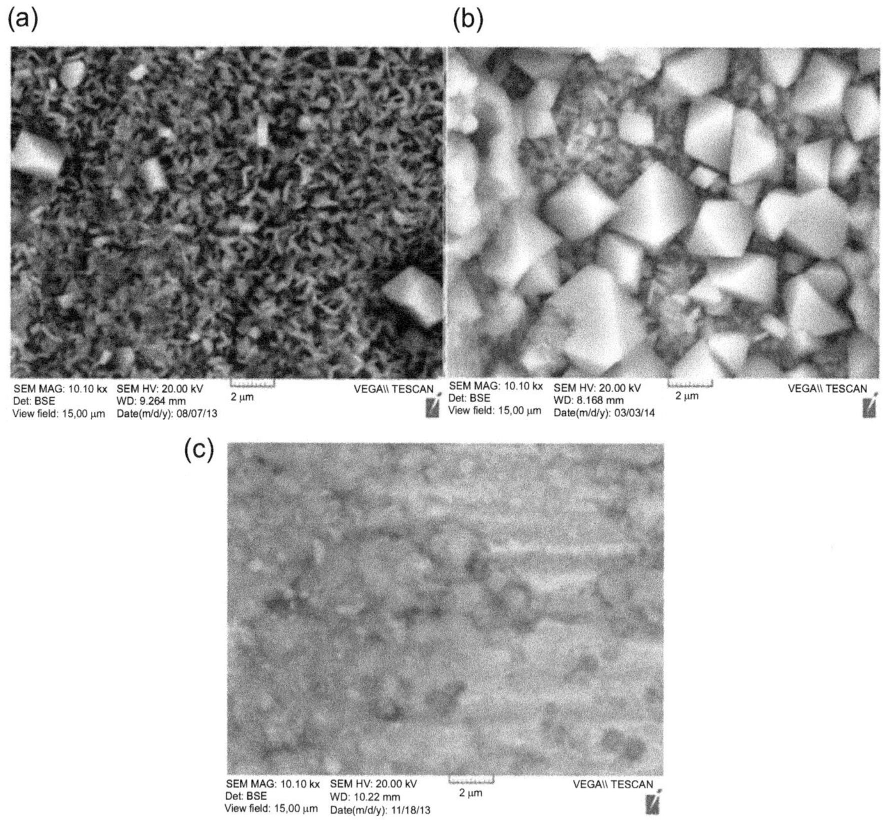

Figure 5.33 Morphology of the surface layers formed on 310 SS at (a) 29, (b) 8 and (c) 0.1 MPa after 1000 h exposure at 625 °C (Li et al., 2015).

Table 5.1 Surface composition and structure on 310 SS at different pressures (Li et al., 2015)

	Surface oxide layer			Recrystallized grain layer	
		Thickness (μm)			
Pressure (MPa)	**Structure**	**Outer**	**Inner**	**Structure**	**Thickness (μm)**
29	Spinel	0.5	0.4	γ w/t $M_{23}C_6$	2
8	Spinel	4.2	3.3	γ w/t $M_{23}C_6$	8
0.1	Spinel	0.2[a]		γ w/t $M_{23}C_6$	1

[a]Does not include the thin continuous layer of silica.

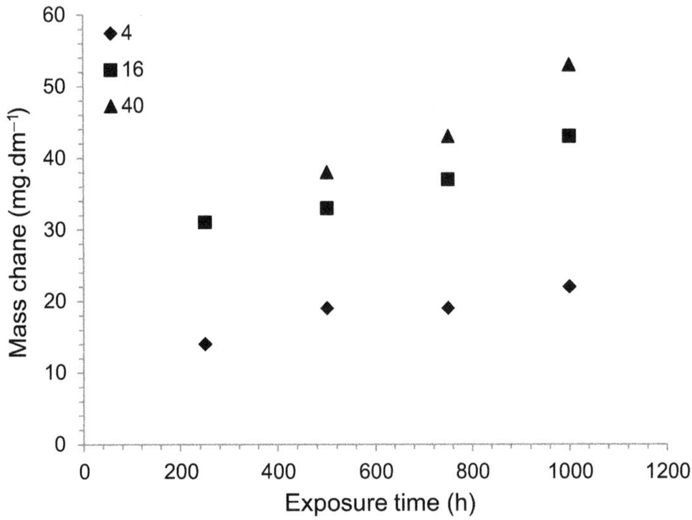

Figure 5.34 Dependence of the weight change of Super 304 in superheated steam at 700 °C on exposure time at three flow rates (values given in the legend).
Adapted from Lujaszewicz, M., Simms, N.J., Dudziak, T., Nicholls, J.R., 2013. Effect of steam flow rate and sample orientation on steam oxidation of ferritic and austenitic steels at 650 and 700 °C. Oxid. Met. 79, 473–483.

the depletion of Cr in the alloy due to the higher rate of oxide growth at the higher flow rates. Knödler and Straub (2014) compared the parabolic rate constants obtained for P92, FB2, Waspalloy and Alloy 625 exposed to SHS from 5000 h laboratory corrosion tests (650 °C for the F/M steels, 750 °C for the Ni-based alloys) with values obtained from samples placed in the bypass lines of an operating power plant (19 MPa) at 630 °C (F/M steels) and 725 °C (Ni-based alloys). There was little difference in rate constants of the F/M steels exposed in the laboratory and the plant, but the rate constants for the Ni-based alloys exposed in the plant were much higher than those measured in the laboratory.

Zhang et al. (2016) studied the effect of flow and static conditions on the oxidation of T22 and P92 in SCW at 550 and 600 °C, 25 MPa, 2 $mg \cdot kg^{-1}$ DO, inlet conductivity of 0.1 $\mu S \cdot cm^{-1}$. It was found that the rate constants were close to parabolic in the flowing system but closer to cubic for the static system. It was also suggested, based on a review of other literature data, that flow promoted the formation of hematite on these alloys. The authors attributed the differences in behaviour to the removal of H_2 generated by corrosion by the flowing coolant, which changed the surface oxygen potential. As part of an international interlaboratory comparison exercise, Guzonas et al. (2016b) examined the dependence of weight gain on autoclave refresh time (Fig. 2.5). Several of the test loops participating in the exercise monitored the DO concentration before and after the test section; one laboratory reported a systematic loss of about 600 $\mu g \cdot kg^{-1}$ between inlet and outlet, assumed to be due to corrosion of the corrosion coupons and the high-temperature sections of the loop including the autoclave.

Comparison of data from descaled corrosion coupons exposed in a static autoclave and in a flow loop suggested that transport of corrosion products in the static autoclave was not a limiting factor, since the weight losses for each alloy tested in the static autoclave were much higher than those tested in the flowing loop. It was postulated that oxide exfoliation may be higher at the higher flow rates in loops resulting in lower weight gains. However, while examination of the coupons exposed to flow showed evidence for oxide exfoliation, the weight gain calculated from the descaled weight loss was **lower** than the measured weight gain, inconsistent with a high oxide loss by exfoliation.

Holcomb (2009) studied chromia evaporation in high-temperature steam and proposed an equation of the form:

$$k_e = Sh_{ave} \cdot \frac{D_{AB} M_0}{LRT} \left(P_0 - P_0' \right) \tag{5.6}$$

for the evaporation rate k_e, where Sh_{ave} is the average Sherwood number, D_{AB} the gaseous diffusion coefficient between $CrO_2(OH)_2$ and the solvent gas, M_0 the molecular mass of $CrO_2(OH)_2$, L the length of the surface, R the gas constant, T the temperature (in K), P_0 the partial pressure of $CrO_2(OH)_2$ and P_0' the bulk partial pressure of $CrO_2(OH)_2$. The inclusion of the partial pressure difference term allows for the reduction or suppression of evaporation if sufficient $CrO_2(OH)_2$ is present in the steam entering the test section (or piping of interest in a plant). This approach is conceptually identical to the approach used to model corrosion product deposition (Chapter 4). By considering the rate of solid-state diffusion of Cr to the surface and the rate of evaporation the mass change of an alloy in a flowing system can be determined.

5.3.5 Heat transfer

Although a major application of in-core alloys in an SCWR is as the fuel cladding, there have been few studies of the effects of heat transfer on corrosion rate. Fig. 5.35 shows data from Gaul and Pearl (1961). The heat transfer tests were carried out at temperatures between 593 and 704 °C and the isothermal tests at 621 °C. The tests were performed at 6.9 MPa, 20 $mg \cdot kg^{-1}$ DO and 2.5 ppm H_2; the test samples were pickled before exposure. Little difference was found in descaled weight loss over a 4000 h exposure period, but more oxide was lost to the system (based on comparing weight gain and descaled weight loss) in the heat transfer tests ($\sim$10%–50%) compared to the isothermal tests (very small weight gains). A similar result (little difference in descaled weight loss but higher oxide loss to the system for the heat transfer surfaces) for Alloy 625 was reported by Pearl et al. (1967). The higher oxide loss to the coolant noted for both alloys would have significant implications for activity transport. Tests of Alloy 800 in SHS at temperatures up to 738 °C under irradiation (Comprelli et al., 1969) showed generally uniform corrosion penetration consistent with out-of-reactor data but higher than predicted oxide loss.

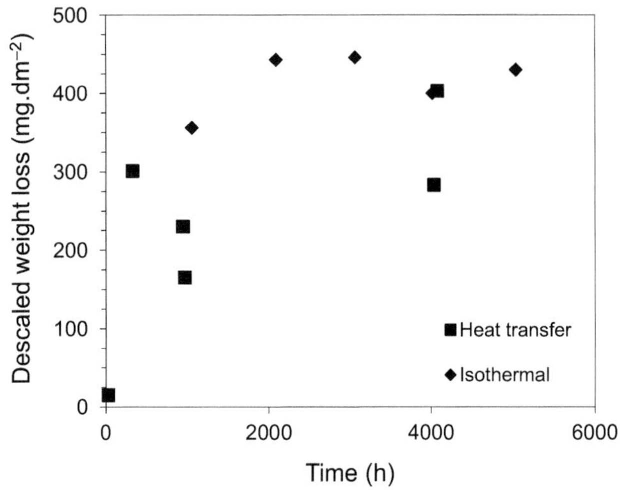

Figure 5.35 Descaled weight loss versus exposure time for 304 SS in supercritical water at 593–704 °C (heat transfer) and at 621 °C (isothermal), 6.9 MPa, 20 mg·kg^{-1} dissolved oxygen and 2.5 ppm H_2.
Data from Gaul, G.G., Pearl, W.L., 1961. Corrosion of Type 304 Stainless Steel in Simulated Superheat Reactor Environments. General Electric Company Report GEAP-3779.

5.3.6 Ageing

A fundamental challenge for SCWR concepts is that the high peak fuel cladding temperature (up to 850 °C for the Canadian SCWR concept) results in slow microstructural evolution due to thermal ageing (diffusion and secondary phase precipitation, particularly involving grain boundaries), as discussed in Chapter 3. Much of the testing to date has used as-fabricated materials in relatively short duration tests, and the influence of intermetallic phases requiring longer ageing times for their formation (e.g., χ, η and σ phases) has received little attention. In an early study for SCWO development, Daigo et al. (2006) studied the effects of thermal ageing on Ni-based alloys in SCW containing 0.01 mol·kg^{-1} H_2SO_4 at 400 °C, 30 MPa. A reduction in Charpy impact toughness and in corrosion resistance was noted, the degree of degradation being significant for high Cr alloys. Jiao et al. (2014) studied the effects of thermal ageing on the corrosion resistance of mill-annealed (MA) and thermally treated (TT) 316L and 310S in SCW (550 °C, 25 MPa, 8 mg·kg^{-1} DO). The treatments produced the expected intermetallic phases ($M_{23}C_6$, χ, η and σ phases). It was concluded that thermal ageing did not have a major effect on corrosion resistance, likely due to the discontinuous nature of the precipitates formed on grain boundaries and within grains.

5.3.7 Irradiation

Because corrosion is an interfacial phenomenon, it is difficult to disentangle the effects of water radiolysis from the effects of irradiation of the metal surface and oxide layers.

Alloys exposed to SCW at temperatures up to 700 °C after in-reactor irradiation (not in SCW) (Higuchi et al., 2007) showed no increase in general corrosion; however, the effects of coolant irradiation (water radiolysis) could not be assessed in this work. Zr−1%Nb and Ni−Cr-based alloys irradiated by a 9.76 MeV/6.23 kW_e electron beam for 497 h in water in the vicinity of the pseudocritical point showed enhanced oxidation rates under irradiation (Bakai et al., 2015). As noted in Section 5.3.5, tests of Alloy 800 in SHS at temperatures up to 738 °C under irradiation (Comprelli et al., 1969) showed generally uniform corrosion penetration consistent with out-of-reactor data but higher than predicted oxide loss. As discussed in Chapter 4, Ishigure et al. (1980) reported that γ-irradiation increased the release of dissolved Fe species from 304 SS in subcritical water.

5.4 Oxide morphology

The morphology and composition of the surface oxides formed in SCW can be complex, and these characteristics are affected by all of the key variables discussed in Section 5.3. Generically, in water the oxide structure depends on the thermodynamic stability of the oxides (or hydroxides) of the metallic constituents of the alloy (as described, for example, by a Pourbaix or Ellingham diagram) and the rates of solid-state diffusion of metal ions and oxidants in the oxide layers formed. Kofstad (1988) provides a good overview of the thermodynamic and kinetic aspects of oxide growth in high-temperature gases. Under the conditions expected in an SCWR core, the expected thermodynamic stabilities of the oxides formed is predicted to be $Al_2O_3 > TiO_2 > SiO_2 > MnO > Cr_2O_3 > Fe_3O_4 > NiO$ for low DO concentrations. At higher DO concentrations (DO > $\sim 2\ mg \cdot kg^{-1}$ based on the experimental data presented in Section 5.3), Cr(VI) hydroxide appears to be formed and is relatively soluble in SCW, and formation of hematite is favoured over magnetite. Metal diffusion rates are in the order Fe > Ni > Cr.

5.4.1 *Ferritic–martensitic steels*

A large number of very detailed studies of the oxide formed on F/M steels exposed to SCW now exist (Jang et al., 2005b; Ren et al., 2006; Ampornrat and Was, 2007; Hwang et al., 2008; Ampornrat, 2011; Tan et al., 2010; Bischoff, 2011; Liu, 2013; Li, 2012), and the results for different alloys are in general agreement. The oxide has a layered structure with an outer Fe_3O_4 layer, an inner $(FeCr)_3O_4$ layer with the spinel structure and an innermost diffusion layer. Cr_2O_3 is sometimes observed in the inner layer, depending on the temperature and Cr content of the alloy, and may or may not form a continuous layer. The outer layer predominantly consists of coarse columnar grains, while the inner layer is composed of very fine equiaxed grains. The diffusion layer can be quite complex in structure and thickness depending on the temperature. Bischoff (2011) noted that oxide precipitates in NF616 tended to form along lath boundaries. He noted that while the oxide thickness increased with exposure time at 500 °C, the oxide morphology remained relatively constant. At 500 °C, the inner oxide was observed to be quite porous and was divided into a layer near the inner oxide/outer oxide interface that was relatively compact, and a layer near the inner layer/diffusion layer interface that was quite porous. In 500 °C steam, there was very little porosity but similar dual layer morphology. For HCM12A at 600 °C the diffusion layer was not as uniform as that found on NF616 and appeared to be determined by the microstructure. The outer layer was more porous and the porosity of this layer increased with time, whereas the inner oxide porosity decreased with time. At 500 °C the oxides were more uniform and were relatively constant in morphology over the entire exposure time. A similar division of the inner oxide into a porous and nonporous layer was found. A 14Cr ODS sample exposed at 500 °C had a large diffusion layer (slightly larger than the inner layer) with many oxide precipitates forming long dendrites in the growth direction along ferrite grain boundaries. Both the inner and outer layers were very porous.

Marker experiments have shown that the inner oxide/outer oxide interphase corresponds to the original alloy surface, i.e., the oxide grows from both the inner oxide–metal interface and the outer oxide–SCW interface. Bischoff (2011) and Bischoff et al. (2012) described the results of experiments in which Pd markers were placed on corrosion coupons to study the location of the original water–metal interface using a photolithographic method. As the surface roughness affects the focus and illumination during the photolithographic exposure, all samples, including ones without markers, were polished to a 0.05 μm finish. The marked and unmarked coupons were then exposed to SCW (500 °C, 25 MPa, ~1000 h) and SHS (500 °C, 10.8 MPa, ~1350 h), removed, cross-sectioned and the marker location determined by SEM. At both pressures the markers were observed at the outer–inner layer interface (Fig. 5.36), confirming that oxidation of F/M steels in SCW occurs by simultaneous by outward migration of iron to form the outer layer and inward migration of oxygen to form the inner layer.

The steady-state oxidation process is depicted in Fig. 5.37, which shows an outer oxide composed of columnar grains containing pores, an inner, more fine-grained oxide containing pores and voids, and oxidation at grain boundaries in the diffusion region. Inward diffusion of O_2 and H_2O, and some outward diffusion of iron (as ions or various hydrolysis products, see Fig. 4.9), is via pores, and inward diffusion of O^{2-} and OH^-, and outward diffusion of some iron (as ions), is via solid-state diffusion. The relative importance of these various processes is not yet clear and likely varies depending on factors such as temperature, chromium concentration, surface finish, grain size, SCW density and concentrations of minor alloying elements.

A significant feature of all corrosion studies of F/M steels in SCW is the presence of porosity (Fig. 5.38), the size and density of the pores changing with temperature. Tomlinson and Cory (1989) noted the presence of porosity in the oxides formed on 9Cr–1Mo steel after 10–20 h exposure to a mixture of Ar and steam at 501 °C (flow rate 50 $mL \cdot min^{-1}$). They stated that the pores appeared to be largely situated at grain boundaries and grain boundary triple points. Cook and Miles (2012) noted that increasing the DO concentration from 5 to 200 $\mu g \cdot kg^{-1}$ increased the porosity. Zhang et al. (2016) found that the porosity 'healed' (was no longer observed in

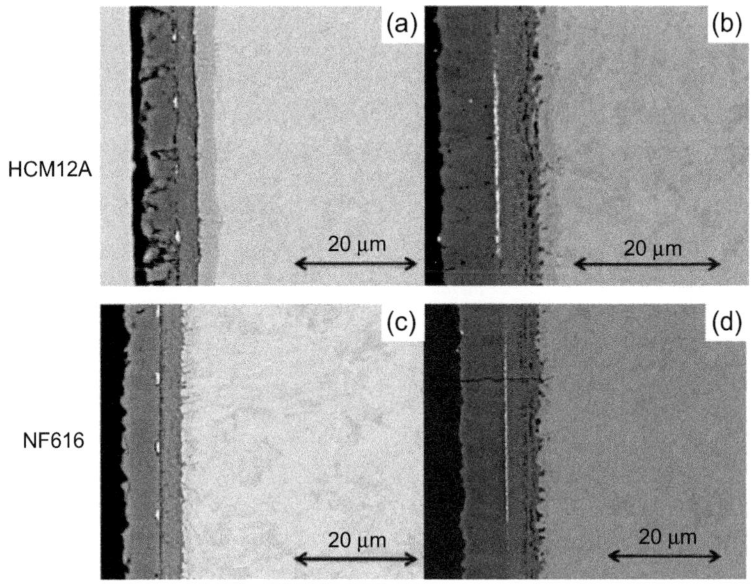

Figure 5.36 SEM images of the marked oxide layers formed on HCM12A (a and b) and NF616 (c and d) in 500 °C for 6 weeks in both steam (a and c) and SCW (b and d) (Bischoff et al., 2012). The marker can be seen as the thin, bright white features located roughly in the middle of the oxide layers.

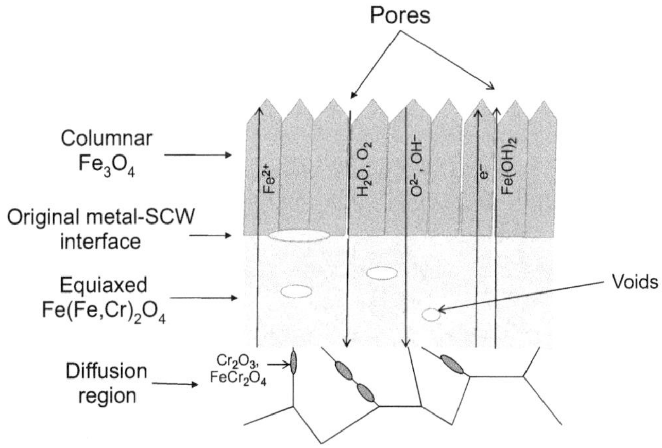

Figure 5.37 Schematic of the oxide layers formed on F/M steels in supercritical water (SCW). The pores have been shown as direct, although an interconnected network is more plausible.

plan-view SEM images) after 1000 h exposure to SCW at 550 and 600 °C. They also observed that fewer pores were formed at higher temperatures. Li (2012) noted that the porosity formed on model Fe−9Cr and Fe 9Cr 1.5X (X = Si, Al, Mn, V, or Ti) alloys showed a surface finish dependence, being the lowest on diamond-blade cut surfaces. It was noted that the oxides formed on surfaces with a finish finer than 1 μm were

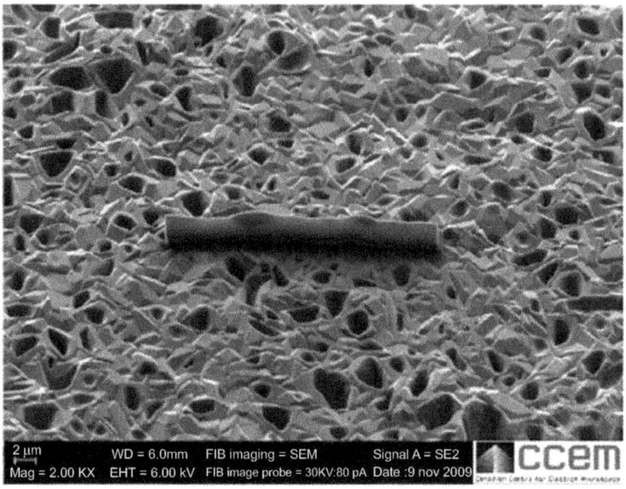

Figure 5.38 Surface of the oxide on a model 14Cr alloy exposed to supercritical water, showing the large amount of surface porosity. The elongated strip on the centre of the image is a thin protective coating used for the protection of the topmost surface for the TEM sample preparation. Scale bar: 2 μm (Artymowizc et al., 2010).

highly porous, suggesting that the rate of formation of the protective Cr-rich inner layer is important. It was also found that the oxide formed on martensite grains on F/M steels was more porous than on ferrite grains, possibly due to differences in grain boundary defect content. These data suggest that the porosity is associated with the initial formation of the outer oxide, disappearing as the oxide reaches a steady-state structure.

Cook et al. (2010) suggested the pores were of hydrothermal origin[10]. One possible hydrothermal mechanism would be the rapid initial formation of highly defective magnetite crystals on the surface, which then dissolve and reprecipitate in the hydrated surface layer resulting in the observed 'healing'. As the density of water at the surface decreases with increasing temperature, fewer pores would be expected at higher temperatures as observed. Zhu et al. (2016) suggested that the formation and healing of the pores was related to the differences in rates of diffusion along grain boundaries and the bulk grain. Tan and Allen (2009) proposed that the porosity was due to localized corrosion resulting from the adsorption of Cu_2S particles transported to the surface from other loop surfaces. Chemical analysis of the loop water found 6 $\mu g \cdot kg^{-1}$ Cu, as well as 36 $\mu g \cdot kg^{-1}$ Mn and measurable concentrations of Ni and Mo. They found that the pores formed on HCM12A polished to a 1 μm finish and exposed to SCW (500 °C, 25 MPa, flow rate ~1 $m \cdot s^{-1}$) for up to ~1000 had a depth of ~1.5 μm with some as deep as ~4 μm (40% of the surface magnetite film). They noted that if the pores were interconnected then the corrosion rate could be significantly increased.

[10] A. Anderson (St. Francis Xavier University) has made similar comments concerning the hydrothermal nature of oxide morphologies on surfaces exposed to SCW.

5.4.2 Austenitic steels

Most studies of the oxides formed on austenitic alloys exposed to SCW show a layered structure consisting of an outer magnetite or mixed magnetite–hematite layer and an inner layer with the spinal structure (Fe, Cr, Ni, M)$_3$O$_4$ (M: other minor elements) or a corundum-type oxide M_2O_3 (M = Fe, Cr, Mn). Payet (2011) studied the growth of the oxide layers using water containing isotopically labelled oxygen. 316L SS was exposed to SCW (600 °C, 25 MPa) for two periods: an initial oxidation period in $H_2^{16}O$, and the second in $H_2^{18}O$. After the SCW exposures the coupons were analysed in cross-section using SIMS depth profiling to determine the ^{16}O and ^{18}O profiles. ^{18}O was observed at both the metal/oxide and oxide/SCW interfaces, demonstrating that, as with F/M steels, the oxide layer grows outward and inward simultaneously, the inner oxide growing at the metal/oxide interface and the outer layer growing at the oxide/SCW interface (Fig. 5.39).

Alloy microstructure plays a large role in determining the oxide morphology. It was found that for the 20Cr–25Ni alloy NF709, the near-surface austenite grains were

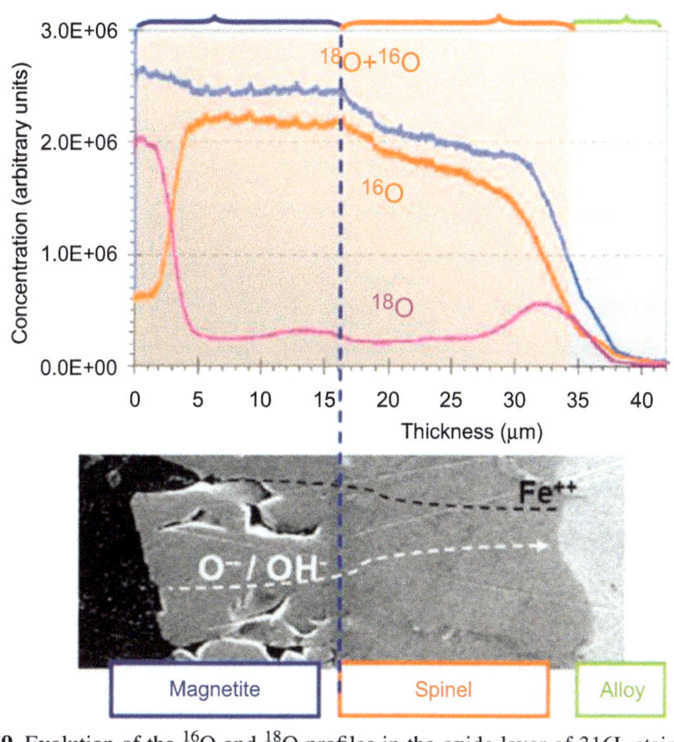

Figure 5.39 Evolution of the ^{16}O and ^{18}O profiles in the oxide layer of 316L stainless steel exposed to a first oxidation period (760 h) at 600 °C, 25 MPa in $H_2^{16}O$ followed by a second period (305 h) in $H_2^{18}O$ at 600 °C, 25 MPa

From Sarrade, S., Féron, D., Rouillard, F., Perrina, S., Robin, R., Ruizc, J.C., Turc, H.-A., 2017. Overview on corrosion in supercritical fluids. J. Supercrit. Fluids 120, 335–344.

readily oxidized intragranularly to form the inner oxide scale in which the oxide grains display preferential orientations imposed by the original austenite grain orientations (Chen et al., 2008). It is worth noting that the extensive porosity observed in the outer oxide formed on F/M steels in SCW (Fig. 5.38) is not observed on austenitic alloys. Nodular oxides are often observed on austenitic steels exposed to SCW; Chen et al. (2005, 2007a) reported that the oxide layer formed on D9 SS exhibited a nodular structure with an uneven surface after SCW exposure at 500 °C. The spacing of the nodules was 5−10 μm, comparable to the size of the austenite grains in the alloy. The oxide formed on Alloy 800HT in SCW (625 °C, 25 MPa, 500 h, static autoclave) was found to consist of oxide nodules on top of a base oxide layer (Mahboubi, 2014; Mahboubi et al., 2015) (Fig. 5.40). The nodules were enriched in Cr, Fe, Mn and O, while the underlying oxide pit was enriched in Cr, Mn and O. The base oxide layer (130 nm thick) was enriched in Cr, Mn and O with a small amount of Fe. The area beneath

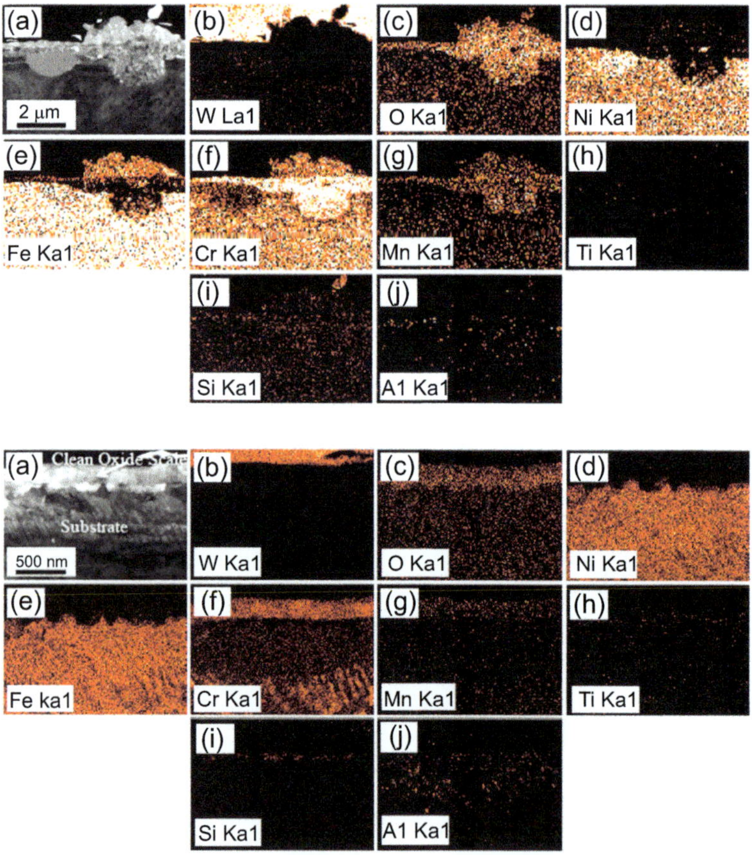

Figure 5.40 Low magnification BF-STEM images of Alloy 800HT coupon exposed to SCW (625 °C, 25 MPa, 500 h, static autoclave) in cross-section (a) and associated STEM/EDS maps (b)−(j). Top − typical island: bottom − clean oxide scale (Mahboubi, 2014).

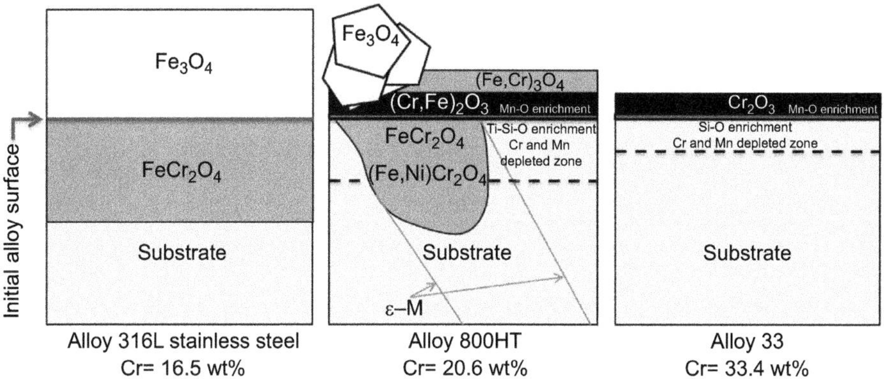

Figure 5.41 Summary of oxide scale structure and composition formed on austenitic Fe–Cr–Ni alloys after exposure in 25 MPa SCW at 550 °C for 500 h in a static autoclave (Mahboubi, 2014).

the base oxide layer (633 nm thick) was depleted in Cr and Mn and enriched in Fe and Ni. A thin layer (62 nm thick) immediately below the base oxide layer was enriched in Ti, Si and O.

Fig. 5.41 depicts schematically the oxide layer structures forming on Type 316L SS (16 wt% Cr), Alloy 800HT (20.6 wt% Cr) and Alloy 33 (33.4 wt% Cr) in 25 MPa SCW at 550 °C in a static autoclave. Higher bulk Cr content promotes formation of a corundum-type M_2O_3 (M = Fe, Cr and Mn) external base layer. The composition of the M_2O_3 layer has been shown to depend on the Cr content of the alloy and the temperature: the layer formed on Alloy 33 contains significantly less Fe than that formed on Alloy 800HT (Mahboubi, 2014).

Fig. 5.42 is a schematic representation of a mechanism of nodule formation proposed by Zhang et al. (2009a). Owing to the large difference in diffusion rates of Cr

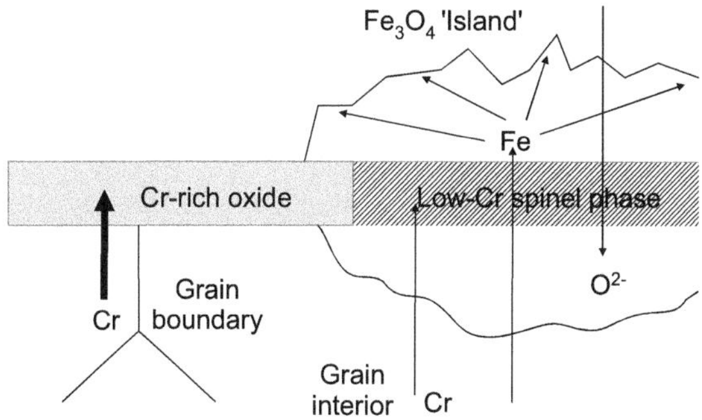

Figure 5.42 Schematic representation of the mechanism of nodule formation on austenitic steels.

Adapted from Zhang, L., Zhu, F., Tang, R., 2009a. Corrosion mechanisms of candidate structural materials for supercritical water-cooled reactor. Front. Energy Power Eng. China 3 (2), 233–240.

atoms along a grain boundary and through the grain interior (as discussed in Section 5.3.2 (Fig. 5.17)), the faster diffusion of Cr atoms on grain boundaries can supply sufficient Cr to the metal-oxide interface to compensate for Cr consumption in the oxide film above grain boundary, whereas Cr diffusion to the surface above a grain interior will be slower. As a result, the oxide film above the grain interior can become depleted in Cr and nonprotective, facilitating diffusion of oxidants to the base metal forming a mirror image nodule beneath the film. Iron atoms from the base metal diffuse across the film and react with the oxygen or water to form an oxide island of Fe_3O_4 just above the film as seen in Fig. 5.40.

There have been few detailed long-term studies of oxide growth in SCW. Behnamian et al. (2017) presented the results of a long-term (20,000 h) capsule test of 316L SS at 500 °C. After 500 h exposure the outer layer oxide was roughly 5 μm thick and composed of crystals of roughly the same diameter as the thickness. Increasing the exposure time to 5000 h led to an increase in oxide layer thickness and an increase in crystallite diameter (Fig. 5.43) After 20,000 h exposure, the thickness of the outer oxide layer was 20–25 μm and the thickness of the inner oxide layer was 13–53 μm. The outer oxide layer exhibited a continuous but porous morphology after 10,000 h exposure. A few microcracks were detected on the inner oxide layer, possibly due to oxidation and increased stress intensity on the inner surface of the capsule during SCW exposure. The oxide thickness and weight change were both found to follow a parabolic rate law ($n = 0.427$ from thickness measurements, $n = 0.455$ from weight change measurements). The observed porosity was attributed to the collapse of vacancies into pores when the vacancy concentration in the Fe_3O_4 layer becomes high enough. Fig. 5.44 illustrates schematically the long-term evolution of the oxidation of 316L SS.

5.4.3 Ni-based alloys

With the exception of Alloy 625, there have been fewer studies of Ni-based alloys under SCWR conditions than those of the iron-based alloys. In most studies, this alloy shows considerably higher corrosion resistance compared with austenitic steels. The oxides found on Ni-based alloys are thin compared with those found on the other alloy classes; oxide thicknesses on Alloy 625 and Alloy 690 were ~50 nm after 483 h exposure to SCW at 450 °C and 23 MPa (Guzonas et al., 2007). Weight changes are typically small even at 600 °C, making measurement difficult and leading to significant scatter in the data. Weight losses are often observed. For Alloy 625, the surface oxide has been found to consist of oxide particles with a typical size of about 1 μm on a uniform oxide layer composed of very fine oxide particles in the size range of tens of nanometres. Pits of varying diameters are typically observed on the sample surface (Was et al., 2006; Fujisawa et al., 2006; Guzonas et al., 2007; Chang et al., 2012; Ren et al., 2007) possibly initiated at metal carbide inclusions (Ren et al., 2007). Pitting was found to be suppressed when the samples were exposed to relatively higher

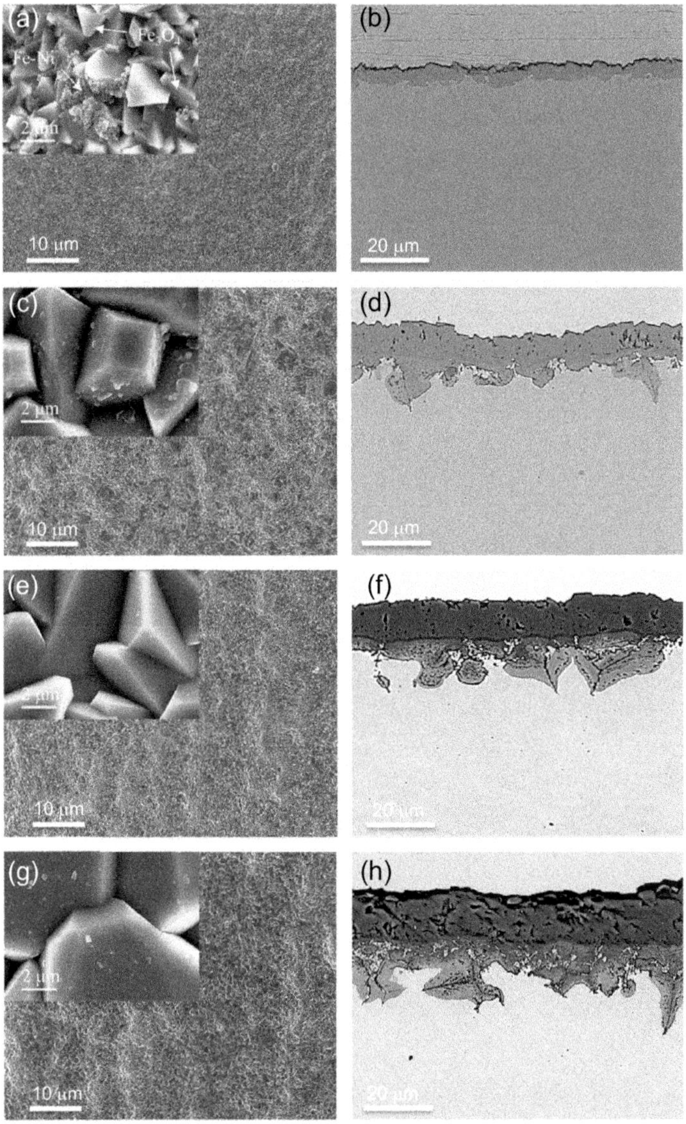

Figure 5.43 Top-view and cross-sectional SEM micrographs of the oxide films on alloy 316L exposed to the supercritical water at 500 °C and 25 MPa for (a, b) 500 h, (c, d) 5000 h, (e, f) 10,000 h, and (g, h) 20,000 h.
From Behnamian, Y., Mostafaei, A., Kohandehghan, A., Zahiri, B., Zheng, W., Guzonas, D., Chmielus, M., Chen, W., Luo, J.L., 2017. Corrosion behavior of alloy 316L stainless steel after exposure to supercritical water at 500 °C for 20,000 h. J. Supercrit. Fluids 127, 191–199.

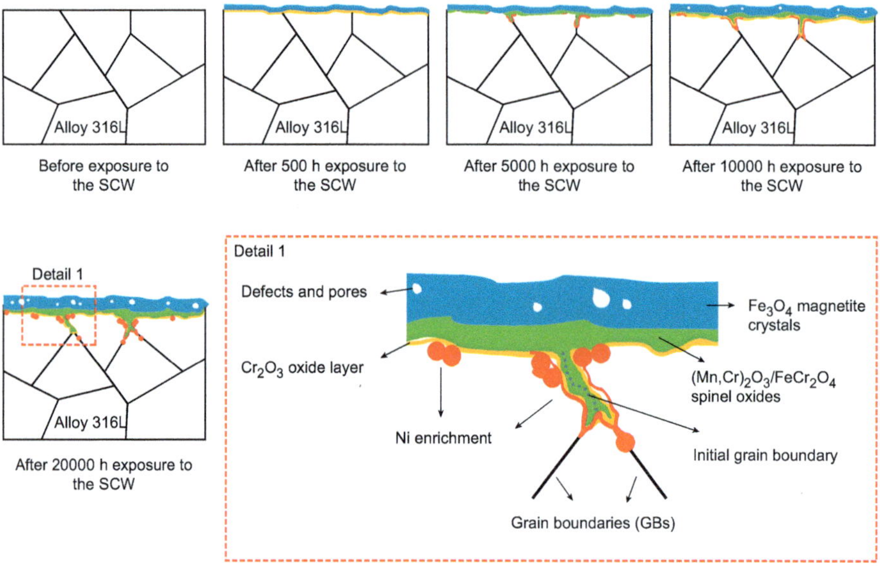

Figure 5.44 Schematic representation of the oxidation of 316 SS in supercritical water (SCW) (500 °C, 25 MPa, capsule specimens) at various times up to 20,000 h.
From Behnamian, Y., Mostafaei, A., Kohandehghan, A., Zahiric, B., Zheng, W., Guzonas, D., Chmielus, M., Chen, W., Luo, J.L., 2017. Corrosion behavior of alloy 316L stainless steel after exposure to supercritical water at 500 °C for 20,000 h. J. Supercrit. Fluids 127, 191–199.

oxygen content or higher temperature SCW. Intergranular corrosion was reported at T > 600 °C after prolonged exposures (Ren et al., 2007).

The surface oxide formed on Alloy 625 by corrosion in SCW has a double-layer structure consisting of a nickel/iron-rich outer layer and a chromium-rich inner layer. Li et al. (2009) reported that after 483 h exposure to SCW at 450 °C, 23 MPa a thin (~50 nm) Cr-rich surface oxide formed whose thickness was independent of SCW density, although the weight gain observed at lower SCW density became weight loss as the SCW density increased from 125 to 200 $kg \cdot m^3$ (Fig. 5.27). Raman spectra of the oxides formed on Alloy 625 and Alloy 690 after exposure to SCW at 450 °C for 483 h (Guzonas et al., 2007) indicated mainly $NiFe_2O_4$. Zhang et al. (2009b) exposed Alloy 625 to SCW (500 °C, 25 MPa) for 1000 h and found a coarse outer $NiFe_2O_4$ layer and an inner fine-grained layer consisting of $NiCr_2O_4$. Nodular corrosion was also observed, attributed to localized attack of γ' clusters in the matrix. After SCW exposure of Alloy 625 at 600 °C for 1026 h, the major phase identified was a spinel phase with a stoichiometry of $Ni(Fe,Cr)_2O_4$, and the other two oxides were identified as $Cr_{1.3}Fe_{0.7}O_3$ and NiO (Ren et al., 2007). Auger electron spectroscopy of the surface of Alloy 625 showed a duplex oxide structure after SCW exposure at both 500 and 600 °C (Ren et al., 2007), consisting of a nickel/iron-rich outer layer and a chromium-rich inner spinel layer with an underlying diffusion layer. Increasing the temperature to 600 °C resulted in the formation of a uniform oxide surface with fine particulates, while the grain boundaries were outlined by a topographically elevated oxide. Zhang et al. (2009b) noted that with longer exposure times of C-276 to SCW the precipitated phases (present in the alloy for strengthening) can be oxidized, breaking the outer protective oxide layer and forming microcracks on the surface, which can increase corrosion.

5.4.4 Coatings

The use of coatings to improve material performance in an SCWR has been investigated by a number of groups (Khatamian, 2013; Guzonas et al., 2005; Wills et al., 2007; Zheng et al., 2008; Hui et al., 2011; Huang and Guzonas, 2013; van Nieuwenhove et al., 2013), especially for the Canadian SCWR concept with its much higher peak cladding temperatures. For example, a substrate alloy with good resistance to creep and radiation damage could be alloyed with a surface layer that has high resistance to general corrosion and SCC (Zheng et al., 2008). If the surface layer could be produced after fabrication of the in-core components, i.e., after the forming and joining operations, then formability and weldability issues could be avoided. There are also components that could possibly benefit from the application of a thermal-barrier coating.

Khatamian (2013) explored Cr coatings of the zirconium alloys Zircaloy-2, Zircaloy-4, Zr-2.5Nb, Zr−1Nb and Zr−Cr−Fe, while Hui et al. (2011) demonstrated the successful deposition of ZrO_2 coatings on Zircaloy at room temperature. In both cases the corrosion resistance of the substrate in SCW was significantly improved. Guzonas et al. (2005) showed that reasonably robust zirconia films could be deposited on carbon steel and stainless steel surfaces using a sol−gel-based technique. Wills et al. (2007) examined the use of an atmospheric pressure plasma jet to form coatings on a variety of surfaces for use in an SCWR, although suitably robust coatings could not be produced. A number of MCrAlY-type coatings have been explored for use in the Canadian SCWR concept (Huang and Guzonas, 2013); these types of materials are needed as a bond layer between the substrate and an upper zirconia coating should an YSZ corrosion or thermal-barrier coating be needed, especially for in-core components other than the fuel cladding. Long-term testing of NiCrAlY and FeCrAlY in SCW showed very minimal weight change for both materials, the weight change for Ni−20 Cr−5Al being only 0.009 $mg \cdot dm^{-2}$ after 6300 h exposure to SCW at 550 °C (Huang and Guzonas, 2014). Coatings of TiAlN, CrN and ZrO_2 were deposited on 316L stainless steel and Alloy 600 substrates by physical vapour deposition (Van Nieuwenhove et al., 2013); while the performance of the TiAlN and ZrO2 coatings in SCW was not satisfactory, the CrN coating provided a stable and protective corrosion barrier. Coatings for application in an SCWR remain an active area of research, but significant development work would be needed to qualify such hybrid materials for use as a fuel cladding. As a result, coatings were excluded from consideration during the final selection of candidate fuel cladding materials for the 'Mark I' version of the Canadian SCWR concept.

Other forms of surface composition modification have been tested. Ytrrium surface treatment was used to modify the surface of a 9CrODS alloy. An yttrium-rich layer formed on the Y-surface-treated 9CrODS alloy, dividing the magnetite layer into two parts (Tan et al., 2007; Chen et al., 2007c). The outer magnetite layer was porous, while the inner magnetite layer was fine grained. The yttrium was concentrated mainly in one central layer, suggesting that this layer formed during the initial stages of exposure due to the high thermodynamic stability of Y oxide. This layer was subsequently driven outward due to the growth of the fine Fe_3O_4 grains underneath. XRD analyses indicated that the yttrium was predominantly present as $YFeO_3$. The authors concluded that while the $YFeO_3$ particulates were an effective local diffusion barrier for cation diffusion, the interparticle regions provided additional diffusion paths for cations.

5.5 Oxide growth kinetics

From an SCWR designer's perspective, a key output required of any program to study corrosion is a means to predict the total metal loss (or metal penetration) at the end of in-service life of the component and the uncertainties associated with that prediction.

Typically, the kinetics of corrosion (change in metal loss, weight gain, oxide thickness or other parameter with time) are classified as logarithmic, parabolic or linear. These classifications arise from theoretical considerations derived from idealized models. In real alloys, these classifications do not always fit the data, and the rate law may change with time or temperature. Kofstad (1988) and Hauffe (1976) provide thorough discussions of the various rate laws that can be encountered and their mechanistic bases.

The logarithmic rate law has the form:

$$x = k\,\log \cdot \log(t + t_0) + C \tag{5.7}$$

An inverse logarithmic rate law is also found:

$$x^{-1} = C - kilog \cdot \log(t) \tag{5.8}$$

Parabolic kinetics are often assumed as the default in the SCWR and SCFPP corrosion literature, being the form expected if the diffusion of a reactant through a growing oxide film is the rate-limiting step:

$$x^2 = k_p \cdot t + C \tag{5.9}$$

Linear kinetics can result when the rate determining step is a surface reaction limited by supply of reactant, the existence of a porous oxide that allows rapid diffusion of reactant to the reaction front, or diffusion through a barrier layer of constant thickness:

$$x = k_{lin} \cdot t + C \tag{5.10}$$

Although care must be taken, the use of double logarithmic plots (log x vs. log t) can help determine the type of kinetics being followed and any changes in rate law with time:

$$\log(x) = \frac{1}{m}\cdot\log(t) + C \tag{5.11}$$

The value of m (1, 2 or 3) indicates linear, parabolic or cubic kinetics, respectively. Noninteger values of m are often found in real systems.

Models of oxide growth typically assume that the oxide layer structure is already established, ignoring the initial period of oxide formation from a bare surface. Choudhry et al. (2016) studied the evolution of oxygen, hydrogen and dissolved metal concentrations during the initial stages of corrosion of Alloy 800H in SCW containing 20 ppm DO. Fig. 5.45 shows the concentrations of O_2 and Fe, Cr and Ni (the major constituents of Alloy 800H), as well as the concentrations of O_2, Mn, Al, Cu and Ti in the collected samples as a function of exposure time at 650 °C. Similar data were obtained at 550 °C. At both temperatures the time profile for each species could be divided into four stages (I, II, III and IV) defined by the DO concentration measured at the exit of the reactor. The first stage corresponded to an initial period during which the O_2 concentration measured at the exit was less than 300 ppb. The second stage was a brief period of very rapid increase in O_2 concentration at the exit. During the third stage the O_2 concentration at the exit gradually increased. During the fourth stage the O_2 concentration at the exit reached that of the feedwater and remained constant, indicative of a low, steady-state rate of oxidation.

During Stage I at 650 °C, relatively high concentrations of Fe, Ni and Al and a very low concentration of dissolved O_2 ($\sim$300 ppb) were observed in the effluent. Almost all of the added O_2 was consumed by oxidation of the alloy. The amounts of Fe, Ni and Al released were only a small fraction (about 1/500) of the total amount of oxide formed as calculated from the oxygen consumption. The concentration of Fe in the coolant during this period was about 6×10^{-10} m in reasonable agreement with the calculated solubility of magnetite (1×10^{-11} m) reported by Olive (2012). During Stage I, Fe was released at roughly the same proportion as its content in the alloy and there was slightly less Ni release and significantly more Mn and Al release than that expected based on alloy composition. Stage I lasted about 90 h at 550 °C and 300 h at 650 °C.

Stage II lasted less than 24 h at 650 °C and about 4 h at 550 °C. During this time the oxygen concentration at the outlet increased rapidly to about 17 ppm at both temperatures. At 650 °C there was an abrupt decrease in the Fe release and a rapid increase in Cr release during Stage II. Cr, Al, Cu and Ti releases were much larger than expected based on their concentrations in the alloy. At the same time an increase in Cr release is observed, a large reduction in Mn release is also seen. During Stage III, the relative amounts of the various elements released were similar to those observed in Stage II. The concentrations of Al, Mn, Ti and Cu slowly decreased with exposure time. The rate at which the O_2 concentration at the outlet increased with time during Stage III was roughly the same at 500, 550 and 650 °C. The behaviour of Cr during Stage III

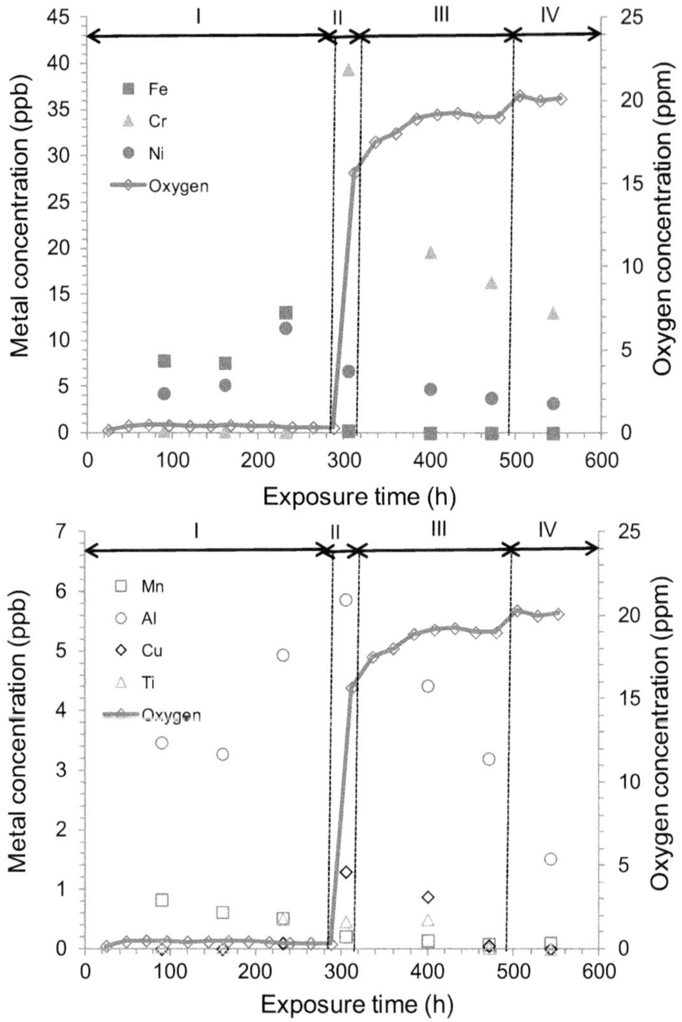

Figure 5.45 Concentrations of dissolved Fe, Cr, Ni (top) and Mn, Al, Cu and Ti (bottom), in ppb (μg L^{-1}), and the oxygen concentration in ppm (mg L^{-1}) in the effluent water sample versus exposure time. Temperature, pressure and flow rate were held constant at 650 °C, 25 MPa and 0.1 mL·min^{-1}, respectively (Choudhry et al., 2016).

differed at 550 and 650 °C, increasing with time at 550 °C and decreasing with time at 650 °C.

During Stage IV, Cr and Al release remained higher than expected based on their concentrations in the alloy. The fraction release at steady-state is the same at 550 and 650 °C. Based on these observations a model of the initial stages of oxide growth was developed (Fig. 5.46).

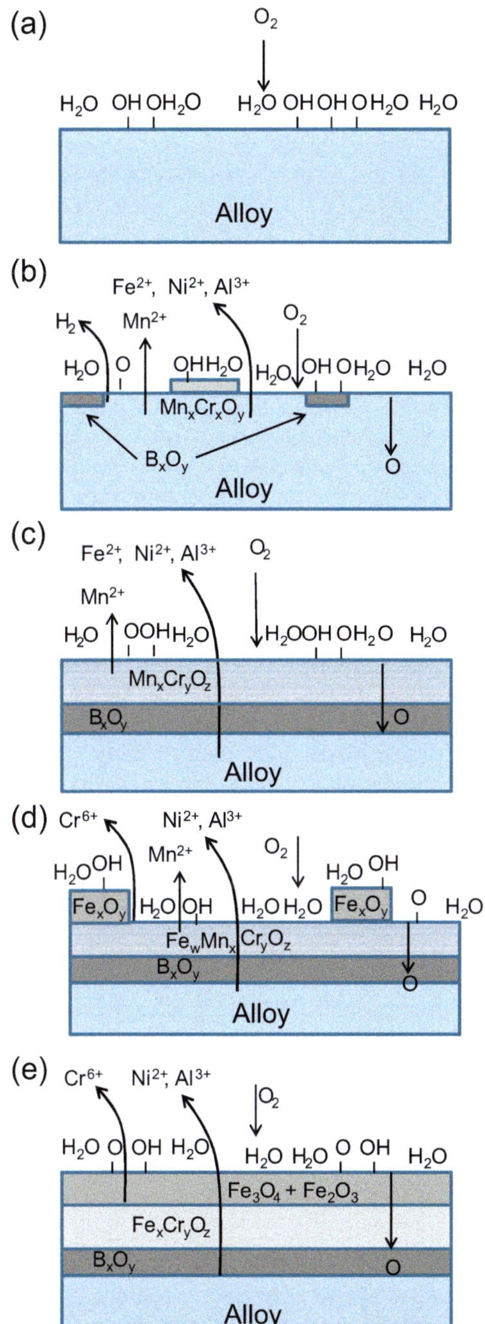

Figure 5.46 Model of oxide formation on Alloy 800H in oxygenated water. Oxide layer thicknesses are not to scale. Based on the metal release data, B is likely Ti (i.e., the inner layer is TiO_2). Note that while the metal species released into solution are written as ions, they are most likely to be uncharged ion pairs (e.g., $Fe(OH)_2$, $Al(OH)_3$) (Choudhry et al., 2016).

Novotny et al. (2013) reported a similar transient behaviour of DO in tests at $T = 500$ and 550 °C, 25 MPa, flow rate 15–17 $L \cdot h^{-1}$ (autoclave refresh time ~6 min), inlet DO concentration of 1985–2020 $\mu g \cdot kg^{-1}$ and inlet conductivity <0.1 $\mu S \cdot cm^{-1}$. The outlet water conductivity slowly decreased from 0.4 $\mu S \cdot cm^{-1}$ at the beginning of the test to 0.12 $\mu S \cdot cm^{-1}$, and the autoclave outlet DO concentration steadily increased during the initial stage of the test from 0 to ~1800 $\mu g \cdot kg^{-1}$, after which it remained almost constant at 1800 ± 10 $\mu g \cdot kg^{-1}$ for the rest of the heating sequence.

Phenomenologically, the corrosion rate (CR) is expected to have the following functional form:

$$CR = k \cdot \exp\left(-\frac{E_a}{RT}\right) \cdot t^x \tag{5.12}$$

where k is the rate constant, E_a is the activation energy, T is the absolute temperature, t is the exposure time, R is the gas constant, and x (= 1/m in Eq. 5.11) is usually between 0 and 1. Measuring the time and temperature dependencies in independent experiments gives the values for x, E_a and k needed to perform the required double extrapolation (time and temperature) to obtain end-of-life corrosion penetration. Limitations in available data sets can lead to considerable uncertainties associated with such extrapolations.

While parabolic kinetics are often assumed, this assumption must be carefully assessed. The use of parabolic rate constants facilitates the comparison of different data sets, but data presented previously in this chapter suggest kinetics can range from cubic to almost linear depending on the alloy and test conditions. As Fig. 5.45 shows, establishment of a 'steady-state' oxide growth can take hundreds of hours, so that kinetic parameters derived from short-term tests (e.g., 500 h) may be of limited value. A significant amount of long-term corrosion data suggests that corrosion of Alloy 800H in SHS follows linear kinetics, e.g., Pearl et al. (1965). Brush (1965) developed a model for the corrosion kinetics that consisted of both a temperature dependent term that was independent of time term and a term that was linear in time. Guzonas et al. (2016a) assessed a large data set on Alloy 800H corrosion from a variety of sources and found that linear kinetics using an expression similar in form to that of Brush (1965) better fit the long-term corrosion data and resulted in a more conservative value for end-of-life metal penetration. The following equation was proposed for predicting the in-service lifetime weight gain ΔW of Alloy 800H:

$$\Delta W = (0.6 \cdot T - 302) + 17 \exp(51,000/R \cdot T) \cdot t \tag{5.13}$$

where R is the gas constant, T is the temperature, and t is the exposure time. Similar expressions were derived for 347 SS, 310 SS and Alloy 625.

Steeves et al. (2015) used data obtained over several years and the corrosion model of Guzonas and Cook (2012) to derive kinetic parameters for Alloy 625 corrosion in

SCW. Data were obtained at 400, 500, 550 and 600 °C; weight losses were observed, the weight loss being essentially linear with time at each temperature:

$$WL = 323.4 \cdot \exp(-8684 \cdot T^{-1}) \cdot t. \tag{5.14}$$

For comparison, Pearl et al. (1967) reported a rate law for Alloy 625 of the form:

$$WL = 2427 \cdot \exp(-3073 \cdot T^{-1}) + 1.07 \times 106 \cdot \exp(-9944 \cdot T^{-1}) \cdot t \tag{5.15}$$

derived from loop tests in SHS at a flow rate of 174–217 $kg \cdot h^{-1}$ and a pressure of 6.9 MPa. To simulate water radiolysis, 20 $mg \cdot kg^{-1}$ oxygen and 2.5 $mg \cdot kg^{-1}$ dissolved hydrogen were maintained in the loop. The activation energies obtained by Steeves et al. (2015) (72.2 $kJ \cdot mol^{-1}$) and by Pearl et al. (1967) (81.7 $kJ \cdot mol^{-1}$) are in relatively good agreement. The model of Pearl et al. (1967) includes data obtained at temperatures below 500 °C; a reanalysis omitting the lower temperature data gives a value for E_a of 74.4 $kJ \cdot mol^{-1}$. The revised rate law derived from Pearl's data:

$$WL = 260 \cdot \exp(-8950 \cdot T^{-1}) \cdot t. \tag{5.16}$$

is in excellent agreement with Eq. (5.14).

5.6 Mechanisms and modelling

Improved materials for SCWR development may be required because of the limitations in materials properties, e.g., loading capacity, as well as high-temperature corrosion and EAC resistance. However, new material options are only accepted and introduced to routine operation after having been proven to perform in the desired environment and this is a very slow process for nuclear-grade materials. To successfully introduce a new material for a given component, the candidate material must fulfil a range of criteria on properties related to cost, fabrication, manufacturing, etc. and then show sufficient improvement compared with existing materials to earn acceptance. Although the criteria may be clear and competition from the existing solutions is well defined, it can be challenging to achieve significant improvement; improving a specific property may require a disproportionately high level of additional expensive alloying, or because of conflicting simultaneously required properties, i.e., adverse effect on neutron economy via higher Ni content in bulk alloy to improve general corrosion resistance. Advancing SCWR material development therefore requires a combination of theory, modelling and reliable experimental data.

SCWR developers require models to predict critical parameters such as metal penetration or oxide film thickness at various times during component life. Models to predict the general corrosion behaviour of alloys vary in sophistication from purely empirical, based on fitting an assumed rate law to measured corrosion rate data, to deterministic models based on mathematical formulations of corrosion mechanisms. Ultimately, all models must be validated by experimental measurements and must be able to predict the time and temperature dependencies, the effects of water chemistry, the oxide structure and layer composition and the effect of minor alloying elements.

Some modelling options of varying degrees of sophistication have been considered for predicting the estimated service life of candidate SCWR materials in terms of corrosion resistance. These approaches for predicting the corrosion rate in SCW are discussed in the following sections.

5.6.1 *Empirical and phenomenological models*

Corrosion in subcritical water and high-density SCW is an electrochemical process involving distinct oxidation and reduction half-reactions that can be separated as long as there is both electronic and electrolytic contact between anodic and cathodic sites. The low ionic conductivity of low-density SCW does not favour such separation and the corrosion mechanism becomes analogous to gas-phase oxidation.

To account for changes in physical properties of water in the vicinity of the critical point, Guzonas and Cook (2012) extended a phenomenological model of Guan and Macdonald (2009) by explicitly[11] adding direct CO as a parallel process to EO. The relative corrosion rate, R, is given by Eq. (5.17):

$$R = k_{EO} \exp\left(-\frac{E_{EO}}{RT}\right) C_H^m C_{O_2}^n + k_{CO} \exp\left(-\frac{E_{CO}}{RT}\right) C_{O_2}^P \tag{5.17}$$

where k_{EO} (E_{EO}) and k_{CO} (E_{CO}) are heterogeneous rate constants (activation energies) for the EO and CO mechanisms, respectively; T is the absolute temperature; $C_{O_2}^n$ and $C_{O_2}^p$ are the oxygen concentrations; C_{H+}^m is the hydrogen ion concentration; and m, n and p are reaction orders with respect to each component. Fig. 5.47 shows the EO and CO components and the total relative corrosion as a function of temperature. The Arrhenius temperature dependence of the EO component causes R to increase with increasing temperature up to T_c after which the rapid drop in SCW density decreases reactant availability resulting in a drop in R. Well above T_c, the CO mechanism becomes dominant and R again exhibits Arrhenius temperature behaviour. A **local** maximum in R near T_c is predicted.

The model (Fig. 5.47) predictions can be compared with the data for corrosion of 304 SS in subcritical water and superheated steam (Maekawa et al., 1968), which shows a similar inflection near T_c (Fig. 5.47); changes in slope above and below T_c suggest differences in activation energy. Steeves et al. (2015) reported local maxima near T_c for both Alloy 800H and Alloy 625 (Fig. 5.48) in SCW and used Eq. (5.17) to extract activation energies for the EO and CO components. While this

[11] Guan and Macdonald (2009) introduced the concept of parallel EO and CO mechanisms but did not develop a model that included both processes.

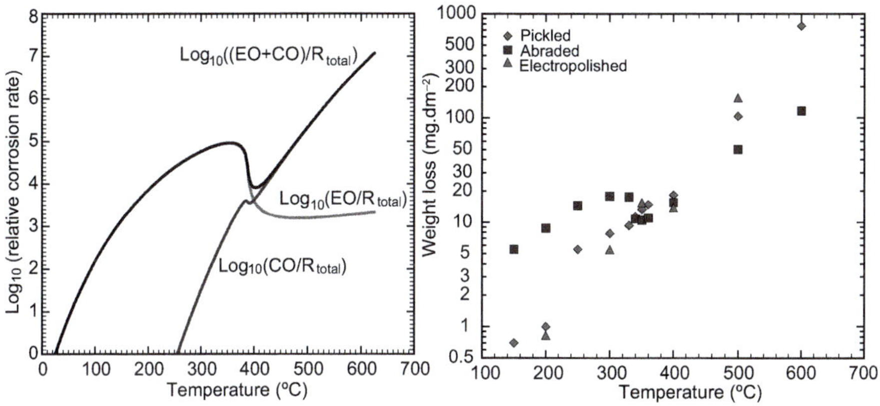

Figure 5.47 Predicted relative corrosion rate using Eq. 5.17 (left) for m, n = 0.5, p= 1, pressure = 25 MPa, E_{EO} = 50 kJ/mol and E_{CO} = 200 kJ/mol, and data for corrosion of 304 SS in subcritical water and superheated steam (Maekawa et al., 1968) (right).
From Guzonas, D.A., Cook, W.G., 2012. Cycle chemistry and its effect on materials in a supercritical water-cooled reactor: a synthesis of current understanding. Corros. Sci. 65, 48–66.

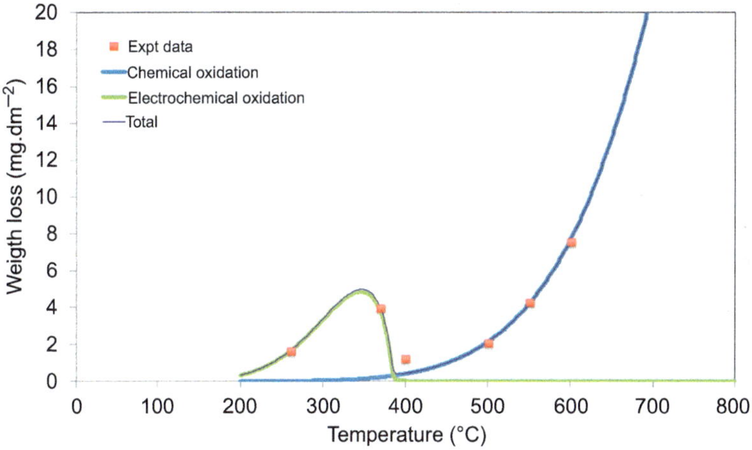

Figure 5.48 Experimental and modelled corrosion rates of Alloy 625 (Steeves et al., 2015).

phenomenological model provides insights into the origins of the temperature dependencies of the corrosion rate above and below T_c and enables the calculation of activation energies, a major shortcoming is the inability to predict the time dependence. This type of model could be coupled with empirical measurements of the time dependence (Eq. 5.12) to address this shortcoming.

Such semiempirical models have good predictive power for a particular alloy if sufficient data are available. However, these models cannot be used to predict the behaviour of new alloys.

5.6.2 Deterministic models

A deterministic model for corrosion under SCW conditions was originally developed by Bojinov and coworkers (Betova et al., 2008, 2009a,b; Penttilä et al., 2011) based on the model for oxide film growth on stainless steels in light-water reactor (LWR) conditions. The mixed-conduction model (MCM) emphasizes the coupling between ionic and electronic defects in quasi-steady-state passive films. The model enables the determination of the main kinetic and transport parameters of oxide layers needed to calculate the oxide film thickness as a function of exposure time on many alloys under SCW conditions and can provide insights into the effects of alloying elements on SCW oxidation resistance.

According to the MCM (Betova et al., 2008, 2009a,b; Penttilä et al., 2011; Bojinov et al., 2001, 2005a,b; 2007) the growth of the inner, compact oxide layer proceeds via generation of normal cation positions and oxygen vacancies at the alloy/oxide interface. Oxygen vacancies are then transported by diffusion–migration to the film/electrolyte interface where they are consumed by reaction with adsorbed water. In parallel, metal cation transmission through the layer, either by generation of cation vacancies at the outer interface and their transport and consumption at the inner interface or by generation, transport and consumption of interstitial cations, is also expected to occur; these cations are exchanged with normal cation sublattice sites (interstitialcy mechanism). The inner layer is assumed to have the spinel structure; X-ray diffraction shows that this assumption is true for many alloys (Penttilä et al., 2011, 2015a,b). It is also assumed that outer oxide layer growth is due to direct reaction of cations transported through the inner layer via interstitial sites with water and/or oxygen at the inner layer/outer layer interface layer (Ehlers et al., 2006; Sun et al., 2009b; Ren et al., 2006). A simplified picture of these processes is presented in Fig. 5.49.

Film growth under free corrosion conditions requires transfer of electrons through the inner layer to couple oxidation of metal constituents with the reduction of water or DO. The MCM assumes that ionic point defects play the role of electron donors or acceptors (Bojinov et al., 2005a), electronic conduction being coupled to the transport of ionic defects. The model does not consider electronic conduction to be rate-limiting since in-situ electric resistance and electrochemical impedance measurements during oxidation of steels in SCW between 500 and 700 °C have shown that the oxide layer–specific conductivity is comparable to that on pure iron (Betova et al., 2006), i.e., electronically conductive.

Similar to what has been observed in LWR conditions at temperatures above 150 °C and in SCW (Betova et al., 2008, 2009a, 2012; Bojinov et al., 2005a; 2007), the large number of defects in the oxide is unlikely to support high electric field conditions during oxidation over the temperature range 500–700 °C. Thus the low-field approximation of the generalized transport equation of Fromhold and Cook (1967) was used for the flux of point defects of type j ($j = I$, O and V, i.e., interstitial cations, oxygen and cation vacancies):

$$J_j(x,t) = -D_j \frac{\delta c_j(x,t)}{\delta x} - \frac{XF\vec{E}}{RT} D_j c_j(x,t) \tag{5.18}$$

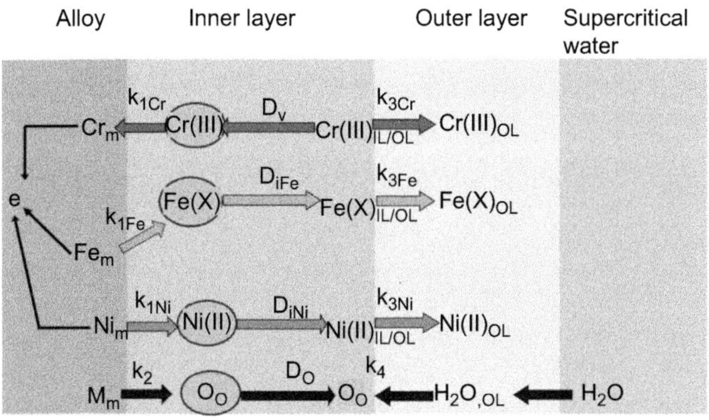

Figure 5.49 Simplified scheme of the growth of the inner and outer oxide layers formed on an Fe–Cr–Ni alloy. D_O = diffusion coefficient of oxygen vacancies in the inner layer ($cm^2\ s^{-1}$), D_V = diffusion coefficient of cation vacancies in the inner layer ($cm^2\ s^{-1}$), D_I = diffusion coefficient of interstitial cations in the inner layer ($cm^2\ s^{-1}$), $k_{1,j}$ = (j = Fe, Ni, Mn, Si, Cu, Nb, Ti, Mo) rate constant of interstitial cation formation at the alloy/inner layer interface ($mol\ cm^{-2}\ s^{-1}$), k_2 = rate constant of oxide formation via oxidation and injection of oxygen vacancies ($mol\ cm^{-2}\ s^{-1}$) and $k_{3i,j}$ = (j = Fe, Cr, Ni, Mn, Si, Cu, Nb, Ti, Mo) rate constant of formation of outer layer at the inner layer/outer layer interface ($cm\ s^{-1}$).

where c_j = (j = Fe, Cr, Ni, Mn, Si, Cu, Nb, Ti, Mo) molar concentration ($mol\ cm^{-3}$), D_j = diffusion coefficient in the inner layer ($cm^2\ s^{-1}$), X = nominal valence of a cation in the oxide (X = 2 for Ni and Mn, X = 3 for Cr and X = 2.67 for Fe), F = Faraday constant, $\overrightarrow{E}$ = the electric field strength in the inner layer ($V\ cm^{-1}$), R = gas constant, and T = temperature (K).

The electric field strength in the oxide was considered to be independent of distance within the oxide (Bojinov et al., 2005a; Macdonald, 2012) due to band-to-band tunnelling between electronic states that produces a separation of charge to compensate variations in field strength.

The concentration of a metallic oxide constituent can be expressed as its atomic fraction, where $V_{m,MO}$ is the molar volume of the phase in the layer, assumed to be a spinel oxide ($47\ cm^3\ mol^{-1}$). While formation of other phases (e.g., Fe_2O_3 or Cr_2O_3) has been reported during oxidation of steels in SCW, the differences in molar volumes of these phases is on the order of 10%, and the assumption introduces little error.

The nonstationary transport equations result from the differentiation of the flux with respect to the variation of concentration over time for a given component of the film. To obtain the compositional profiles in the inner layer, the system of equations is solved subject to initial and boundary conditions using a Crank–Nicolson method (Crank and Nicolson, 1947), which takes into account growth mechanisms via oxygen anion and chromium cation vacancies. A system of equations analogous to the inner layer treatment was employed to calculate outer layer compositional profiles.

To predict inner layer oxide growth as a function of exposure time, a growth law (Eq. 5.19) derived from both the PDM (point defect model) (Macdonald, 2011) and MCM (Betova et al., 2008, 2009a; Bojinov et al., 2007) was adopted:

$$L_{in}(t) = L_{in}(t=0) + \frac{1}{b}\ln\left[1 + V_{m,MO}k_2 b e^{-bL_{in}(t=0)}t\right], \quad b = \frac{3\alpha_2 F\vec{E}}{RT} \tag{5.19}$$

where L_{in} = thickness of the inner layer of oxide (cm), $L_{in}(t=0)$ = initial thickness of the inner layer of oxide (cm), $V_{m,MO}$ = molar volume of the phase in the oxide layer ($cm^3 \cdot mol^{-1}$), k_2 = rate constant of oxide formation via oxidation and injection of oxygen vacancies ($mol \cdot cm^{-2} \cdot s^{-1}$) and t = time (seconds). For outer layer growth in SCW (Penttilä et al., 2011) a parabolic law derived from the hypothesis that it can be approximated by a diffusion-like mechanism was found to be inadequate, especially for highly alloyed materials, as significant concentrations of Cr and other constituents, e.g., Ni, Al, etc., were observed in the outer oxide. It was also assumed that a dissolution–precipitation mechanism was unlikely due to the negligible solubility of oxides. Growth of the outer layer was therefore assumed to occur via reaction with water of interstitial cations transported through the inner layer. A model for bilayer film growth (Sloppy et al., 2013) was used as a starting point. Outer layer growth is assumed to be due to the flux of interstitial cations through the inner layer:

$$\frac{dL_{out}(t)}{dt} = -V_{m,MO}J_I \tag{5.20}$$

where J_I = flux of interstitial cations in the inner layer ($mol \cdot cm^{-2} \cdot s^{-1}$). The flux of cations via interstitial lattice sites is equal to the rate constant of oxidation of a metallic constituent j with injection of interstitial cations into the oxide,

$$J_I = -\sum_j k_{1j} y_{j,a} e^{-bL_{in}(t)} \tag{5.21}$$

leading to the following differential equation:

$$\frac{dL_{out}(t)}{dt} = V_{m,MO}\sum_j k_{1j} y_{j,a} e^{-bL_{in}(t)} \tag{5.22}$$

This equation is valid subject to the assumption that transfer coefficients for all reactions at the alloy/inner layer interface are equal. Integration from 0 to t (oxidation time) and 0 to $L_{out}(t)$ gives the following growth law:

$$L_{out}(t) = \frac{\sum_j k_{1j} y_{j,a}}{k_2}\left(L_{in}(t) - L_{in}(t=0)\right) \tag{5.23}$$

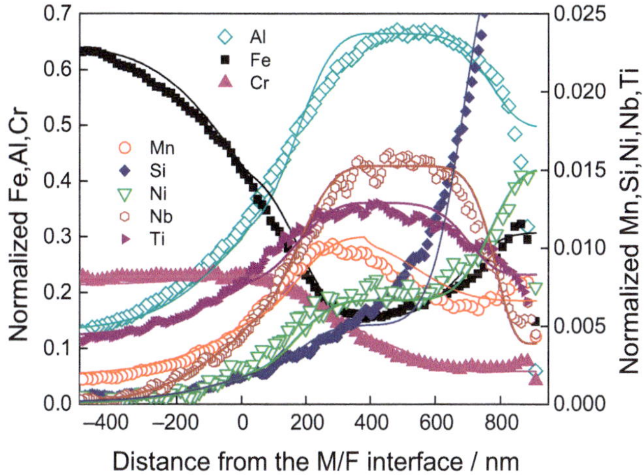

Figure 5.50 Comparison between measured (points) and calculated (solid lines) fractions of metallic constituents in the oxide formed on MA956 as a function of distance from the alloy/oxide interface after 600 h exposure at 650 °C (Penttilä et al., 2015a,b). The fractions of metallic constituents are normalized to the total concentration of metallic constituents.

The model validity was tested by comparing model predictions with elemental profiles of oxide constituents measured using glow discharge optical emission spectroscopy (GDOES) (Fig. 5.50). With the chosen values of the interfacial rate constants and diffusion coefficients, good agreement was obtained for the magnitude and depth distribution of the fractions of major (Fe, Cr ,Al) and minor (Mn, Si, Nb, Ti, etc.) oxide constituents.

A sensitivity study performed for each depth profile demonstrated that the confidence intervals for the rate constants and diffusion coefficients for the main alloy constituents are close to ±10%. The sensitivity study identified the following parameters as most strongly affect the compositional profiles: the rate constants at the alloy/inner layer interface; the rate constant of generation of chromium vacancies at the inner layer/water interface; and the diffusion coefficients of point defects in the inner layer.

The rate constants obtained were used to predict oxide film thickness as a function of exposure time. Good correspondence between calculated and measured oxide thicknesses on PM2000 and MA956 (left), and Alloy 690 and Sanicro 28 (right) as a function of exposure times was found (Fig. 5.51), providing additional proof for the validity of the proposed approach.

The model outlined above does not consider the physicochemical properties of the solvent. Guzonas (2013) suggested that the expression proposed by Cook and Olive (2011) for the precipitated oxide thickness (see Chapter 4) be used to introduce such a dependence:

$$\frac{dL_O}{dt} = k_{OLFe}\left(C_{\frac{Q}{S}} - C_{sat}\right) \tag{5.24}$$

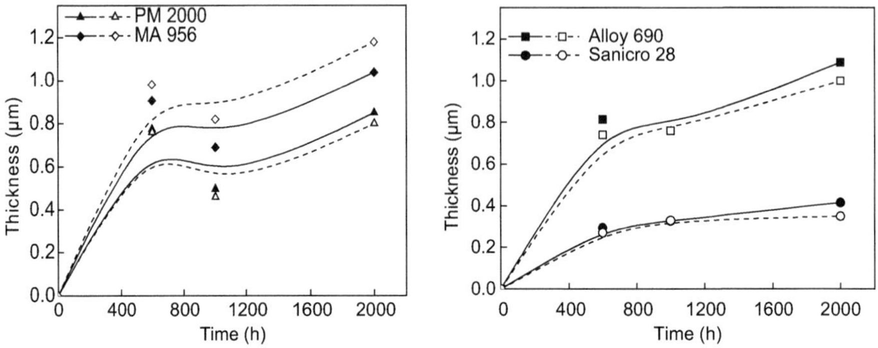

Figure 5.51 Comparison of oxide thicknesses calculated from the model (*open symbols*) and estimated from glow discharge optical emission spectroscopy depth profiles (*closed symbols*) on PM2000 and MA956 (left) and Alloy 690 and Sanicro 28 (right) at 650 °C in supercritical water (Penttilä et al., 2015a,b).

where dL_0/dt = change in outer layer thickness, k_{OLFe} = kinetic deposition (dissolution) constant (notation of Penttilä et al. (2011)), and $C_{O/S}$ and C_{sat} = oxide solution and saturation concentrations, respectively. By combine Eq. (5.24) with the expression for the outer oxide thickness and using $C_{sat} = K_s \cdot \rho^n$ the following expression for the change in oxide thickness as a function of SCW density was obtained:

$$\frac{dL_0}{dt} = D_{0,Fe} \cdot \frac{\Delta y_{Fe}}{L_0} + k_{OLFe}\left(C_{\frac{O}{S}} - K_s \cdot \rho^n\right) \tag{5.25}$$

This approach is similar to that used by Macdonald (1999) to model the effect of oxide dissolution on inner layer thickness. When $C_{O/S}$ is small, e.g., low impurity concentrations at the test section inlet, the oxide thickness will depend on SCW density. In low-density SCW, the oxide thickness will be dominated by the first term in Eqs. (5.5)−(5.24) unless the oxide solubility is high, and weight gain will be a reasonable measure of corrosion rate unless there is oxide exfoliation. Eq. (5.25) also introduce the possibility of adding a flow rate dependence.

References

Abe, F., Yoshida, H., 1985. Corrosion behaviours of heat-resisting alloys in steam at 800 °C and 40 atm pressure. Z. fur Metllkd. 76, 219−225.

Allen, T.R., Chen, Y., Ren, X., Sridharan, K., Tan, L., Was, G.S., West, E., Guzonas, D.A., 2012. Material performance in supercritical water. In: Konings, R.J.M. (Ed.), Comprehensive Nuclear Materials, vol. 5. Elsevier, Amsterdam, pp. 279−326.

Ampornrat, P., Was, G.S., 2007. Oxidation of ferritic-martensitic alloys T91, HCM12A and HT-9 in supercritical water. J. Nucl. Mater. 371 (1−3), 1−17.

Ampornrat, P., 2011. Determination of Oxidation Mechanisms of Ferritic/martensitic Alloys in Supercritical Water (Ph.D. thesis). Nuclear Engineering and Radiological Sciences, University of Michigan.

Artymowicz, D., Andrei, C., Huang, J., Miles, J., Cook, W.G., Botton, G.A., Newman, R.C., 2010. Preliminary study of oxidation mechanisms of high Cr steels in supercritical water. In: 2nd Canada-China Joint Workshop on Supercritical Water-cooled Reactors (CCSC-2010) Toronto, Ontario, Canada, April 25–28, 2010.

Artymowicz, D., Andrei, C., Cook, W.G., Newman, R.C., 2014. Influence of the cold work on the supercritical water oxidation of Fe14Cr1.5Si model alloy. In: Canada-China Conference on Advanced Reactor Development (CCCARD-2014), Niagara Falls, Canada, April 27–30, 2014.

Asay, R.H., 1990. Plant Tests of Surface Preconditioning for Steam Generators. EPRI Report NP-6616.

Bakai, O.S., Guzonas, D.A., Boriskin, V.M., Dovbnya, A.M., Dyuldya, S.V., 2015. Combined effect of irradiation, temperature, and water coolant flow on corrosion of Zr–, Ni–Cr-,and Fe–Cr-based alloys. In: 7th International Symposium on Supercritical Water-cooled Reactors (ISSCWR-7), March 15–18, 2015, Helsinki, Finland. Paper ISSCWR7-2012.

Behnamian, Y., Mostafaei, A., Kohandehghan, A., Zahiri, B., Zheng, W., Guzonas, D., Chmielus, M., Chen, W., Luo, J.L., 2017. Corrosion behavior of alloy 316L stainless steel after exposure to supercritical water at 500 °C for 20,000 h. J. Supercrit. Fluids 127, 191–199.

Betova, M.I., Bojinov, M., Kinnunen, P., Lehtovuori, V., Peltonen, S., Penttilä, S., Saario, T., 2006. Composition, structure, and properties of corrosion layers on ferritic and austenitic steels in ultrasupercritical water. J. Electrochem. Soc. 153 (11), B464–B473.

Betova, I., Bojinov, M., Karastoyanov, V., Kinnunen, P., Saario, T., May 2012. Effect of water chemistry on the oxide film on alloy 690 during simulated hot functional testing of a pressurised water reactor. Corros. Sci. 58, 20–32.

Betova, I., Bojinov, M., Kinnunen, P., Lehtovuori, V., Penttila, S., Saario, T., 2007. Surface film electrochemistry of AISI316 stainless steel and its constituents in supercritical water. In: Proc. ICAPP 2007, Nice, France, May 13–18, 2007, Paper 7123.

Betova, I., Bojinov, M., Kinnunen, P., Lundgren, K., Saario, T., 2008. Mixed-conduction model for stainless steel in a high-temperature electrolyte: estimation of kinetic parameters of inner layer constituents. J. Electrochem. Soc. 155 (2), C81–C92.

Betova, I., Bojinov, M., Kinnunen, P., Lundgren, K., Saario, T., 2009a. Influence of Zn on the oxide layer on AISI 316L(NG) stainless steel in simulated pressurised water reactor coolant. Electrochim. Acta 54 (3), 1056–1069.

Betova, I., Bojinov, M., Kinnunen, P., Lundgren, K., Saario, T., 2009b. A kinetic model of the oxide growth and restructuring on structural materials in nuclear power plants. In: Hagy, F.H. (Ed.), Structural Materials and Engineering. Nova Science Publishers, , New York, pp. 91–133.

Berge, P., 1997. Importance of surface preparation for corrosion control in nuclear power stations. Mater. Perform. 36 (11), 56–62.

Bischoff, J., 2011. Oxidation Behavior of Ferritic-martensitic and ODS Steels in Supercritical Water (Ph.D. dissertation in Nuclear Engineering). The Pennsylvania State University, Department of Mechanical and Nuclear Engineering.

Bischoff, J., Motta, A.T., Eichfeld, C., Comstock, R.J., Cao, G., Allen, T.R., 2012. Corrosion of ferritic-martensitic steels in steam and supercritical water. J. Nucl. Mater. xx, xx.

Bojinov, M., Fabricius, G., Kinnunen, P., Laitinen, T., Mäkelä, K., Saario, T., Sundholm, G., 2001. Electrochemical study of the passive behaviour of Ni-Cr alloys in a borate solution – a mixed conduction model approach. J. Electroanal. Chem. 504 (1), 29–44.

Bojinov, M., Kinnunen, P., Lundgren, K., Wikmark, G., 2005a. A mixed-conduction model for the oxidation of stainless steel in a high-temperature electrolyte. Estimation of kinetic parameters of oxide layer growth and restructuring. J. Electrochem. Soc. 152 (7), B250–B261.

Bojinov, M., Heikinheimo, L., Saario, T., Tuurna, S., 2005b. Characterisation of corrosion films on steels after long-term exposure to simulated supercritical water conditions. In: Proceedings of the American Nuclear Society – International Congress on Advances in Nuclear Power Plants 2005, ICAPP'05, pp. 1799–1807.

Bojinov, M., Galtayries, A., Kinnunen, P., Machet, A., Marcus, P., 2007. Estimation of the parameters of oxide film growth on nickel-based alloys in high temperature water electrolytes. Electrochim. Acta 52 (26), 7475–7483.

Boyd, W.K., Pray, H.A., 1957. Corrosion of stainless steels in supercritical water. Corrosion 13 (6), 33–42.

Brar, C., Davis, W., Mann, E., Ridley, C., Wirachowsky, B., Rangacharyulu, C., Guzonas, D., Leung, L., 2015. GEN IV SCWR cladding analysis project: nickel content in SCWR cladding material. In: Proceedings 5th Int. Youth Conference on Energy, Pisa, Italy, May 27–30, 2015, pp. 103–110.

Brush, E.G., 1965. Corrosion rate law considerations in superheated steam. Nucl. Appl. 1 (3), 246–251.

Chang, K.-H., Huang, J.-H., Yan, C.-B., Yeh, T.-K., Chen, F.-R. Kai, J.-J., 2012. Corrosion behavior of alloy 625 in supercritical water environments. Prog. Nucl. Energy 1–12.

Chen, Y., Sridharan, K., Allen, T.R., 2005. Corrosion Behavior of NF616 and D9 as Candidate Alloys for Supercritical Water Reactors. Corrosion 2005. NACE International, Houston, TX. Paper 05391.

Chen, Y., Sridharan, K., Allen, T.R., 2007a. Corrosion of Candidate Austenitic Stainless Steels for Supercritical Water Reactors. Corrosion 2007, Nashville, TN. NACE International, Houston, TX. Paper 07408.

Chen, Y., Sridharan, K., Allen, T.R., 2007b. Corrosion of 9Cr oxide dispersion strengthened steel in supercritical water environment. J. Nucl. Mater. 371, 118–128.

Chen, Y., Sridharan, K., Kruizenga, A., Tan, L., Allen, T.R., 2008. Corrosion behavior of austenitic alloys in supercritical water. Corros. Sci.

Chen, Y., Sridharan, K., Allen, T.R., 2007c. Effect of yttrium thin film surface modification on corrosion layers formed on 9Cr ODS steel in supercritical water". In: Allen, T.R., Busby, J.T., King, P.J. (Eds.), 13th Environmental Degradation of Materials in Nuclear Power Systems 2007. Canadian Nuclear Society.

Cho, H.S., Kimura, A., Ukai, S., Fujiwara, M., 2004. Corrosion properties of oxide dispersion strengthened steels in supercritical water environment. J. Nucl. Mater. 329–333, 387–391.

Choudhry, K.I., Mahboubi, S., Botton, G.A., Kish, J.R., Svishchev, I.M., 2015. Corrosion of engineering materials in a supercritical water cooled reactor: characterization of oxide scales on Alloy 800H and stainless steel 316. Corros. Sci. 100, 222–230.

Choudhry, K.I., Guzonas, D.A., Kallikragas, D.T., Svishchev, I.M., 2016. On-line monitoring of oxide formation and dissolution on alloy 800H in supercritical water. Corros. Sci. 111, 574–582.

Comprelli, F.A., Busboom, H.J., Spalaris, C.N., 1969. Comparison of radiation damage studies and fuel cladding performance for Incoloy-800. In: Irradiation Effects in Structural Alloys for Thermal and Fast Reactors, ASTM STP 457. American Society for Testing and Materials, pp. 400–413.

Cook, W., Miles, J., Li, J., Kuyucak, S., Zheng, W., 2010. Preliminary analysis of candidate alloys for use in the CANDU-SCWR. In: 2nd Canada-China Joint Workshop on Supercritical Water-cooled Reactors (CCSC-2010) Toronto, Ontario, Canada, 2010 April 25–28.

Cook, W.G., Olive, R.P., 2011. Prediction of crud deposition in a CANDU-SCWR core through corrosion product solubility and transport modelling. In: Proceedings of the 5th International Symposium on Supercritical Water-Cooled Reactors (ISSCWR5), March 13–16, 2011. Canadian Nuclear Society, Vancouver, Canada. Toronto, Ontario. Paper P83.

Cook, W.G., Olive, R.P., 2012a. Pourbaix diagrams for the iron-water system extended to high-subcritical and low-supercritical conditions. Corros. Sci. 55, 326–331.

Cook, W.G., Olive, R.P., 2012b. Pourbaix diagrams for chromium, aluminum and titanium extended to high-subcritical and low-supercritical conditions. Corros. Sci. 58, 291–298.

Cook, W.G., Olive, R.P., 2012c. Pourbaix diagrams for the nickel-water system extended to high-subcritical and low-supercritical conditions. Corros. Sci. 58, 284–290.

Cook, W., Miles, J., 2012. Effect of dissolved oxygen on the corrosion and oxide morphology of materials in SCW. In: 3rd Canada-China Joint Workshop on Supercritical Water-cooled Reactors (CCSC-2012), Xi'an, China, April 18–20, 2012. Paper 12063.

Cowen, H.C., Longton, P.B., Hand, K., 1966. Corrosion of Stainless Steels and Nickel-Based Alloys in Supercritical Steam. U.K. Atomic Energy Authority TRG Memorandum 3399 (C).

Cox, B., 1973. Accelerated Oxidation of Zircaloy-2 in Supercritical Steam. Atomic Energy of Canada Limited Report AECL 4448.

Crank, J., Nicolson, P., 1947. A practical method for numerical evaluation of solutions of partial differential equations of the heat conduction type. Math. Proc. Camb. Philos. Soc. 43 (1), 50–67.

Daigo, Y., Watanabe, Y., Sugahara, K., Isobe, T., 2006. Compatibility of nickel-based alloys with supercritical water applications: aging effects on corrosion resistance and mechanical properties. Corrosion 62 (2), 174–181.

Dong, Z., Chen, W., Zheng, W., Guzonas, 2012. Corrosion behavior of chromium oxide based ceramics in supercritical water (SCW) environments. Corros. Sci. 65, 461–471.

Dong, Z., Liu, Z., Li, M., Luo, J.-L., Chen, W., Zheng, W., Guzonas, D., 2015. Effect of ultrasonic impact peening on the corrosion of ferritic-martensitic steels in supercritical water. J. Nucl. Mater. 457, 266–272.

Dong, Z., Li, M., Behnamian, Y., Liu, Z., Luo, J.-L., Chen, W., Zheng, W., Guzonas, D., 2014 June. Corrosion Behavior of Ferritic-martensitic Steels in High Temperature Hydrothermal Environments. NACE Northern Area Western Conference. Alberta, Edmonton, pp. 27–30.

Eberle, F., Kitterman, J.H., 1968. Scale Formation on Superheater Alloys Exposed to High Temperature Steam. In: Lien, G.E. (Ed.), Behavior of Superheater Alloys in High Temperature, High Pressure Steam. The American Society of Mechanical Engineers, New York.

Ehlers, J., Young, D.J., Smaardijk, E.J., Tyagi, A.K., Penkalla, H.J., Singheiser, L., Quadakkers, W.J., 2006. Enhanced oxidation of the 9%Cr Steel P91 in water vapour containing environments. Corros. Sci. 48 (11), 3428–3454.

Fromhold Jr., A.T., Cook, E.L., 1967. Diffusion currents in large electric fields for discrete lattices. J. Appl. Phys. 38, 1546–1551.

Fry, A., Osgerby, S., Wright, M., 2002. Oxidation of Alloys in Steam Environments – A Review. National Physical Laboratory Report MATC(A)90.

Fujisawa, R., Nishimura, K., Nishida, T., Sakaihara, M., Kurata, Y., Watanabe, Y., 2006. Corrosion behavior of nickel-based alloys and type 316 stainless steel in slightly oxidizing or reducing supercritical water. Corrosion 62 (3), 270–274.

Fujiwara, K., Watanabe, K., Domae, M., Katsumura, Y., 2007. Stability of Chromium Oxide Film Formed by Metal Organic Chemical Vapor Deposition in High Temperature Water up to Supercritical Region. Corrosion 2007, March 11–15, Nashville, Tennessee.

Galerie, A., Wouters, Y., Caillet, M., 2001. The kinetic behavior of metals in water vapour at high temperatures: can general rules be proposed. Mater. Sci. Forum 369–372, 231–238.

Gaul, G.G., Pearl, W.L., 1961. Corrosion of Type 304 Stainless Steel in Simulated Superheat Reactor Environments. General Electric Company Report GEAP-3779.

Gilewicz-Wolter, J., Żurek, Z., Dudała, J., Lis, J., Homa, M., Wolter, M., 2006. Diffusion rates of ^{51}Cr, ^{54}Mn and ^{59}Fe in MnC_2O_4 and $FeCr_2O_4$ spinels. Adv. Sci. Technol. 46, 27–31.

Glebov, V.P., Antikain, P.A., Taratuta, V.A., Mamet, A.P., Kurilenko, E.P., 1976. Investigation of the influence of pH of the medium at supercritical pressure and of hydrogen concentration in it on the intensity of oxidation of different types of steel. Teploenerggetika 23, 77–81.

Gómez-Briceño, D., Blázquez, F., Sáez-Maderuelo, A., 2013. Oxidation of austenitic and ferritic/martensitic alloys in supercritical water. J. Supercrit. Fluids 78, 103–113.

Gong, M., Zhou, Z., Hu, H., Penttilä, S., Toivonen, A., 2015. Corrosion behavior of ^{14}Cr ODS ferritic steels in a supercritical water. In: 7th International Symposium on Supercritical Water-Cooled Reactors (ISSCWR-7), 2015 March 15–18, Helsinki, Finland. Paper ISSCWR7-2059.

Gu, G.P., Zheng, W., Guzonas, D., 2010. Corrosion database for SCWR development. In: 2nd Canada-China Joint Workshop on Supercritical Water-cooled Reactors (CCSC-2010) Toronto, Ontario, Canada, 2010 April 25–28.

Guan, X., Macdonald, D.D., 2009. Determination of corrosion mechanisms and estimation of electrochemical kinetics of metal corrosion in high subcritical and supercritical aqueous systems. Corrosion 65 (6), 376–387.

Guzonas, D.A., Wills, J.S., McRae, G.A., Sullivan, S., Chu, K., Heaslip, K., Stone, M., 2005. Corrosion-resistant coatings for use in a supercritical water CANDU reactor. In: Proceedings of the 12th Int. Conf. on Environmental Degradation of Materials in Nuclear Power Systems – Water Reactors, Salt Lake City, August 14–18, 2005. TMS (The Minerals Metals and Materials Society).

Guzonas, D.A., Wills, J., Do, T., Michel, J., 2007. Corrosion of candidate materials for use in a supercritical water CANDU® reactor. In: 13th International Conference on Environmental Degradation of Materials in Nuclear Power Systems, pp. 1250–1261.

Guzonas, D., Wills, J., Dole, H., Michel, J., Jang, S., Haycock, M., Chutumstid, M., 2010. Steel corrosion in supercritical water: an assessment of the key parameters. In: 2nd Canada-China Joint Workshop on Supercritical Water-cooled Reactors (CCSC-2010), Toronto, Ontario, Canada, April 25–28, 2010.

Guzonas, D.A., Cook, W.G., 2012. Cycle chemistry and its effect on materials in a supercritical water-cooled reactor: a synthesis of current understanding. Corros. Sci. 65, 48–66.

Guzonas, D.A., 2013. The physical chemistry of corrosion in a supercritical water-cooled reactor. In: 16th International Conference on the Properties of Water and Steam (ICPWS16), University of Greenwich, London, UK, September 1-5, 2013.

Guzonas, D.A., Bissonette, K., Deschenes, L., Dole, H., Cook, W., 2013. Mechanistic aspects of corrosion in a supercritical water-cooled reactor. In: ISSCWR-6: The 6th International Symposium on Supercritical Water-Cooled Reactors, March 03–07, 2013, Shenzhen, Guangdong, China.

Guzonas, D.A., Edwards, M.E., Zheng, W., 2016a. "Assessment of candidate fuel cladding alloys for the Canadian supercritical water-cooled reactor concept. J. Nucl. Eng. Rad. Sci. 2, 011016.

Guzonas, D., Penttilä, S., Cook, W., Zheng, W., Novotny, R., Sáez-Maderuelo, A., Kaneda, J., 2016b. The reproducibility of corrosion testing in supercritical water – results of an international round robin exercise. Corros. Sci. 106, 147–156.

Guzonas, D., Novotny, R., Penttilä, S., 2017. Corrosion phenomena induced by supercritical water in generation IV nuclear Reactors. In: Yvon, P. (Ed.), Structural Materials for Generation IV Nuclear Reactors. Woodhead Publishing.
Hauffe, K., 1976. Gas-solid reactions – oxidation. In: Hannay, N.B. (Ed.), "Treatise on Solid-state Chemistry". Plenum Press, New York (Chapter 8).
Higuchi, S., Sakurai, S., Ishida, T., 2007. A study of fuel behavior in an SCWR core with high power density. In: Proc. ICAPP'07, Nice, France, May 13–18, 2007. Paper 7206.
Holcomb, G.R., Alman, D.E., 2006. The effect of manganese additions on the reactive evaporation of chromium in Ni–Cr alloys. Scr. Mater. 54, 1821–1825.
Holcomb, G.R., 2009. Steam oxidation and chromia evaporation in ultrasupercritical steam boilers and turbines. J. Electrochem. Soc. 156, C292–C297.
Holcomb, G.R., 2014. High Pressure Steam Oxidation of Alloys for Ultrasupercritical Conditions, vol. 82, pp. 271–295.
Holcomb, G.R., Carney, C., Doğan, Ö.M., 2016. Oxidation of alloys for energy applications in supercritical CO_2 and H_2O. Corros. Sci. 109, 22–35.
Hsu, R.H., Rankin, W.N., Summer, M.E., 1994. Characterizing Pre-polished Type 304L Stainless Steel (U). Westinghouse Savannah River Company Report WSRC-MS-94-0437.
Hudson, M.J.B., Ocken, H., 1996. Dose reduction measures at the Millstone Point – 2 PWR. In: Water Chemistry of Nuclear Reactor Systems 7, vol. 2. BNES, p. 9.
Huang, X., Guzonas, D., 2013. Summary of Recent Coating Material Development and Testing in SCW. In: 6th International Symposium on Supercritical Water-cooled Reactors ISSCWR-6, March 03–07, 2013, Shenzhen, Guangdong, China. Paper ISSCWR6-13012.
Huang, X., Guzonas, D., 2014. Characterization of Ni–20Cr–5Al model alloy in supercritical water. J. Nucl. Mater. 445, 298–307.
Huang, X., Guzonas, D., Li, J., 2015a. Characterization of Fe–20Cr–6Al-Y model alloy in supercritical water. Corros. Eng. Sci. Technol. 50, 137–148.
Huang, X., Li, J., Amirkhiz, B.S., Liu, P., 2015b. "Effect of water density on the oxidation behavior of alloy A-286 at 625°C- A TEM study. J.Nucl. Mater. 467, 758–769.
Hwang, S.S., Lee, B.H., Kim, J.-G., Jang, J., 2005. Corrosion behavior of F/M steels and high Ni alloys in supercritical water. In: Proceedings of Global 2005, Tsukuba, Japan, October 9–13, 2005. Paper 043.
Hwang, S.S., Lee, B.H., Kim, J.G., Jang, J., 2008. SCC and corrosion evaluations of the F/M steels for a supercritical water reactor. J. Nucl. Mater. 372, 177–181.
Hui, R., Cook, W., Sun, C., Xie, Y., Yao, P., Miles, J., Olive, R., Li, J., Zheng, W., Zhang, L., 2011. Deposition, characterization and performance evaluation of ceramic coatings on metallic substrates for supercritical water-cooled reactors. Surf. Coatings Technol. 205, 3512–3519.
Ishigure, K., Kawaguchi, M., Oshoma, K., 1980. The effect of radiation on the corrosion product release from metals in high temperature water. In: Proceedings of the 4th BNES Conference on Water Chemistry of Nuclear Reactor Systems, Bournemouth, UK. Paper 43.
Jacob, Y.P., Haanappel, V.A.C., Stroosnijder, M.F., Buscail, H., Fielitz, P., Borchardt, G., 2002. The effect of gas composition on the isothermal oxidation behavior of PM chromium. Corros. Sci. 44, 2027–2039.
Jiao, Y., Kish, J., Zheng, W., Guzonas, D., Cook, W., 2014. Effect of thermal ageing on the corrosion resistance of stainless steel type 316L exposed in supercritical water. In: Canada-China Conference on Advanced Reactor Development (CCCARD-2014), Niagara Falls, Canada, April 27–30, 2014.

Jeong, Y.H., Park, J.Y., Kim, H.G., Busby, J.T., Gartner, E., Atzmon, M., Was, G.S., Comstock, R.J., Chu, Y.S., Gomes da Silva, M., Yilmazbayhan, A., Motta, A.T., 2005. Corrosion of zirconium-based fuel cladding alloys in supercritical water. In: Proceedings of the 12th International Conference on Environmental Degradation of Materials in Nuclear Power Systems-Water Reactors, pp. 1369–1378.

Jang, J., Lee, Y., Han, C.H., Yi, Y.S., Hwang, S.S., Sik, S., 2005a. Effect of Cr content on supercritical water corrosion of high Cr alloys. Mater. Sci. Forum 475–479, 1483–1486.

Jang, J., Han, C.H., Lee, B.H., Yi, Y.S., Hwang, S.S., 2005b. Corrosion behavior of 9Cr F/M steels in supercritical water. In: Proceeding of ICAPP 05, Seoul, Korea, No. 5136.

Kaneda, J., Kasahara, S., Kuniya, J., Moriya, K., Kano, F., Saito, N., Shioiri, A., Shibayama, T., Takahashi, H., 2005. General corrosion properties of titanium based alloys for the fuel claddings in the supercritical water-cooled reactor. In: Proc. 12th International Conference on Environmental Degradation of Materials in Nuclear Power Systems-Water Reactors, pp. 1409–1419.

Karasawa, H., Fuse, M., Kiuchi, K., Katsumura, Y., 2004. Radiolysis of supercritical water. In: 5th Int. Workshop on LWR Coolant Water Radiolysis and Electrochemistry, 2004 October 15, San Francisco, USA.

Kasahara, S., Kuniyo, J., Moriya, K., Saito, N., Shiga, S., September 2003. General Corrosion of Iron, Nickel and Titanium Alloys as Candidate Materials for the Fuel Cladding of Supercritical-Water Cooled Power Reactor, GENES4/ANP2003, 2003, Kyoto, Japan, Paper 1132.

Khatamian, D., 2013. Corrosion and deuterium uptake of Zr-based alloys in supercritical water. J. Supercrit. Fluids 78, 132–142.

Kim, J.-H., Kim, B.K., Kim, D.-I., Choi, P.-P., Raabe, D., Yi, K.-W., 2015. The role of grain boundaries in the initial oxidation behaviour of austenitic stainless steel containing alloyed Cu at 700 °C for advanced thermal power plant applications. Corros. Sci. 96, 52–66.

Knodler, R., Straub, S., 2014. Steam oxidation of martensitic steels and Ni-base alloys: comparison of lab tests with operating power plants. Oxid. Met. 82, 113–122.

Kofstad, P., 1988. High Temperature Corrosion. Elsevier Applied Science, London.

Kritzer, P., 2004. Corrosion in high-temperature and supercritical water and aqueous solutions: a review. J. Supercrit. Fluids 29, 1.

Kritzer, P., Boukos, N., Dinjus, E., 1998. Corrosion of Alloy 625 in high temperature sulfate solutions. Corrosion 54, 689–699.

Kritzer, P., Boukos, N., Dinjus, E., 1999. Factors controlling corrosion in high-temperature aqueous solutions: a contribution to the dissociation and solubility data influencing corrosion processes. J. Supercrit. Fluids 15, 205–227.

Kriksunov, L.B., Macdonald, D.D., 1995a. Corrosion in supercritical water oxidation systems: a phenomenological analysis. J. Electrochem. Soc. 142, 4069.

Kriksunov, L.B., Macdonald, D.D., 1995b. Corrosion testing and prediction in SCWO environments. HTD. In: Proc. of the ASME Heat Transfer Division, vols. 317–2, p. 281.

Kriksunov, L.B., Macdonald, D.D., 1997. Potential-pH diagrams for iron in supercritical water. Corros. Sci. 53, 605–611.

Leistikow, S., 1972. Isothermal steam corrosion of commercial grade austenitic stainless steels and nickel base alloys in two technical surface conditions. In: Proceedings of the Fourth International Congress on Metallic Corrosion, Amsterdam, Netherlands, September 7–14, 1969. National Association of Corrosion Engineers, Houston, TX, pp. 278–290.

Li, J., Guzonas, D., Zheng, W., Wills, J., Dole, H., Michel, J., Woo, O.T., 2009. Property of passive films form on alloys tested in SCW water. In: 4th Int. Symposium on Supercritical Water-cooled Reactors", Heidelberg, Germany, March 8–11, 2009. Paper 56.

Li, W., Woo, O.T., Guzonas, D., Li, J., Huang, X., Sanchez, R., Bibby, C.D., 2015. Effect of pressure on the corrosion of materials in high temperature water. In: Characterization of Minerals, Metals, and Materials. John Wiley & Sons Inc., Hoboken, NJ (Chapter 12).

Li, J., Huang, X., Zeng, Y.W., Woo, O.T., Guzonas, D.A., 2014. Microscopy investigation on the corrosion of canadian generation IV SCWR materials. In: 19th Pacific Basin Nuclear Conference (PBNC 2014) Hyatt Regency Hotel, Vancouver, British Columbia, Canada, August 24–28, 2014.

Li, W., 2012. Oxidation Resistance and Nanoscale Oxidation Mechanisms in Model Binary and Ternary Alloys Exposed to Supercritical Water (M.A.Sc. thesis). Dept. of Materials Science and Engineering, University of Toronto.

Li, Q., Zhou, B., Yao, M., Liu, W., Chu, Y., 2007. "Corrosion behavior of zirconium alloys in supercritical water at 550°C and 25 MPa". Rare Metal Mater. Eng. 36, 1358–1361.

Liu, Z., 2013. Corrosion Behavior of Designed Ferritic-martensitic Steels in Supercritical Water (M.Sc. thesis). Department of Chemical and Material Engineering, University of Alberta, Canada.

Longton, P.B., 1966. The Oxidation of Iron- and Nickel-Based Alloys in Supercritical Steam; A Review of the Available Data. United Kingdom Atomic Energy Authority TRG Report 1144.

Lillard, S., 2012. Corrosion and compatibility. In: Konings, R.J.M. (Ed.), Comprehensive Nuclear Materials, vol. 5. Elsevier, Amsterdam, pp. 279–326.

Lujaszewicz, M., Simms, N.J., Dudziak, T., Nicholls, J.R., 2013. Effect of steam flow rate and sample orientation on steam oxidation of ferritic and austenitic steels at 650 and 700°C. Oxid. Met. 79, 473–483.

Macdonald, D.D., 1999. Passivity – the key to our metals-based civilization. Pure Appl. Chem. 71 (6), 951–978.

Macdonald, D.D., 2004. Effect of pressure on the rate of corrosion of metals in high subcritical and supercritical aqueous systems. J. Supercrit. Fluids 30, 375.

Macdonald, D.D., 2011. The history of the point defect model for the passive state: a brief review of film growth aspects. Electrochim. Acta 56, 1761–1772.

Macdonald, D.D., 2012. The passive state in our reactive metals-based civilization. Arabian J. Sci. Eng. 37 (5), 1143–1185.

Maekawa, T., Kagawa, M., Nakajima, N., 1968. Corrosion behaviors of stainless steel in high-temperature water and superheated steam. Trans. Jpn. Inst. Metals 9 (2), 130–136.

Mahboubi, S., September 2014. Effect of Cr Content on Corrosion Resistance of Fe-Cr-Ni Alloys Exposed in Supercritical Water (SCW) (M.A.Sc. thesis). Department of Materials Science and Engineering, McMaster University.

Mahboubi, S., Botton, G.A., Kish, J.R., 2014. Microstructural characterization of oxide scales grown on austenitic Fe-Cr-Ni alloys exposed to supercritical water (SCW). In: 2014 Canada-China Conference on Advanced Reactor Development (CCCARD-2014), Niagara Falls, Canada, April 27–30, 2014. Paper CCCAD2014-017.

Mahboubi, S., Botton, G.A., Kish, J.R., 2015. On the oxidation resistance of alloy 800HT exposed in supercritical water (SCW). Corrosion 71 (8), 992–1002.

Mitton, D.B., Yoon, J.H., Cline, J.A., Kim, H.S., Eliaz, N., Latanision, R.M., 2000. The corrosion behaviour of nickel-base alloys in SCWO systems. Ind. Eng. Chem. Res. 39, 4689–4696.

Montgomery, M., Karlsson, A., 1995. Survey of oxidation in streamside conditions. VGB Kraftw. 75, 235–240.

Motta, A.T., Yilmazbayhan, A., Gomes da Silva, M.J., Comstock, R.J., Was, G.S., Busby, J.T., Gartner, E., Peng, Q., Jeong, Y.H., Park, J.Y., 2007a. Zirconium alloys for supercritical water reactor applications: challenges and possibilities. J. Nucl. Mater. 371 (1–3), 61–75.

Motta, A.T., Siwy, A.D., Kunkle, J.M., Bischoff, J.B., Comstock, R.J., Chen, Y., Allen, T.R., 2007b. Microbeam synchrotron radiation diffraction and fluorescence study of corrosion layers formed on 9Cr ODS steel in supercritical water. In: Allen, T.R., Busby, J.T., King, P.J. (Eds.), 13th International Conference on Environmental Degradation of Materials in Nuclear Power Systems. Canadian Nuclear Society, pp. 1501–1513.

Nezakat, M., Akhiani, H., Penttilä, S., Szpunar, J., 2014. Oxidation Resistance of Thermomechanically Processed Austenitic Stainless Steel 316L in Super Critical Water. PBNC2014-350.

Novotny, R., Janik, P., Nilsson, K.F., Siegl, J., Hausild, P., 2013. Pre-qualification of cladding materials for SCWR fuel qualification testing facility. In: 6th International Symposium on Supercritical Water-Cooled Reactors, March 3–7, 2013, Shenzhen, Guangdong, China. China: CGNPC, 2013. Paper ISSCWR6-13069.

Olive, R.P., 2012. Pourbaix Diagrams, Solubility Predictions and Corrosion-product Deposition Modelling for the Supercritical Water Reactor (Ph.D. thesis). University of New Brunswick, Department of Chemical Engineering.

Otsuka, N., Shida, Y., Fujikawa, H., 1989. Internal-external transition for the oxidation of Fe-Cr-Ni austenitic stainless steels in steam. Oxid. Met. 32, 13.

Payet, M., 2011. Corrosion en Eau Supercritique – Apport à la Compréhension des Mécanismes pour des Alliages Fe-Cr-Ni de Structure c.f.c. (Doctoral thesis) Ecole Doctorale Arts et Metiers ED 415, Paris.

Pearl, W.L., Brush, E.G., Gaul, G.G., Wozadlo, G.P., 1965. General corrosion of Incoloy 800 in simulated superheat reactor environment. Nucl. Technol. 1 (3), 235–245.

Pearl, W.L., Brush, E.G., Gaul, G.G., Leistikow, S., 1967. General corrosion of Inconel 625 in simulated superheat reactor environment. Nucl. Appl. 3, 418–432.

Peng, Q., Gartner, E., Busby, J.T., Motta, A.T., Was, G.S., 2007. Corrosion behavior of model zirconium alloys in deaerated supercritical water at 500°C. Corrosion 63, 577–590.

Penttilä, S., Betova, I., Bojinov, M., Kinnunen, P., Toivonen, A., 2011. Estimation of kinetic parameters of the corrosion layer constituents on steels in supercritical water coolant conditions. Corros. Sci. 53, 4193–4203.

Penttila, S., Toivonen, A., 2013. Oxidation and SCC behaviour of austenitic and ODS steels in supercritical water. In: ISSCWR-6: The 6th International Symposium on Supercritical Water-cooled Reactors, March 03–07, 2013, Shenzhen, Guangdong, China. CGNPC, China. Paper ISSCWR6-13029.

Penttilä, S., Betova, I., Bojinov, M., Kinnunen, P., Toivonen, A., 2015a. Oxidation model for construction materials in supercritical water – estimation of kinetic and transport parameters. In: 7th Int. Symp. on Supercritical Water-cooled Reactors (ISSCWR-7), March 15–18, 2015, Helsinki, Finland. Paper #2064.

Penttilä, S., Betova, I., Bojinov, M., Kinnunen, P., Toivonen, A., 2015b. Oxidation model for construction materials in supercritical water – estimation of kinetic and transport parameters. Corros. Sci. 100, 36–46.

Peraldi, R., Pint, B.A., 2004. Effect of Cr and Ni contents on the oxidation behavior of ferritic and austenitic model alloys in air with water vapor. Oxid. Met. 61 (5/6), 463–483.

Potter, E.C., Mann, G.M.W., 1963. Mechanism of magnetite growth on low-carbon steel in steam and aqueous solutions up to 550 degrees C. In: 2nd Int. Congress on Metallic Corrosion, National Association of Corrosion Engineers, Houston.

Qiu, L., Guzonas, D.A., 2012. Prediction of metal oxide stability in supercritical water reactors–Pourbaix Versus Ellingham. In: The 3rd Canada-China Joint Workshop on Supercritical Water-cooled Reactors (CCSC-2012), Xian, China, April 25–28, 2012.

Ren, X., Sridharan, K., Allen, T.R., 2006. Corrosion of ferritic–martensitic steel HT9 in supercritical water. J. Nucl. Mater. 358 (2–3), 227–234.

Ren, X., Sridharan, K., Allen, T.R., 2007. Corrosion behavior of alloys 625 and 718 in supercritical water. Corrosion 63, 603.

Ren, X., Sridharan, K., Allen, T.R., 2008. Effect of Shot Peening on the Oxidation of Ferritic-Martensitic Steels in Supercritical Water. Corrosion, Paper 08425.

Ren, X., Sridharan, K., Allen, T.R., 2010. Effect of grain refinement on corrosion of ferritic–martensitic steels in supercritical water environment. Mater. Corros. 61, 748–755.

Robertson, J., 1991. The mechanism of high temperature aqueous corrosion of stainless steels. Corros. Sci. 32, 443–465.

Rodriguez, D., Merwin, A., Chidambaram, D., 2014. On the oxidation of stainless steel alloy 304 in subcritical and supercritical water. J. Nucl. Mater. 452, 440–445.

Rodriguez, D., Chidambaram, D., 2015. Oxidation of stainless steel 216 and Nitronic 50 in supercritical and ultrasupercritical water. Appl. Surf. Sci. 347, 10–16.

Ru, X., Staehle, R.W., 2013a. Historical experience providing bases for predicting corrosion and stress corrosion in emerging supercritical water nuclear technology: Part 1 – review. Corrosion 69 (3), 211–229.

Ru, X., Staehle, R.W., 2013b. Historical experience providing bases for predicting corrosion and stress corrosion in emerging supercritical water nuclear technology: Part 2 – review. Corrosion 69 (4), 319–334.

Ru, X., Staehle, R.W., 2013c. Historical experience providing bases for predicting corrosion and stress corrosion in emerging supercritical water nuclear technology: Part 3 – review. Corrosion 69 (5), 423–447.

Ruther, W.E., Greenberg, S., 1964. Corrosion of steels and nickel alloys in superheated steam. J. Electrochem. Soc. 111, 1116–1121.

Ruther, W.E., Schlueter, R.R., Lee, R.H., Hart, R.K., 1966. Corrosion behavior of steels and nickel alloys in superheated steam. Corrosion 22 (5), 147–155.

Saito, N., Tsuchiya, Y., Yamamoto, S., Akai, Y., Yotsuyanaji, T., Domae, M., Katsumura, Y., 2006. Chemical thermodynamic considerations on corrosion products in supercritical water-cooled reactor coolant. Nucl. Technol. 155, 105–112.

Sanchez, R.G., 2014. Corrosion Performance of Candidate Materials for Canadian Gen IV Supercritical Water Cooled Reactor (M.A.Sc. thesis). Dept. of Mechanical Engineering, Carleton University, Ottawa, Canada.

Sarver, J., 2009. The oxidation behavior of candidate materials for advanced energy systems in steam at temperatures between 650 °C and 800 °C. In: 14th Int. Conf. on Environmental Degradation of Materials in Nuclear Power Systems, Virginia Beach, VA, August 23–27, 2009.

Sarver, J.M., Tanzosh, J.M., 2013. Effect of temperature, alloy composition and surface treatment on the steamside oxidation/oxide exfoliation behavior of candidate A-USC boiler materials. Babcock & Wilcox Technical Paper BR-1898. In: Proceedings of the Seventh International Conference on Advances in Materials Technology for Fossil Power Plants, Waikoloa, Hawaii, USA, October 22–25, 2013.

Sawochka, S., Leonard, M., 2006. Source Term Reduction: Impact of Plant Design and Chemistry on PWR Shutdown Releases and Dose Rates. Electric Power Research Institute Report 1013507.

Sarrade, S., Féron, D., Rouillard, F., Perrina, S., Robin, R., Ruizc, J.C., Turc, H.-A., 2017. Overview on corrosion in supercritical fluids. J. Supercrit. Fluids 120, 335–344.

Saunders, S.M., Monteiro, M., Rizzo, F., 2008. The oxidation behavior of metals and alloys at high temperatures in atmospheres containing water vapour: a review. Prog. Mater. Sci. 53, 775–837.

Schulenberg, T., 2013. Material requirements of the high performance light water reactor. J. Supercrit. Fluids 77, 127–133.

Sloppy, J.D., Lu, Z., Dickey, E.C., Macdonald, D.D., 2013. Growth mechanism of anodic tantalum pentoxide formed in phosphoric acid. Electrochim. Acta 87 (1), 82–91.

Spalaris, C.N., 1963. Finding a corrosion resistant cladding for superheater fuels. Nucleonics 21.

Steeves, G., Cook, W., Guzonas, D., 2015. Development of kinetic models for the long-term corrosion behaviour of candidate alloys for the Canadian SCWR. In: 7th International Symposium on Supercritical Water-cooled Reactors (ISSCWR-7), March 15–18, 2015, Helsinki, Finland. Paper-2076.

Sun, M., Wu, X., Zhang, Z., Han, E.-H., 2009a. Oxidation of 316 stainless steel in supercritical water. Corros. Sci. 51 (5), 1069–1072.

Sun, M., Wu, X., Han, E.-H., Rao, J., 2009b. Microstructural characteristics of oxide scales grown on stainless steel exposed to supercritical water. Scr. Mater. 61 (10), 996–999.

Sun, C., Hui, R., Qu, W., Yick, S., 2009c. Progress in corrosion resistant materials for supercritical water reactors. Corros. Sci. 51, 2508–2523.

Svishchev, I.M., Carvajal-Ortiz, R.A., Choudhry, K.I., Guzonas, D.A., 2013. Corrosion behavior of stainless steel 316 in sub- and supercritical aqueous environments: effect of LiOH additions. Corros. Sci. 72, 20–25.

Tan, L., Allen, T.R., 2005. An EBSD study of grain boundary engineered Incoloy alloy 800H. Metall. Mater. Trans. A 36 (7), 1921–1925.

Tan, L., Allen, T.R., 2009. Localized corrosion of magnetite on ferritic-martensitic steels exposed to supercritical water. Corros. Sci. 51, 2503–2507.

Tan, L., Sridharan, K., Allen, T.R., 2006a. The Effect of grain boundary engineering on the oxidation behavior of Incoloy alloy 800H in supercritical water. J. Nucl. Mater. 348, 263.

Tan, L., Yang, Y., Allen, T.R., 2006b. Oxidation behavior of alloy HCM12A exposed in supercritical water. Corros. Sci. 48, 3123.

Tan, L., Machut, M.T., Sridharan, K., Allen, T.R., 2007. Oxidation behavior of HCM12A exposed in harsh environments. J. Nucl. Mater. 371, 171–175.

Tan, L., Ren, X., Sridharan, K., Allen, T.R., 2008a. Effect of shot-peening on the oxidation of alloy 800H exposed to supercritical water and cyclic oxidation. Corros. Sci. 50, 2040–2046.

Tan, L., Ren, X., Sridharan, K., Allen, T.R., 2008b. Corrosion behavior of Ni-base alloys for advanced high temperature water-cooled nuclear plants. Corros. Sci. 50, 3056.

Tan, L., Ren, X., Allen, T.R., 2010. Corrosion behavior of 9–12% Cr Ferritic–martensitic steels in supercritical water. Corros. Sci. 52, 1520–1528.

Tomlinson, L., Cory, N.J., 1989. Hydrogen emission during the steam oxidation of ferritic steels: kinetics and mechanism. Corros. Sci. 29 (8), 939–965.

Tsuchiya, Y., Saito, N., Kano, F., Ookawa, M., Kaneda, J., Hara, N., 2007. Corrosion and SCC Properties of Fine Grain Stainless Steel in Subcritical and Supercritical Pure Water. NACE 2007. NACE, Houston.

Van Nieuwenhove, R., Balak, J., Toivonen, A., Penttilä, S., Ehrnsten, U., 2013. Investigation of coatings, applied by PVD, for the corrosion protection of materials in supercritical water. In: 6th International Symposium on Supercritical Water-cooled Reactors (ISSCWR-6), March 03–07, 2013, Shenzhen, Guangdong, China. Paper ISSCWR6-13024.

Wills, J.S.C., Chiu, A., Guzonas, D., 2007. Deposition of anti-corrosion coatings by atmospheric pressure plasma jet. In: 28th Annual Conference of the Canadian Nuclear Society, June 3–6, 2007, St. John, New Brunswick.

Was, G.S., Teysseyre, S., McKinley, J., 2004. Corrosion and Stress Corrosion Cracking of Iron- and Nickel-base Austenitic Alloys in Supercritical Water. NACE Corrosion 2004, Paper 04492.

Was, G.S., Allen, T.R., 2005. Time, temperature, and dissolved oxygen dependence of oxidation of austenitic and ferritic-martensitic alloys in supercritical water. In: Proc. International Congress on Advances in Nuclear Power Plants (ICAPP'05), Seoul, Korea, Paper 5690, May 15–19, 2005. Paper 5690.

Was, G.S., Ampornrat, P., Gupta, G., Teysseyre, S., West, E.A., Allen, T.R., Sridharan, K., Tan, L., Chen, Y., Ren, X., Pister, C., 2007. Corrosion and stress corrosion cracking in supercritical water. J. Nucl. Mater. 371, 176–201.

Was, G.S., Teysseyre, S., Jiao, Z., 2006. Corrosion of austenitic alloys in supercritical water. Corrosion 62, 989–1005.

Was, G.S., Teysseyre, S., 2005. Challenges and recent progress in corrosion and stress corrosion of austenitic alloys for supercritical water reactor core components. In: Proceedings of the 12th International Conference on Environmental Degradation of Materials in Nuclear Power Systems-water Reactors. The Minerals, Materials and Metals Society, Warrendale, PA, pp. 1343–1357.

Watanabe, Y., Kobayashi, T., Adschiri, T., 2001. Significance of Water Density on Corrosion Behavior of Alloys in Supercritical Water. NACE Corrosion 2001, Paper 01369.

Watanabe, Y., Shoji, K., Adschiri, T., Sue, K., 2002. Effects of Oxygen Concentration on Corrosion Behavior of Alloys in Acidic Supercritical Water. NACE Corrosion 2002, Paper 02355.

Watanabe, Y., Daigo, Y., 2005. Corrosion rate and oxide scale characteristics of austenitic alloys in supercritical water. Materials Science Forum, v 522-523, High-Temperature Oxidation and Corrosion 2005. In: Proceedings of the International Symposium on High-Temperature Oxidation and Corrosion 2005, 2006, pp. 213–220.

Wright, I.G., Tortorelli, P.F., 2007. Program on Technology Innovation: Oxide Growth and Exfoliation on Alloys Exposed to Steam. Electric Power Research Institute Report 1013666.

Wright, I.G., Dooley, R.B., 2010. A review of the oxidation behavior of structural alloys in steam. Int. Mater. Rev. 55, 129–167.

Xu, R., Yetisir, M., Hamilton, H., 2014. Thermal-mechanical behavior of fuel element in SCWR design. In: 2014 Canada-China Conference on Advanced Reactor Development, Niagara Falls, Ontario Canada, April 27–30, 2014. Paper CCCARD2014-005.

Yi, Y., Lee, B., Kim, S., Jang, J., 2006. Corrosion and corrosion fatigue behaviors of 9Cr steel in a supercritical water condition. Mater. Sci. Eng. A Struct. Mater. Prop. Microstruct. Process. 429, 161–168.

Yi, Y.-S., Watanabe, Y., Kondo, T., Kimura, H., Saito, M., 2001. Oxidation rate of advanced heat-resistant steels for ultra-supercritical boilers in pressurized superheated steam. J. Press. Vessel Technol. 123, 391–397.

Yamamura, T., Okuyama, N., Shiokawa, Y., Oku, M., Tomiyasu, H., Sugiyama, W., 2005. Chemical states in oxide films on stainless steel treated in supercritical water: factor analysis of X-ray photoelectron spectra. J. Electrochem. Soc. 152 (12), B540–B546.

Yuan, J., Wu, X., Wang, W., Zhu, S., Wang, F., 2013. The effect of surface finish on the scaling behavior of stainless steel in steam and supercritical water. Oxid. Met. 79, 541–551.

Zemniak, S.E., Hanson, M., Sunder, P.C., 2008. Electropolishing effects on corrosion behavior of 304 stainless steel in high temperature, hydrogenated water. Corros. Sci. 50, 2465–2477.

Zhang, L., Zhu, F., Tang, R., 2009a. Corrosion mechanisms of candidate structural materials for supercritical water-cooled reactor. Front. Energy Power Eng. China 3 (2), 233–240.

Zhang, Q., Tang, R., Li, C., Luo, X., Long, C., Yin, K., 2009b. Corrosion behavior of Ni-base alloys in supercritical water. Nucl. Eng. Technol. 41, 107–112.
Zhang, L., Bao, Y., Tang, R., 2012. Selection and corrosion evaluation tests of candidate SCWR fuel cladding materials. Nucl. Eng. Des. 249, 180–187.
Zhang, N.Q., Zhu, Z.-L., Xu, H., Mao, X.-P., Li, J., 2016. Oxidation of ferritic and ferritic-martensitic steels in flowing and static supercritical water. Corros. Sci. 103, 124–131.
Zheng, W., Guzonas, D., Li, J., 2008. Novel approach to the development of in-core materials for a supercritical water cooled reactor. In: 16th Pacific Basin Nuclear Conference (16 PBNC), Aomori, Japan, October 13–18, 2008. Paper ID P16P1413.
Zhou, Z.Y., Lvov, S.N., Wei, X.J., Benning, L.G., Macdonald, D.D., 2002. Quantitative evaluation of general corrosion of Type 304 stainless steel in subcritical and supercritical aqueous solutions via electrochemical noise analysis. Corros. Sci. 44, 841–860.
Zhu, Z., Xu, H., Jiang, D., Mao, X., Zhang, N., 2016. Influence of temperature on the oxidation behavior of a ferritic-martensitic steel in supercritical water. Corros. Sci. 113, 172–179.
Zhu, Z., Xu, H., Jiang, D., Yue, G., Li, B., Zhang, N., 2016. The role of dissolved oxygen in supercritical water in the oxidation of ferritic-martensitic steel. J. Supercrit. Fluids 108, 56–60.
Zurek, J., Young, D.J., Essuman, E., Hänsel, M., Penkalla, H.J., Niewolak, L., Quadakkers, W.J., 2008. Growth and adherence of chromia-based surfaces scales on Ni-base alloys in high and low pO_2 gases. Mater. Sci. Eng. A 477, 259–270.

Environmentally assisted cracking 6

6.1 Introduction

Environmentally assisted cracking (EAC) is a complex phenomenon driven by the synergistic interaction of mechanical, chemical and metallurgical factors as summarized in Fig. 6.1. EAC includes environment-affected damage mechanisms such as stress corrosion cracking (SCC), strain-induced corrosion cracking (SICC) and corrosion fatigue (CF). These degradation modes can be distinguished (for example) in terms of the nature of the stress; typically, SCC is caused by constant stress, SICC by monotonically increasing strain and CF by cycling stress. The complex interplay between causative factors makes experimental measurements of EAC difficult. As a result, our state of knowledge on EAC under supercritical water-cooled reactor (SCWR) conditions is not as well advanced as that of general corrosion. The requirement for all three causative factors to be present makes it possible in some circumstances to compensate for suboptimal values of one factor through careful

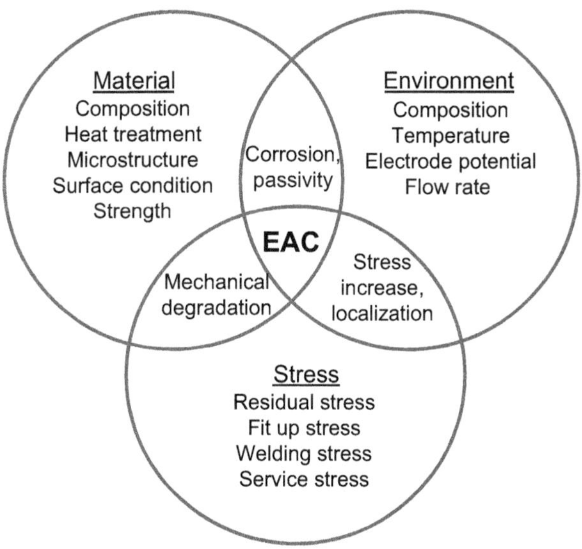

Figure 6.1 Venn diagram illustrating how the simultaneous occurrence of tensile stress, susceptible material conditions and a corrosive environment can lead to environmentally assisted cracking (EAC).
After Speidel, M.O., 1984. Stress corrosion and corrosion fatigue mechanics. In: Speidel, M.O., Atrens, A. (Eds.), Corrosion in Power Generating Equipment. Plenum Press, New York.

Materials and Water Chemistry for Supercritical Water-cooled Reactors
https://doi.org/10.1016/B978-0-08-102049-4.00006-4

optimization of the other two, e.g., optimizing water chemistry, design and residual stresses to make up for material deficiencies.

Fig. 6.1 somewhat over-simplifies matters since, as Andresen and Ford (2011) stressed, parameters such as cold work or irradiation can affect more than one circle of influence independently (cf. Fig. 1.3).

The danger of EAC is in its stochastic nature and the difficulty of its detection at early stages by conventional nondestructive techniques, since cracks propagate with little or no macroscopic plastic deformation. An alloy affected by EAC does not usually display abnormal mechanical properties (yield strength and tensile strength). This is especially important for pressure boundary components, as sudden failure can have catastrophic consequences. Failure of the fuel cladding is also undesirable since this releases radioactive material into the coolant (Chapter 4). However, the presence of a through-wall crack in the fuel cladding can be readily detected on line and monitored by sampling and measuring the coolant fission gas activity, allowing the reactor to be shutdown if required to remove the defective fuel rod(s).

EAC-induced cracks usually propagate perpendicular to the principal tensile stress. As a result, intergranular (interdendritic in weld metals) and transgranular crack propagation can be observed for the same alloy, depending on the environment, microstructure or stress/strain state, and the modes of EAC can be mixed or switch from one to the other. The degree of cracking susceptibility for a particular alloy may change in any given alloy/environment system with relatively subtle changes in material, stress and environment conditions.

EAC phenomena began to gain attention in the nuclear industry in the mid-1970s with a series of intergranular stress corrosion cracking (IGSCC) incidents, specifically IGSCC of sensitized and cold-worked stainless steels in boiling water reactors (BWRs) and steam generator tube cracking in pressurized water reactors (PWRs) (Combrade et al., 2010; Staehle and Gorman, 2003). Most of these incidents were the result of improper selection of structural materials, residual stresses and inadequate water treatment. EAC remains an important issue for water-cooled reactors (WCRs), in particular due to the increased average age of existing nuclear power plants (NPPs). Therefore EAC was recognized at the beginning of SCWR development as a key life-limiting degradation mechanism for SCWR components in the core and downstream of the core[1].

Stainless steels and nickel-base alloys have been extensively used in WCRs, and stainless steel fuel cladding has been widely used in LWRs. The La Crosse BWR operated for 20 years using 348 SS fuel cladding; more than 50,000 fuel rods passed through the core during its lifetime (Strasser et al., 1982). During the first five cycles of operation, about 1.3% of the rods failed, essentially all at higher than core-average burn-up. The failure rates during subsequent cycles were reduced through changes

[1] Various in-core components have different in-service lifetimes and hence the relative importance of EAC causative factors may not be the same between components; for example, fuel cladding versus core internals.

such as a limit on maximum burn-up and tighter operating restrictions (e.g., on control rod movements). The failure mechanism was ascribed to a combination of internal stresses from pellet-cladding interaction and outside diameter cladding corrosion resulting from the high concentration of oxidants in the coolant due to water radiolysis. Type 347NG stainless steel has been used extensively in German BWRs for core internals with no cracking being observed unless the material was inadvertently sensitized and/or there was surface cold work. Scott (2013) noted that 'Strain-hardened SSs, mainly Type 316 and to a lesser extent niobium-stabilized Type 347, have been used successfully for many decades in PWRs for bolting and other purposes where moderate strength is required, and without apparent problems, except when highly irradiated'. The Russian alloy 1Kh18Ni10T was used successfully in the superheaters in Beloyarsk at 8 MPa and 550 °C (Emel'yanov et al., 1972). Stainless steels have also been used extensively in the superheater region of supercritical fossil power plants (SCFPPs). German once-through boilers have operated for periods on the order of 10^5 h using Nb-stabilized steels (DIN 8CrNiNb163, X8CrNiMoNb1616 or X8CrNiMoVNb1613), and Russian SCFPPs have significant experience using Ti-stabilized 12Cr18Ni12Ti. EAC has been a significant issue for steam turbines (blades, discs, rotors) due to the transport of impurities in the feedwater and their deposition in the turbine (cf. Fig. 4.15) (Zhou and Turnbull, 2002). The occurrence of EAC was strongly correlated with the presence of oxygen in the steam. Sodium, chloride and sulphate ions have been found to increase the risk of EAC of turbine components.

Allen et al. (2012), Ru and Staehle (2013a,b,c), Was et al. (2007), and Zheng et al. (2011, 2014) have reviewed the available SCC data obtained in support of SCWR development, and the reader is encouraged to consult these reviews for further information. The published proceedings of the on-going symposium series 'International Symposium on Environmental Degradation of Materials in Nuclear Power Systems – Water Reactors' provide a snapshot of current R&D in EAC for WCRs (including on occasion SCWRs), as well as historical perspective.

6.2 Effects of key variables

Chronologically EAC processes in many alloy/environment systems consist of the following steps (Staehle, 2007):

- Engineering crack initiation' – includes the precursor period (specific stress, metallurgical and environmental conditions may develop at metal/solution interface, crack initiation, coalescence and short crack growth.
- Propagation of a single dominating crack' – crack growth rate (CGR) is rarely constant and can accelerate or decelerate depending on plant operating conditions.

The following section discusses the effect of key variables (Fig. 6.1) on EAC in SCW based on results collected to support SCWR development and other WCR experience. The focus will be on aspects relevant to SCWR development.

6.2.1 Environmental factors

As discussed in Chapter 4, SCWR in-core water chemistry is expected to be similar to that of a BWR[2] operating with no added H_2 (normal water chemistry, NWC) or added H_2 (hydrogen water chemistry, HWC). In BWRs, the major environmental factors controlling EAC propagation rate for austenitic stainless steels and Ni-based alloys are the electrochemical corrosion potential (ECP), determined by the concentrations of oxidizing and reducing species, and concentration of anionic impurities, reflected in the conductivity. EAC initiation and crack growth of austenitic stainless steels and Ni-based alloys in high temperature subcritical water have been studied for many years (Andresen and Ford, 2011; Raja and Shoji, 2011), and although outstanding questions remain, a significant fundamental and practical knowledge base now exists. This knowledge base has informed and guided the SCWR R&D effort related to EAC as discussed below.

6.2.1.1 Temperature

EAC susceptibilities are thermally activated (Andresen and Ford, 2011), but in BWRs, temperature is generally only a minor factor except during startup or shutdown. In plants, cracking is largely confined to components operating at temperatures >250 °C (Kilian et al., 2005). However, given the very wide range of temperatures in an SCWR core and the large variation in SCW properties in the near-critical region, temperature is expected to play a much larger role in EAC in an SCWR. As with general corrosion, the EAC susceptibility of materials must be evaluated for subcritical water ($280 < T < T_c$), near-critical and low temperature SCW ($374 < T < 450$ °C) and high temperature SCW ($T \gg T_c$, up to the peak cladding temperature).

The EAC susceptibility/temperature dependence for stainless steels in subcritical water is complicated and does not necessarily follow an Arrhenius-type relationship because in many cases more than one rate-controlling process is present over the range of temperatures (Fig. 6.2). Given the discussions in previous chapters regarding the complex temperature dependencies of the rates of various chemical processes (corrosion product deposition and water radiolysis (Chapter 4) and general corrosion (Chapter 5)), it seems reasonable to expect similar complex temperature dependencies for EAC in an SCWR. Allen et al. (2012) concluded from data obtained in constant extension rate tensile (CERT) tests of a number of austenitic stainless steels over the temperature range 400–650 °C that there is a general decrease in maximum stress with increasing temperature but no clear trend in the failure strain.

6.2.1.1.1 Near-critical region – the effect of supercritical water density

In the near-critical region, the effects of temperature are intermixed with rapid changes in other physical and chemical properties. A number of studies of EAC in SCW in support of supercritical water oxidation (SCWO) development have been reported, but most used high concentrations of aggressive anions (chloride, sulphate) well outside

[2] However, there will be no stripping of hydrogen into the steam phase in an SCWR.

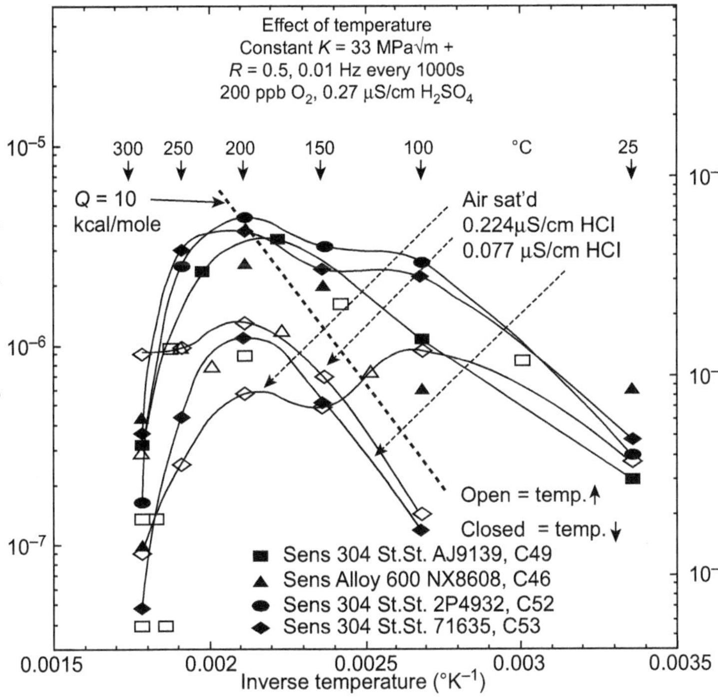

Figure 6.2 Crack propagation rate as a function of test temperature for sensitized stainless steel under constant stress intensity factor in water containing various impurities (Andresen, 1992).

the range of concentration expected in an SCWR under all but the worst out-of-specification conditions. Tsuchiya et al. (2003) carried out CERT tests on sensitized 304 and 316 stainless steels tensile specimens with a strain rate of $4 \times 10^{-7}\ s^{-1}$ at temperatures from 290 to 550 °C at 25 MPa in high purity water with 8 ppm dissolved oxygen (DO). Fig. 6.3 plots IGSCC ratio as a function of SCW density, illustrating the strong correlation between SCW density and the occurrence of SCC. This dependence is expected if, as in BWRs, the SCC mechanism is driven by anodic dissolution. At the low SCW density at the higher temperatures, the extent of ion dissociation is low (cf. Fig. 4.12) and the solubility of metal oxides is low (Fig. 4.10), reducing the aggressiveness of the solution, lowering the solution conductivity and reducing metal dissolution. No signs of IGSCC were observed at temperatures above 400 °C (Fig. 6.3), but both alloys had small cracks on the side surface after exposure to SCW. The change in IGSCC ratio correlated well with the SCW density and the trend in metal oxide solubilities. No SCC was observed for 316L SS.

A similar dependence on SCW density was reported by Watanabe et al. (2003), who measured the effect of SCW pressure on cracking of sensitized 316 SS in pure water containing 8 $mg \cdot kg^{-1}$ DO at a constant temperature of 400 °C. The strain to failure and maximum stress decreased with increasing pressure (increasing density and dielectric constant) and there was an increase in the degree of intergranular fracture. Fig. 6.4

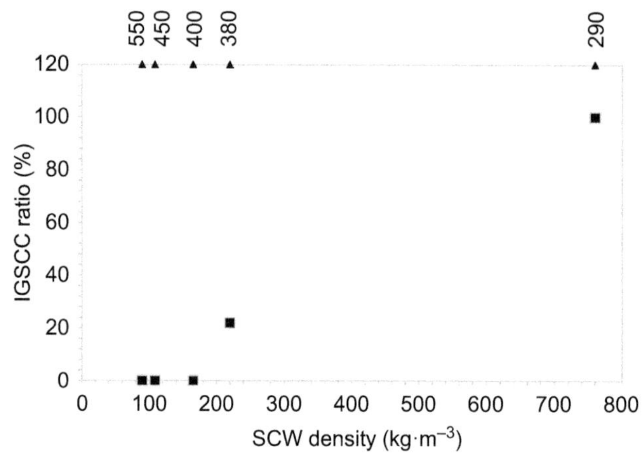

Figure 6.3 Intergranular stress corrosion cracking (IGSCC) susceptibility of sensitized 304SS in supercritical water (SCW) (25 MPa, 8 ppm dissolved oxygen, conductivity 0.06 $\mu S \cdot cm^{-1}$, strain rate $4 \times 10^{-7}\ s^{-1}$) as a function of SCW density. Corresponding temperatures are shown at the top of the figure (in °C).
Data from Tsuchiya, Y., Kano, F., Saito, N., Shioiri, A., Kasahara, S., Moriya, K., Takahashi, H., September 15–19, 2003. SCC and Irradiation Properties of Metals Under Supercritical-Water Cooled Power Reactor Conditions. GENES4/ANP2003, Kyoto, Japan. Paper 1096.

shows the dependence of failure strain on the dielectric constant of water and includes a datum obtained at 360 °C, 25 MPa, where the dielectric constant of water is essentially the same as that at 400 °C, 60 MPa. Fig. 6.4 suggests that under these conditions the variation in dielectric constant (and thus in ion association, oxide solubility, etc.)

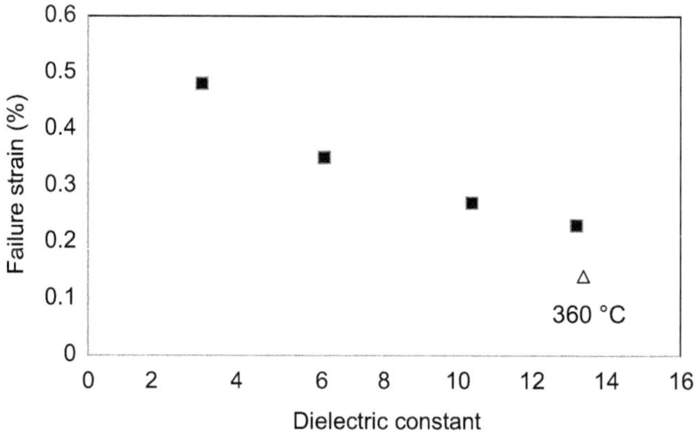

Figure 6.4 Dependence of failure strain on dielectric constant for 316 SS strained at $2.8 \times 10^{-6}\ s^{-1}$ in 400 °C supercritical water containing 8 $mg \cdot kg^{-1}$ oxygen. The open triangle datum was obtained at 360 °C, 25 MPa.
Data from Watanabe, Y., Abe, H., Daigo, Y., 2003. Environmentally-assisted cracking of sensitized stainless steel in supercritical water. In: Proceedings of GENES4/NP2003, Kyoto, Japan, September 15–19, 2003, Paper 1183.

can account for the majority of the observed pressure dependence. They attributed their observations to increased ion stability and oxide solubility, and an increase in water conductivity with pressure, which increased anodic dissolution. The authors noted that solution-annealed 316L SS did not show SCC at 400 °C, even at 60 MPa.

Peng et al. (2007) performed CGR tests in SCW (25 MPa, 2000 $\mu g \cdot kg^{-1}$ DO) with 20% CW 316L at stress intensity factor K_i of 27.6 $MPa \cdot m^{1/2}$. They monitored the CGR from 288 to 500 °C (across the subcritical – supercritical transition) to determine the temperature and DO dependence of the CGR in SCW in the near-critical region (Fig. 6.5). At $T < T_c$, the CGR increased with increasing temperature with an activation energy of 26 $kJ \cdot mol^{-1}$, within the range of values reported for stainless steels in water over the temperature range 25–288 °C. They concluded that the crack growth in subcritical water followed the behaviour expected from the existing database. At $T > T_c$, the CGR decreased with increasing temperature and mirrored the decrease in water density. They attributed the reduction in CGR to crack blunting due to increases in oxidation rate with increasing temperature rather than a change in oxide solubility and accompanying decrease in transport of ions from the crack. Watanabe et al. (2004a,b) reported that cracking severity was higher at 400 °C than at 360 °C in 0.01 $mol \cdot kg^{-1}$ $H_2SO_4 + 800\ mg \cdot kg^{-1}$ O_2. This suggested that a mechanism involving dissolution was unlikely and that increased grain boundary oxidation was important under these very oxidizing conditions, and proposed that different cracking mechanisms were operative in pure water (anodic dissolution) and under highly oxidizing conditions (oxidation cracking). This conclusion is consistent with the

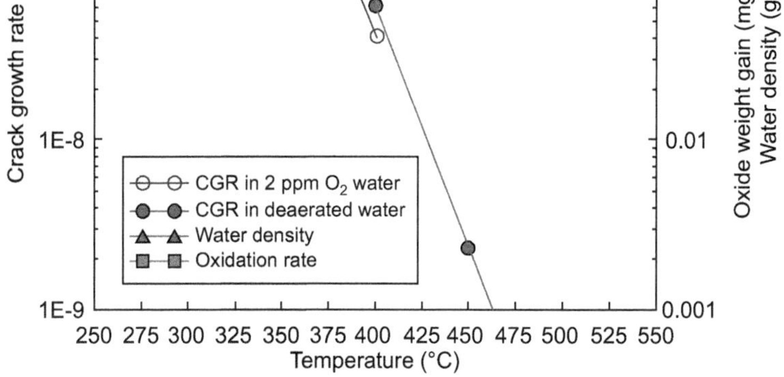

Figure 6.5 Crack growth rate (CGR) versus temperature across the subcritical–supercritical line for a 0.5T compact tension specimen of unsensitized Type 316L stainless steel in pure water. The water density and rate of oxide weight gain over approximately 500 h are also shown (Peng et al. 2007).

results of Watanabe et al. (2003) measured at constant temperature (which eliminates thermally activated changes in oxidation rate).

Was et al. (2007) discussed the effect of temperature on SCC susceptibility of selected austenitic stainless steels in SCW. In the near-critical region where the SCW density is high, it was suggested that the higher dielectric constant and higher oxide solubility lead to higher IGSCC susceptibility. Both the percentage of SCC observed on the fracture surface and the density and depths of secondary cracks along the tensile specimens were considered as SCC indicators.

6.2.1.1.2 High Temperature Behaviour

Most measurements of the temperature dependence of EAC for SCWR development have been made at $T > T_C$. Teysseyre and Was (2006) observed that secondary crack density for 304 SS, 316L SS, Alloy 625 and Alloy 690 decreased with rising temperature while the highest maximum crack length was observed at the highest test temperatures (550 °C). The calculated CGRs in these slow strain rate testing (SSRT) increased with increasing temperature. Instead of SCC/ductile fracture ratio on the fracture surface area, they suggested using a parameter called cracking severity, which incorporates both the density and length of the cracks found along the gauge section. They reported indications of transgranular stress corrosion cracking (TGSCC) on fracture surfaces as well as for secondary cracks. However, they also observed IGSCC indications for secondary cracks after straining 316L in an inert Ar atmosphere. Janik et al. (2013) and Penttilä and Toivonen (2013) evaluated the SCC resistance of 08Cr18Ni10Ti, 347H and 316L at 500 and 550 °C, 25 MPa with 150 $\mu g \cdot kg^{-1}$ DO. At both temperatures, only 316L showed no indications of IGSCC, even when exposed to SCW with higher DO content. On the other hand, in a later test series (Toivonen et al., 2015) on another heat of 316L, a ~8 × 0.5 mm tube showed very high IGSCC susceptibility. Double-loop electrochemical potentiokinetic reactivation (DL-EPR) test comparison of unexposed and exposed samples revealed that the SSRT specimens were thermally sensitized very quickly during the test at 550 °C in SCW with 150 $\mu g \cdot kg^{-1}$ DO. Ru et al. (2015) used U-bend samples of 310S SS to determine the time-to-crack at 290, 380, 550 and 650 °C in water (pH 6.5–7.5, inlet conductivity $\leq \mu S \cdot cm^{-1}$, DO < 10 $\mu g \cdot kg^{-1}$, 25 MPa except at 280 °C). The U-bend samples showed no SCC after exposure for 2000 h. SSRT tests of the same material showed a change in fracture characteristics from dimples at 550 °C to IGSCC at 650 °C. An increase in hardness was reported at 650 °C.

Shen et al. (2014) found that the SCC of AL6XN, HR3C and 316Ti had different temperature dependencies and attributed this difference to differences in cracking mechanism. The change in mechanism was linked to the carbon content of the alloy. Saez-Maderuelo and Gómez-Briceño (2016) reported that the SCC susceptibility of 316L increased with temperature from 400 to 500 °C. Chen et al. (2017) reported the CGR of 310S SS in SCW containing Ar, O_2 or H_2 measured using direct current potential drop at 500 and 550 °C. The CGRs increased with increasing temperature (Fig. 6.6) with an apparent activation energy of ~170 $kJ \cdot mol^{-1}$. These data show the opposite temperature dependence from those reported by Peng et al. (2007) for

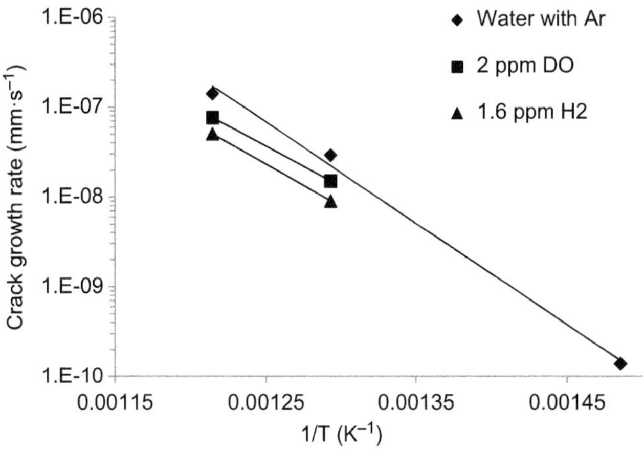

Figure 6.6 Crack growth rate of 310S SS in supercritical water containing Ar, O_2 or H_2 as a function of inverse temperature.
Data from Chen, K., Du, D., Gao, W., Andresen, P.L., Zhang, L., 2017. Investigation of Temperature Effect on Stress Corrosion Crack Growth Behavior of 310S Stainless Steel in Supercritical Water Environment. ISSCWR8.

316L at near-critical temperatures. Chen et al. (2017) suggested that the lower oxidation rate of 310S compared to 316L would lead to less crack blunting of the latter alloy increasing the CGR. An alternate explanation is a change in cracking mechanism analogous to the change in general corrosion mechanism. Chen et al. (2017) proposed a model in which creep played a major role in overall crack growth (see Section 6.3); they also noted that the increase in temperature would lead to sensitization, increasing intergranular corrosion.

6.2.1.2 Water chemistry

WCRs have strict water chemistry monitoring and control regimes to minimize impurity concentrations and control the ECP of the alloys (Hettiarachchi and Weber, 2010; Cowan et al., 2011, 2012). This is especially important in BWRs where it is not possible to completely suppress the net production of oxidizing species (oxygen and hydrogen peroxide) by water radiolysis. The superior performance of stainless steel fuel cladding in PWRs versus BWRs is largely attributed to the more benign water chemistry conditions in PWRs (Strasser et al., 1982), where suppression of the net radiolytic formation of oxidizing species is more easily achieved.

While water conductivity is often used as a measure of the concentrations of impurities, Andresen and Ford (2011) stressed that crack tip anion activity, not bulk water conductivity, is of fundamental importance, and that cracking susceptibility is anion-specific[3]. Strong acid-forming anions (e.g., sulphate or chloride) can have significant

[3] Although the generally accepted mechanism of SCC in BWRs does require a reasonably conductive solution (see Fig. 6.9), at least locally in the vicinity of the crack.

effect on EAC initiation and propagation rate. It is generally thought that SCC requires the transient exposure of bare alloy to the coolant (e.g., film rupture due to tensile stress). If the oxide cannot heal due to the presence of ions that interfere with this process, SCC can initiate and propagate. Thus at $T \gg T_c$, where ion dissociation is not favoured, ionic species can be less detrimental and other factors may become more important.

Fig. 6.7 shows the dependence of SCC of 304 SS in subcritical water on the concentrations of DO and chloride ion for material that was sensitized or solution annealed. Nonsensitized alloy is seen to tolerate a wider range of DO and chloride ion concentrations than sensitized material. The figure also shows the expected normal operating water chemistry range for the SCWR, but unfortunately, there are no data at the expected chloride concentrations or SCWR operating temperatures, and conclusions concerning the risk of SCC in an SCWR based on this diagram are speculative. However, during transients or periods of out-of-specification operation, higher chloride concentrations are possible and Fig. 6.7 would predict an increased risk of

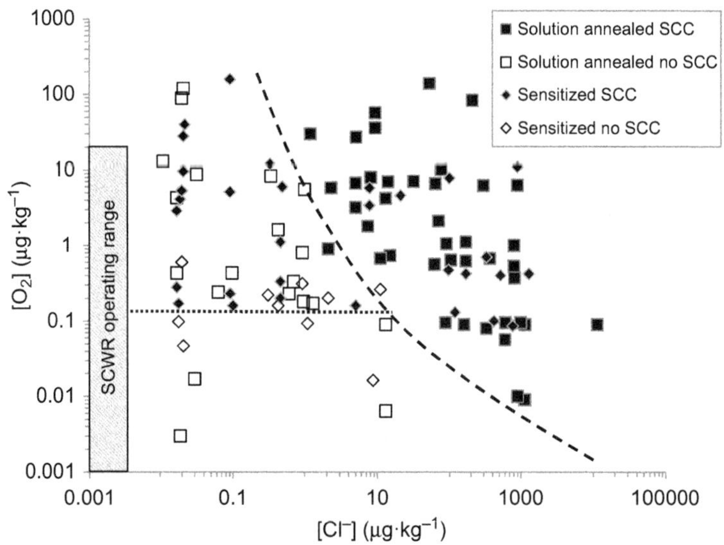

Figure 6.7 Effect of dissolved oxygen and chloride concentrations on stress corrosion cracking (SCC) of 304 SS in subcritical water. The dashed line divides 'Cl^-—O_2' space for nonsensitized 304 into a 'no SCC' region (right of the curve) and an 'SCC' region (left of the curve). Similarly, the dotted horizontal line divides the 'no SCC' region for nonsensitized alloy into 'SCC' (above) and 'no SCC' (below) regions. The shaded region represents the expected normal operating conditions of the supercritical water-cooled reactor (SCWR) (Chapter 4); transient and out-of-specification periods could move the water chemistry outside of this region.

Adapted from Gordon, G.M., Brown, K.S., 1989. Dependence of creviced BWR component IGSCC behavior on coolant chemistry. In: Cubicciotti, D., Simonen, E. (Eds.), Proc. 4th Int. Symp. on Environmental Degradation of Materials in Nuclear Power Systems — Water Reactors, Jekyll Island, August 6—10, 1989. National Association of Corrosion Engineers, pp. 14.46—14.62.

EAC during these periods. A significant fraction of crack growth during BWR operating cycles occurs during start-up, when the water temperature exceeds 93 °C and impurity concentrations are high (Garcia et al., 2012).

Experiments in support of nuclear superheat plant development (Griess et al., 1962; Bevilacqua and Brown, 1963) (Chapter 4) had shown that stainless steel would crack owing to the accumulation of chloride on fuel element surfaces. At the time these tests were performed, it was impossible to produce water with low-ppb concentrations of chloride ions; inevitably, the necessary conditions for cracking of austenitic stainless steels would be fulfilled since oxygen was present in the steam, and during refuelling, start-up and shutdown, the fuel element surfaces would be wetted. Similar results were reported for 1Kh18N10T used for the superheated steam (SHS) channel elements in the Beloyarsk NPP. Test specimens cracked after 144−1100 h of temperature cycling in a chloride-containing environment. It was proposed that moisture deposition on the outer surface and subsequent evaporation may have resulted in high chloride concentrations on the surface (Emel'yanov et al., 1972). As discussed in Chapter 4, at $P > P_c$, the solubility of sodium chloride is sufficiently high that the current best-practice BWR feedwater chloride concentration limit (as low as 0.25 $\mu g \cdot kg^{-1}$) (Stellwag et al., 2011) would be sufficient to prevent in-core chloride precipitation. Bull et al. (2013) reported the results of studies of SCC of 316 SS used as superheater tubes in the advanced gas-cooled reactor (~500 °C at 15 MPa). SCC had been found to be an issue when the required superheat margin was not maintained leading to SCC in dry-out regions. They found that the risk of SCC was minimized by maintaining very low impurity concentrations and a slightly alkaline ionic balance (Fig. 6.8). During normal operation the SCWR coolant will be single-phase, but during start-up and shutdown periods of two-phase operation are likely and in-core deposition of anions is possible. Careful chemistry monitoring and control will be needed during these periods to minimize the EAC risk.

A more serious concern might be chloride enrichment in fuel cladding oxides due to local changes in chemical and physical parameters (e.g., temperature, pH) within the deposit, or feedwater chloride excursions due to chemistry upsets, e.g., condenser tube failure. Staehle and Gorman (2003) have emphasized the importance of crevices and heat transfer surfaces in PWR secondary side SCC. Other impurities are also of concern; for example, lead has been implicated in many of the cracking failures in PWR and pressurized heavy-water reactor steam generator tube failures. Molecular dynamics simulations of chloride ion in SCW at $Fe(OH)_2$ surfaces (Kallikragas et al., 2015) have shown that chloride ion is preferentially found near the surface due to the higher water density there (Chapter 4). Simulations of chloride ion and water confined between surfaces with small gaps (1−8 nm) showed that chloride ion diffused faster when the gap size was smaller, attributed to the smaller hydration shell; at larger gap sizes the larger hydration shell hindered the movement of chloride ion. At subcritical temperatures the opposite behaviour was found suggesting that chloride ion could be more detrimental at supercritical conditions.

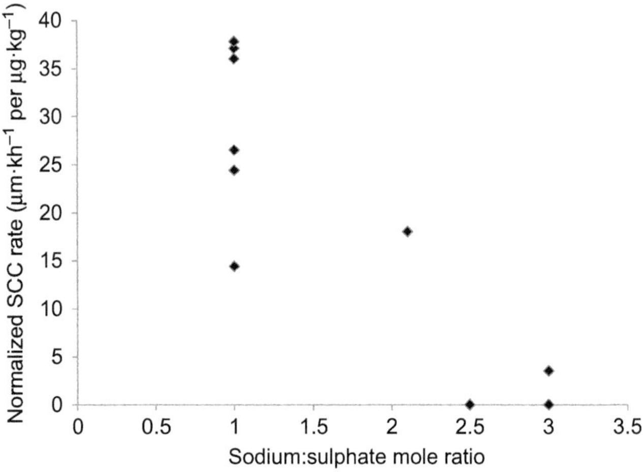

Figure 6.8 Dependence of the stress corrosion cracking rate for 316 SS on the sodium:sulphate mole ratio in superheated steam under conditions representative of the superheater region in the advanced gas-cooled reactor (~500 °C at 15 MPa).
Adapted from Bull, A., Lewis, G., Owen, J., Quirk, G., Rudge, A., Woolsey, I., 2013. Mitigating the risk of stress corrosion of austenitic stainless steels in advanced gas cooled reactor boilers. In: Proceedings of the 16th International Conference on the Properties of Water and Steam, September 1–5, 2013. University of Greenwich, London, UK.

In spite of the importance of anions as a causative factor for SCC in WCRs, anion concentrations are not typically measured in testing carried out in support of SCWR concept development and no systematic study of their effects has been reported. Tan and Allen (2009) reported a chloride concentration of 300 $\mu g \cdot kg^{-1}$ in their loop. Gong et al. (2015) reported that the chloride ion concentration in their loop dropped from 56.8 $\mu g \cdot kg^{-1}$ at 300 °C to 0.7 $\mu g \cdot kg^{-1}$ at 450 °C, and the authors suggested that this might be the result of chloride deposition as the solubility dropped through the critical point. These chloride concentrations are significantly higher than industry best practices (0.25 $\mu g \cdot kg^{-1}$) and, referring to Fig. 6.7, high enough to be of concern.

The second key water chemistry factor in EAC of LWRs is the concentration of oxidants, generally quantified by the ECP. It is now well established that the SCC susceptibility of, and crack propagation in, austenitic steels under BWR coolant conditions is determined by the corrosion potential. Macdonald (2009) outlines the electrochemical basis for SCC in subcritical aqueous environments, the primary phenomenon being the spatial separation of anode (inside the crack) and cathode (on the external alloy surface) (Fig. 6.9). The generation of H_2 by the corrosion reaction and the influx of anions (e.g., Cl^-, SO_4^{2-}) into the crack to compensate for the flow of electrons away from the crack tip through the metal to the cathode can result in an aggressive environment in the crack. The ECP is determined by the surface conditions, the coolant flow rate and the concentrations of oxidants. As O_2 and H_2O_2 give rise to

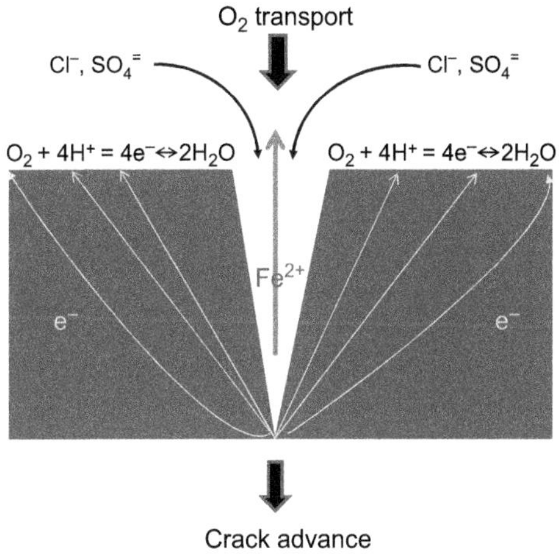

Figure 6.9 Schematic of the electrochemical couples involved in stress corrosion cracking in subcritical aqueous solutions, illustrating the roles played by oxygen and anions in solution. Adapted from Macdonald, D.D., 2009. The electrochemistry of stress corrosion cracking. In: Proceedings of the 14th International Conference on Environmental Degradation of Materials in Nuclear Power Systems Water Reactors. Virginia, USA, 734–744.

different surface oxide structures, the relationship between the concentrations of oxidants and the ECP in a WCR can be complex (Uchida et al., 2009). In addition to the use of nonsensitized materials and minimizing design and residual stresses, minimizing SCC risk in BWRs is achieved by reducing the ECP below −230 mV versus the standard hydrogen electrode. This can be achieved by reducing the concentrations of O_2 and H_2O_2 produced by water radiolysis using HWC, and noble metal addition can be used to improve the recombination rates of hydrogen with O_2 and H_2O_2 at the alloy surface. Deposition of a dielectric (barrier) layer to inhibit transport of oxidant to the metal surface has also been proposed (Zhou et al., 2004).

As with general corrosion (Section 5.3.3.3), the experimental data on the effects of DO concentration on cracking in SCW reported by different authors are not always consistent. Liu and Gong (2017) reported that the maximum crack length increased with increasing DO concentration for 310 SS (polished to a 1000 grit finish) at 620 °C, 25 MPa and a strain rate of $7.5 \times 10^{-7}\ s^{-1}$ (Fig. 6.10). Gong et al. (2015) reported that fracture time for HR3C decreased as the DO concentration increased from 200 to 500 $\mu g \cdot kg^{-1}$ but increased at a DO concentration of 1000 $\mu g \cdot kg^{-1}$. Chromium and manganese depletion was observed at the crack tip. They suggested that at low DO concentrations oxygen enhanced SCC but acted to passivate the surfaces at higher concentrations (crack blunting). They noted an enrichment of C in the oxide, attributed to the high concentration of total organic carbon in the test solution and found evidence for sensitization of the alloy after exposure at the test temperature. Zhang et al. (2015) reported that at 450 °C, increasing the DO concentration from 0 to 2 $mg \cdot kg^{-1}$ increased the SCC susceptibility, while at 550 °C increasing the DO concentration to 2 $mg \cdot kg^{-1}$ decreased the SCC sensitivity. Chen et al. (2017) found that DO

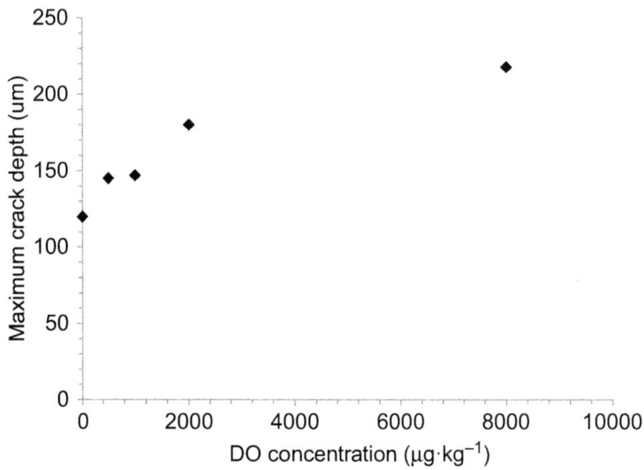

Figure 6.10 Maximum crack length as a function of dissolved oxygen (DO) concentration for 310 SS at 620 °C, 25 MPa and a strain rate of $7.5 \times 10^{-7}\ s^{-1}$ (Liu and Gong, 2017).

concentration had a significant effect on CGRs in SCW; 310S SS exhibited the highest CGRs in water deaerated with Ar and the lowest with 2000 $\mu g \cdot kg^{-1}$ DO content. The lower DO concentration could result in higher SCC susceptibility because of slower oxidation at the crack tip. Fournier et al. (2001) studied the cracking of Alloy 718 and Alloy 690 in aerated SCW at 400 °C, 25 MPa using CERT testing (strain rate $10^{-6}\ s^{-1}$). Alloy 718 was found to be sensitive to SCC while Alloy 690 was not. The authors discussed the role of oxygen embrittlement in the cracking of Ni-base alloys in SCW, suggesting a change of cracking mechanism above T_c.

Woolsey (1989) discussed cracking of the austenitic stainless steels used in the superheaters of SCFPPs in Germany and the former Soviet Union. These plants operate with oxygenated feedwater at oxygen concentrations high enough that oxygen reaches the boiler. No SCC of austenitic steels was reported from German once-through boilers for operating periods on the order of 10^5 h. These plants used Nb-stabilized steels (DIN 8CrNiNb163, X8CrNiMoNb1616 or X8CrNiMoVNb1613). In the former Soviet Union (using Ti-stabilized 12Cr18Ni12Ti), some examples of SCC have been reported for plants using oxygenated chemistry, although other plants operated for longer periods with no reported issues. Woolsey suggested that abnormal operating conditions (e.g., off-load operation) may result in 'wetting' of the surfaces (two-phase conditions) or that feedwater impurities may be the cause (these plants tended to operate with higher concentrations of feedwater impurities than that of German plants). Gruzdev et al. (1970) suggested limiting feedwater O_2 concentrations to $<150\ \mu g \cdot kg^{-1}$ to prevent oxygen from reaching the boilers. Woolsey (1989) suggested operating with a deoxygenated chemistry during start-up and low-load conditions. In SCFPPs, cracking of turbine blades has been a significant issue due to deposition of impurities transported by the steam.

For BWRs, water chemistry operating experience and laboratory data form the basis of SCC mitigation strategies such as HWC and noble metal addition[4] (Cowan et al., 2011, 2012). The EPRI BWR Water Chemistry Guidelines are based on these observations and theoretical predictions of the effects of corrosion potential and coolant purity on EAC susceptibility (EPRI, 2004; Fruzetti and Blok, 2005). Ehrlich et al. (2004) evaluated possible degradation mechanisms of candidate materials for the European SCWR—HPLWR concept and concluded that strict limitations on the concentration of these species through appropriate water chemistry control can substantially reduce the risk (e.g., Fig. 6.7), although this conclusion is only likely to be true at temperatures below ~450 °C where dissociation is favoured and anodic dissolution is the major driving force. It must be emphasized that regardless of the limitations of Fig. 6.7, minimizing impurity concentrations in the SCWR feedwater is still highly recommended.

Ru and Staehle (2013c) compared work reported by Was and Teysseyre (2005) with earlier studies of the same alloys in LWR environments (Coriou et al., 1969) and found some results difficult to explain based on LWR experience. For example, 304L SS and Alloy 625 exhibited IGSCC with crack depths exceeding 300 μm and increasing susceptibility to SCC as the temperature increased from 400 to 550 °C in deaerated SCW in contradiction to results obtained in pure subcritical water at 350 °C. Tests of unsensitized austenitic stainless steels in SCW using chemistry conditions (low DO and low anion concentrations) that should not give rise to SCC nonetheless were found to crack. However, as discussed in Chapter 5, electrochemical processes that involve spatially separated anodic and cathodic sites can no longer occur at $T \gg T_c$ and localized chemical reactions dominate. This suggests that the environmental factors affecting EAC in SCW change at $T \gg T_c$, in particular the role of oxygen and anions. The work of Scenini et al. (2008) on the effects of electropolishing versus mechanical polishing on SCC of Alloy 600 °C-ring and reverse U-bend specimens in hydrogenated steam at 400 °C and 20.7 MPa is of relevance here. Hydrogenated steam was used as a means of accelerating cracking to study primary water SCC. They reported that the electropolished samples showed much greater susceptibility to SCC than did the mechanically polished samples, attributed to the ability of the mechanically polished samples to form an external oxide at all H_2 partial pressures due to the presence of short-circuit diffusion pathways (Chapter 5). On strain-free surfaces at potentials (H_2 partial pressures) below the Ni/NiO equilibrium, no external oxide was formed. The authors proposed the formation of a protective external oxide reduces the transport of O_2 and H_2 to the surface. The observation that surface CW can be beneficial in reducing SCC in hydrogenated SCW is contrary to data obtained under LWR conditions but consistent with the change in corrosion mechanism discussed in Chapter 5; this topic will be revisited in the next section.

[4] When HWC is used in BWRs, the complex in-core geometry leads to nonuniform distribution of H_2 in the core, and inadequate H_2 concentrations in some regions has necessitated the introduction of methods such as noble metal addition to improve effectiveness.

6.2.2 Material factors

The material properties relevant to EAC in high-temperature water include the following (Andresen and Ford, 2011):

- Alloy composition;
- Alloy microstructure, especially the extent of grain boundary sensitization;
- Yield strength;
- Extent of surface damage/cold work.

Historically, the extent of chromium depletion at the grain boundary due to growth of $M_{23}C_6$ carbides has been most associated with IGSCC cases in austenitic stainless steel components in BWRs (Bruemmer et al., 1993). Typically, austenitic stainless steels with more than 0.03% carbon are considered 'sensitized' if they are heated to, or slowly cooled through, the temperature range 520–820 °C; as discussed in Chapter 3, this causes chromium carbides to form at the grain boundaries. As a result, the regions adjacent to the precipitated chromium carbides will be depleted of chromium to less than 12% (Fig. 6.11), resulting in a loss of corrosion resistance in those areas. The most common cause of sensitization in BWRs is welding, where sensitization is contained within the weld heat-affected zone. It is significant in the context of the SCWR that the lower limit of this range is within the temperature range expected in an SCWR core, indicating that sensitization during operation is possible. Indeed, Cao et al. (2011) found that 304 SS with high C content was more susceptible to SCC than the same alloy with low C content. Grain boundary Cr depletion can be prevented by controlling the carbon content to below 0.03% by cooling quickly through the sensitizing temperature range, by solution annealing sensitized material and rapidly

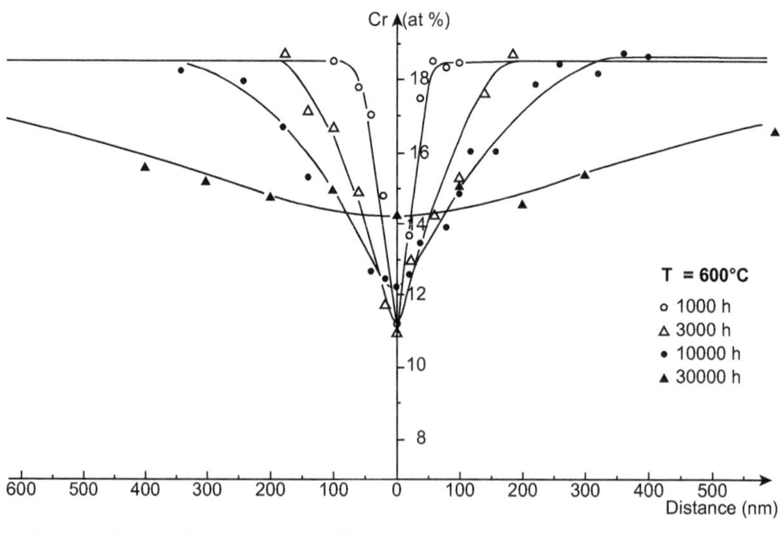

Figure 6.11 Evolution of chromium profiles near grain boundaries during ageing at 550 °C (Sahlaoui et al., 2004).

cooling or by adding strong carbide formers such as titanium or niobium to the steel. In this latter approach, the carbon in the steel preferentially reacts with the titanium or niobium to form titanium or niobium carbides instead of reacting with the chromium. Stainless steels with additions of titanium or niobium are called 'stabilized'. However, IGSCC can still occur in stabilized stainless steels if the carbon content is too high.

Grain boundary sensitization can lead to intergranular crack initiation and propagation, the crack propagation rate depending on other material attributes (yield strength, etc.) as well as water chemistry and loading. Cracking is sensitive to small degrees of grain boundary sensitization and, in particular, the chromium content (Fig. 6.12). This understanding has allowed the nuclear industry to define well-proven mitigation actions based on minimizing the extent of grain boundary sensitization (Andresen and Ford, 2011).

There have been fewer studies of EAC in Ni-base alloys in support of SCWR development, and the results are somewhat inconsistent. Several groups studied the SCC behaviour of Alloy 625 for use in SCWO processes (Kim et al., 2011; Fujisawa et al., 2005; Bosch and Delafosse, 2005) under chemistry conditions considerably more aggressive than those expected in an SCWR. Bosch and Delafosse (2005) performed SCC tests of Alloy 625 using slow strain rate and constant loading methods in SCW at temperatures of 400 and 500 °C. They found that the alloy was sensitive to SCC in SSRT in SCW containing 10 wt% H_2O_2, whereas constant load tests did not show any significant amount of crack propagation. Teysseyre and Was (2006) conducted slow strain rate tests of Alloy 625 in deaerated SCW over the temperature range

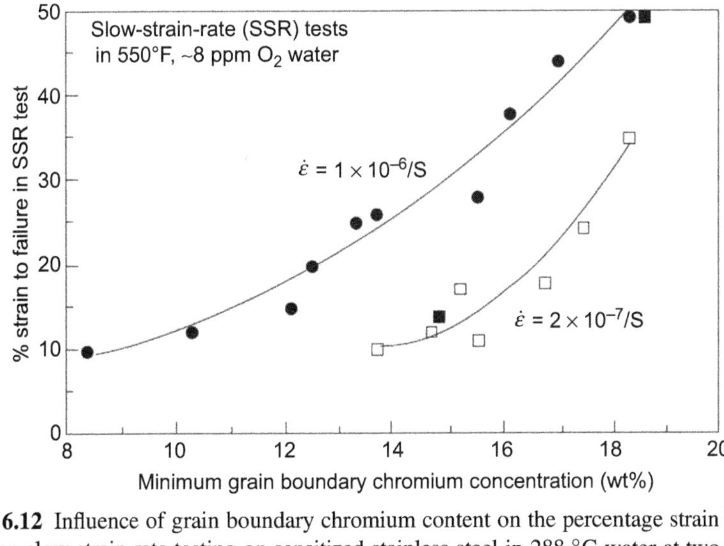

Figure 6.12 Influence of grain boundary chromium content on the percentage strain to failure during an slow strain rate testing on sensitized stainless steel in 288 °C water at two applied strain rates. Note that less than a 1% drop in Cr content adjacent to the grain boundary from the 18% in the bulk material can have a significant effect on intergranular stress corrosion cracking susceptibility (Bruemmer et al., 1993).

of 400–550 °C. Intergranular cracking was observed at all temperatures, the average crack length increasing with increasing temperature, and the cracking was more extensive compared to that found in austenitic alloys under similar conditions. Was et al., (2007) reported SCC growth rates between 3×10^{-9} and 3×10^{-8} $mm \cdot s^{-1}$ for this temperature range based on measurements of the crack depth and the total test time. The activation energy for cracking was reported to be similar to that of 316L.

As discussed in Chapter 3, a fundamental challenge for the SCWR is the slow microstructural evolution due to thermal ageing (diffusion and secondary phase precipitation, particularly involving grain boundaries), leading to long-term increases in SCC susceptibility (Zheng et al., 2011, Jiao et al., 2013). Ageing will be an issue for both structural materials, where the in-service life will be tens of years, and for the fuel cladding, where the in-service life is relatively short but the peak fuel cladding temperature will be high (up to 800 °C for the Canadian SCWR concept). Ru and Staehle (2013a,b,c) predicted that the dynamic thermal range will result in significant ageing of materials (see Chapter 3, Section 3.5), changing grain boundary chemistry and structure and thus changing SCC susceptibility. Until recently, tests have used as-fabricated materials, and the influence of intermetallic phases requiring longer ageing times for their formation (χ, η and σ phases) has received little attention. These phases are of particular concern for stainless steels that contain Nb and Ti to intentionally precipitate carbides (Jiao et al., 2013). With the exception of the work of Udy and Boulger (1953), who carried out SCC tests at 732 °C using static pressure-capsules, the synergistic effect of long-term alloy microstructure evolution at higher temperature on SCC is largely unknown.

Several strategies to mitigate ageing and thus the degree of IGSCC susceptibility have been proposed. Parvathavarthini et al. (2009) proposed using grain boundary engineering to produce a lower effective grain boundary energy[5] (EGBE), which lowers the tendency to sensitisation. Parvathavarthini and Dayal (2002) proposed that the limit of carbon content at which a steel is not susceptible to sensitization is closely connected to the presence of other alloying elements such as chromium, molybdenum, nickel, nitrogen, manganese, boron, silicon as well as titanium and niobium in stabilized steels. In particular, chromium has a pronounced effect on the passivation characteristics of stainless steels. With higher chromium content, the time to reach the resistance limit of chromium depletion at the grain boundaries is shifted to longer time. Thus, alloys with a higher chromium content are more resistant to sensitization.

Deformation type during production (e.g., cross rolling of a plate vs. cold drawing of a tube) results in different microstructures. Nezakat et al. (2014) report that rolling mode had a direct effect on the deformation texture of the remaining austenite. Unidirectional rolling resulted in Brass, Goss and γ-fibre textures, whereas cross rolling formed mainly Brass texture in deformed austenite after 90% reduction in thickness.

The effect of grain size on the intergranular corrosion susceptibility of 316L stainless steel was investigated by Li et al. (2013). Both DL-EPR tests and microstructure inspections of 316L with different grain sizes showed that the susceptibility to

[5] Effective grain boundary energy is a combined factor of several aspects of the nature of the grain boundary.

intergranular corrosion decreased with increasing grain size. Li et al. (2013) suggested that increasing grain size to an optimum level could be an effective way to increase the intergranular corrosion resistance. However, since finer grain size improves corrosion resistance (Chapter 5), it could be a challenge to improve EAC resistance without losing the good mechanical properties and general corrosion resistance resulting from small grain size.

Terada et al. (2008) reported DL-EPR test results of 316L(N) steel creep tested at 600 °C for 7500 and 85,000 h. The DL-EPR tests showed clear sensitization to intergranular corrosion. However, they did not observe $M_{23}C_6$ carbides at grain boundaries, and attributed the sensitization to intergranular corrosion due to intermetallic (mainly σ) phases.

Surface finish and degree of cold working are known to affect the SCC susceptibility of austenitic stainless steels in LWR environments (e.g., core shrouds in Japanese BWRs). Recent studies have shown that cold or warm working increases SCC growth rates of low carbon stainless steels in simulated BWR environments (Shoji et al., 2008). The interactions between various forms of cold work and the crack propagation rate are illustrated in Fig. 6.13 (Andresen and Ford, 2011). As noted by Garud and Ilevbare (2009), the fact that CW affects several of the causative factors for SCC (stress and material) can lead to apparently contradictory conclusions regarding the effects of CW.

The role of cold working in SCC of austenitic stainless steels in PWR and BWR environments has been the subject of much investigation (Andresen, 2002; Garud and Ilevbare, 2009; Hou et al., 2011; Devrient et al., 2011). Under typical PWR conditions, unsensitized austenitic steels can suffer IGSCC and/or TGSCC when the alloys have been subject to a certain amount of permanent plastic deformation, which increases the hardness and yield strength of the microstructure. Strain localization

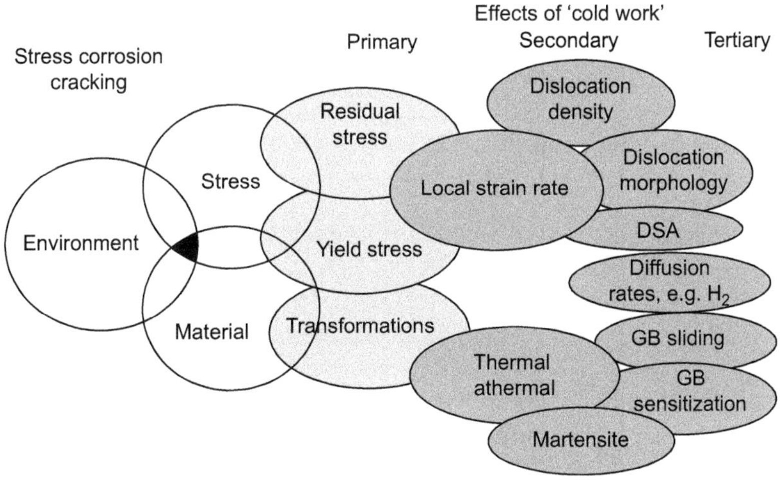

Figure 6.13 Interactions between the parameters associated with 'cold work' and their effect on the conjoint 'material' and 'stress' conditions for stress corrosion crack propagation (Andresen and Ford, 2011).

along grain boundaries due to cold working is the primary reason for crack initiation in these otherwise SCC-immune microstructures. This role of CW is somewhat similar to that of neutron irradiation in that the latter promotes deformation channelling as deformation is concentrated, in the form of deformation bands, along the planes free of obstacles for dislocation movements (Lee et al., 2001). Lozano-Perez et al. (2009) observed deeper oxidation of the deformation bands in stressed samples of cold-worked Alloy 304 than in bands intersecting stress-free surfaces (i.e., unstressed samples), suggesting stress-assisted oxidation of the deformation substructures. The enhancement of oxidation at highly stressed regions around the crack tip was thought to be due to the presence of fast diffusion paths in the highly misoriented and complex plasticity cells (subboundaries) that resulted from prior cold working of the test alloy.

The same effects of CW on SCC of austenitic steels and Ni-based alloys in subcritical water may not be seen under SCW conditions, since at temperatures of 400 °C and higher, thermally activated diffusion will facilitate the microstructural recovery process. At temperatures of $^1/_3\ T_m$ to $^1/_2\ T_m$ recrystallization begins to occur leading to formation of small equiaxed 'soft' grains with low dislocation densities. Indeed, in SCW at 650 °C surfaces with a heavy machining-related deformation layer have shown significant recrystallization with beneficial effects on corrosion resistance (Chapter 5). The possibility of a 'deformation-induced SCC-prone microstructure' in austenitic steels must be considered; such a susceptible microstructure could be produced during the dynamic straining (i.e., tensile cold working) of an SSRT specimen even at SCW temperatures. In this case an ordinarily SCC-immune microstructure could be converted to an 'SCC prone' microstructure by means of excessive plastic straining; this is the situation in a conventional SSRT where a test sample is usually pulled to failure and the presence or absence of SCC on the sample fracture surface or in the gauge length determined during postmortem examination. If the field application does not subject the alloy to such extreme plasticity, this type of laboratory test has no relevance. Unfortunately, most of the reported SSRT or CERT results are based on conditions of heavy, testing-induced, plasticity.

There have been few studies of the effects of these metallurgical factors on SCC in SCW. DL-EPR test results (Oh and Hong, 2000) indicate that CW levels up to 20% enhanced sensitization while 40% CW suppressed sensitization at all ageing times. Also, an increase in the nitrogen content at a similar carbon content accelerated the sensitization at CW levels below 20%.

Novotny et al. (2011) reported the results of SSRT testing of 316L stainless steel in pure 550 °C, 25 MPa SCW with controlled oxygen content. The SSRT specimens were cold-worked unintentionally as a result of the manufacturing process. Strain rate (elongation rate) and oxygen content were varied. The results did not show any significant increase in SCC susceptibility, as characterized by IGSCC crack growth. Fractographic analysis confirmed that failure was due to a combination of TGSCC and ductile fracture. Although CW enhanced SCC susceptibility to a certain extent, faster oxidation in SCW caused crack blunting at later stage of the tests. Sáez-Maderuelo and Gómez-Briceño (2016) measured the SCC susceptibility of annealed and cold-worked 316L SS at 400 and 500 °C in SCW (25 MPa, $< 10\ \mu g \cdot kg^{-1}$ DO). They found that the presence of plastic deformation increased the SCC susceptibility of this alloy. A somewhat contradictory result was reported by Scenini et al. (2008), who found that electropolished samples of Alloy 600 tested in hydrogenated SHS showed much greater susceptibility to SCC than mechanically polished samples, attributed to the ability of the mechanically polished samples to form an external oxide due to the presence of short-circuit diffusion pathways (Chapter 5). On strain-free surfaces at potentials (H_2 partial pressures) below the Ni/NiO equilibrium, no external oxide was formed. Additional studies of the effects of CW are needed to resolve this and other issues. For austenitic stainless steels in SCW, it appears that the positive effect of CW (enhanced corrosion resistance (Section 5.3.2) and higher yield strength) may be counteracted by increased SCC susceptibility (and certainly by lower ductility).

6.2.3 Mechanical factors

Mechanical factors such as stress and strain, residual stress and strain, and strain rate are crucial parameters for EAC and interact strongly with the specific alloy/environment combinations. To date, materials research on EAC in support of SCWR development has focused on assessing susceptibility to SCC of various materials using SSRT and CERT testing, in other words, the propensity for crack initiation in a material and its ability to propagate such cracks. A deeper understanding how various factors affect SCC in SCW does not yet exist nor does a quantitative evaluation of these factors. For instance, quantification of the effect of stress intensity factor K_i was extensively studied for WCR austenitic stainless steel core internals to understand the rate-controlling processes in CGR disposition equations. Generally, for WCRs, the dependence of CGR on K is expressed by an empirical power law equation:

$$\mathrm{CGR}(K_i) = C0 \cdot K_i^n \tag{6.1}$$

Fig. 6.14 (Shoji, 2011) shows an example of experimental CGR versus K_i for non-sensitized austenitic stainless steels in simulated BWR oxygenated water environments.

To date few such studies have been conducted in SCW. Janik (2015) constructed a phenomenological map of SCC susceptibility indication in SCW with respect to DO content and applied strain rate for 316L using available data (Fig. 6.15). He used percentage of IGSCC/TGSCC measured on the fracture surface of a tensile specimen after fracture and ignored eventual secondary cracking on the specimen gauge. Although the results were somewhat inconclusive in so far as the data did not fall neatly into ‘SCC’ and ‘no SCC’ regions, it does appear that at DO concentrations higher than 2000 $\mu g \cdot kg^{-1}$ there is little or no SCC, as noted by some authors (Section 6.2.1.2).

However, in an SSRT test, the crack tip strain rate is known and, when the crack propagation rate is low for the applied strain rate, the localization of the failure may have little to do with the SCC cracks. Therefore the amount of SCC observed on the fracture surface may not be representative of the specimen susceptibility to SCC. Such an approach tends to focus on the propagation of the cracks. Others focus more on the susceptibility to crack initiation, i.e., on the extent of cracking observed on

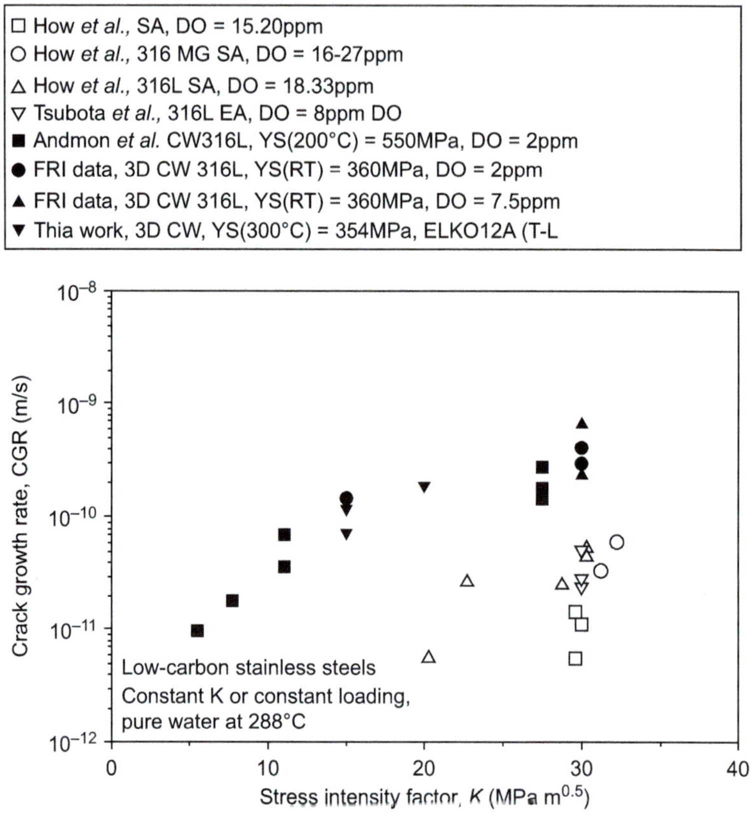

Figure 6.14 Crack growth rate versus stress intensity factor for nonsensitized stainless steels in oxygenated pure water at 288 °C (Shoji, 2011).

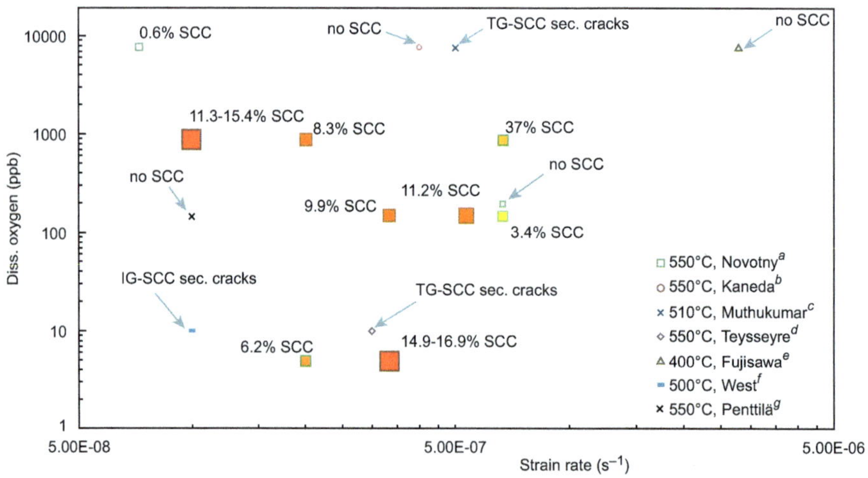

Figure 6.15 Phenomenological map of stress corrosion cracking (SCC) occurrence observed on 316L after slow strain rate testing conducted in supercritical water. The size of the symbol is proportional to the extent of SCC.
After Janik, P., 2015. Acoustic Emission – A Method for Monitoring of Environmentally Assisted Cracking Processes (Doctor thesis). Institute of Chemical Technology Prague, Faculty of Environmental Technology, Department of Power Engineering.

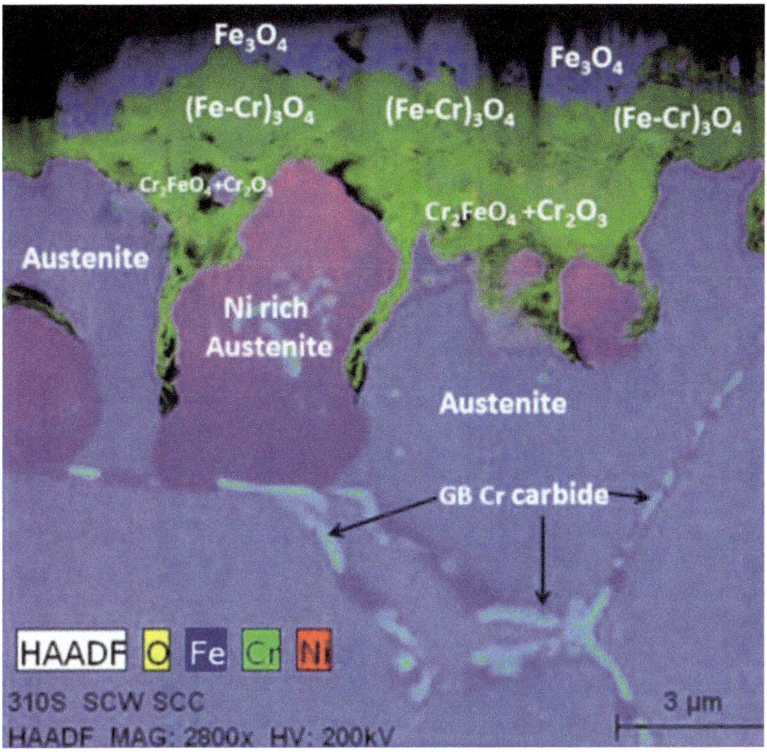

Figure 6.16 TEM-EDX elemental mapping analysis of inner cross section of 310S capsule specimen after 19,000 h exposure to supercritical water at 500 °C, 25 MPa (Zheng et al., 2014).

the gauge surface, using the crack density, average crack length/depth and crack length per unit area as parameters.

Statically–loaded capsule specimens made of the candidate alloys and containing SCW with different initial DO concentrations were tested at 500 °C, 25 MPa for up to 19,000 h with different hoop stresses[6] (Behnamian et al., 2014; Zheng et al., 2014). Fig. 6.16 shows TEM-EDX elemental mapping analysis of oxygen, iron, chromium and nickel of the internal surface of a 310S capsule specimen after 19,000 h exposure. Selective internal oxidation, specifically on grain boundaries, is associated with the intergranular penetration. Some tests were conducted at higher hoop stress, and it was found that the higher the ratio of hoop stress to yield stress the higher the value of maximum and mean crack length (Fig. 6.17). Increasing the hoop stress of the capsule samples from ~62 MPa to approximately the yield strength increased the maximum attack depth in 316L by 39% at 500 °C. For the 310S, the maximum attack depth under the same conditions increased by 35%. A 30% increase in the maximum penetration depth was observed for Alloy 625.

[6] In these pressurized capsules, the loading is biaxial: there are both hoop stress and axial stress with the axial stress at 50% of the hoop stress.

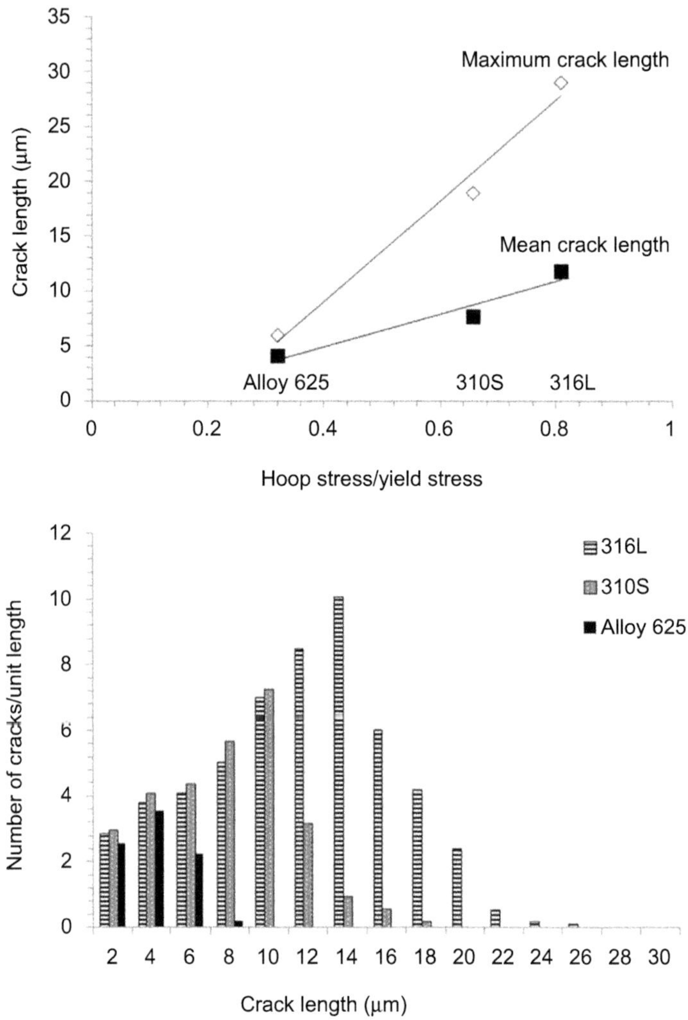

Figure 6.17 As-received, 5000 h, 25 MPa, 500 °C, deionized water, capsule test 8 ppm DO at the start of the test, < 5 ppb at the end.
Adapted from Behnamian, Y., Dong, Z., Kohandehghan, A., Zahiri, R., Mitlin, D., Chen, W., Luo, J.L., Zheng, W., Guzonas D., 2014. Oxidation and stress corrosion cracking of austenitic alloys in neutral pH extreme hydrothermal environments (supercritical water). In: NACE Northern Area Western Conference, Edmonton, Alberta, January 27–30, 2014.

An interesting observation from the long-term (19,000 h) test is that the intergranular attack along the grain boundaries is rather blunt at the very front region, suggesting some kind of dormancy at the time of test termination. Formation of Cr-rich precipitates well ahead of the frontier of the corrosion attack is evident. Given the lack of dynamic straining and the thermal recovery of the microstructure at 500 °C, the corrosion attack, which initially initiated in an intergranular manner, could not continue to grow, and after 19,000 h all intergranular penetrations were less than 15 μm in depth. The IG precipitates could also have improved the resistance to grain boundary slip and sliding in the same manner seen in sensitized Alloy 800H (Janssen et al., 2009).

Teysseyre and Was et al. (109) assessed the SCC susceptibilities of annealed 304 and 316L in pure deaerated SCW using the results of the crack density, average crack length collected at the gauge surface and maximum crack depth from cross sections. They noted that crack density decreased with temperature while the average crack length increased. The measured maximum crack depth significantly increased with increasing temperature. Penttilä and Toivonen (2013) performed SSRT tests on 316NG, 316Ti, 347H and 1.4970 (gauge sections polished to 600 grit) in SCW (550 °C, 25 MPa, 100–150 $\mu g \cdot kg^{-1}$ DO, inlet conductivity < 0.1 $\mu S \cdot cm^{-1}$); indications of transgranular SCC were seen on the fracture surfaces and side surfaces of all materials except 316L, which showed only ductile dimple cracking and no side cracks.

As noted earlier, to date few CGR experiments in SCW have been published. Peng et al. (2007) performed CGR tests in SCW (25 MPa, 2000 $\mu g \cdot kg^{-1}$ DO) with 20% CW 316L at a K_i of 27.6 MPa·m at temperatures between 288 and 500 °C and concluded that CGRs increased with increasing T at $T < T_c$ and decreased with increasing temperature above 400 °C (Fig. 6.5). At 400 °C, decreasing the oxygen concentration in SCW resulted in increased CGR compared to the deaerated case. The apparent contradiction between the temperature dependence of SCC measured by CGR tests versus SSRT tests might be explained by crack blunting by rapid oxidation during CGR testing in contrast to film rupture by the higher strain rate of SSRT tests (Peng et al., 2007).

Chen et al. (2017) measured CGRs of 310S side-grooved compact tension specimen at 500 and 550 °C SCW (25 MPa, Ar, 2000 $\mu g \cdot kg^{-1}$ DO or 1.6 ppm H_2). CGRs increased with increasing temperature, the higher activation energy than in subcritical water attributed to the additional temperature-dependent role of material sensitization and creep. Oxygen had a significant effect on SCC CGRs, 310S SS exhibiting the highest CGRs in water deaerated with Ar and the lowest in SCW with 2000 $\mu g \cdot kg^{-1}$ DO. Based on these results, the authors proposed a revised slip-oxidation–repassivation mechanism to explain the SCC behaviour of 310S SS in SCW.

Arioka et al. (2011) studied the SCC and creep behaviour of cold-worked Alloy 690, 304 SS and 316L SS in a simulated PWR environment as well as in air at temperatures >450 °C. The authors proposed a creep-oxidation mechanism based on the observation of cavities at grain boundaries just ahead of the IG crack tips and crack walls for all test materials. The cavity formation resulted from the diffusion of cold work–induced vacancies driven by the stress gradient at the crack tip.

6.2.3.1 Irradiation factors

In addition to the microstructural defects (point, and linear, i.e., dislocations, and voids discussed in Chapter 3), irradiation also causes changes in material chemistry at the microscopic level such as radiation-induced segregation (RIS) and precipitation, both of which affect SCC susceptibility. The extent of this damage depends on the irradiation dose, flux, irradiation spectrum, as well as the temperature at which the metal is irradiated due to the thermal nature of the diffusional processes involved. Irradiation-assisted SCC of austenitic alloys has been a generic problem in all types of LWR. Major concerns developed in the mid-1970s due to an increasing number of observations of IASCC in highly stressed core components exposed to fast neutron (>1 MeV) irradiation when a 'threshold' fluence level (5×10^{20} $n \cdot cm^{-2}$) was exceeded (Fig. 6.18; Gordon and Brown, 1989).

IASCC was also observed at fluences lower than 5×10^{-20} $n \cdot cm^{-2}$ adjacent to welds in larger stainless steel core structures, the extent of cracking depending on specific combinations of prior thermal sensitization due to the fabrication procedure, fast neutron flux and fluence, weld residual stress and coolant purity (Andresen et al., 1989; Nelson and Andresen, 1991; Andresen and Ford, 1995; Bruemmer et al., 1996; Was et al., 2006).

Irradiation effects include changes to the corrosion potential (due to water radiolysis, Chapter 4), applied tensile stress, grain boundary composition and yield strength. Those changes caused by neutron flux (i.e., corrosion potential) have an immediate effect on cracking susceptibility, whereas fluence-driven changes (RIS, radiation hardening, stress relaxation) have a cumulative effect over an extended time. The specific effects of irradiation on the parameters that control crack propagation in

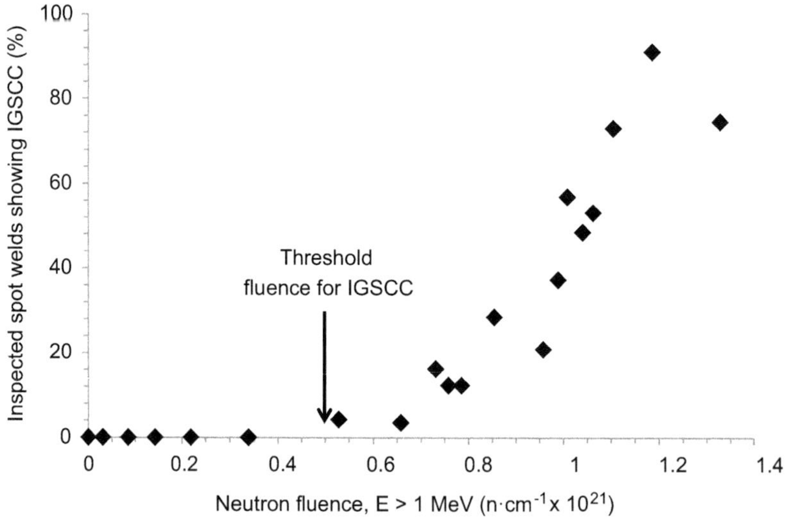

Figure 6.18 Dependence of IASCC on fast neutron fluence for creviced control blade sheath in high-conductivity boiling water reactors (BWRs) where the tensile stress is high due to spot welds and dynamic due to swelling B4C absorber tubes.

Adapted from Gordon, G.M., Brown, K.S., 1989. Dependence of creviced BWR component IGSCC behavior on coolant chemistry. In: Cubicciotti, D., Simonen, E. (Eds.), Proc. 4th Int. Symp. on Environmental Degradation of Materials in Nuclear Power Systems — Water Reactors, Jekyll Island, August 6—10, 1989. National Association of Corrosion Engineers, pp. 14.46—14.62.

stainless steels must consider the following (Andresen et al., 1989; Nelson and Andresen, 1991; Andresen and Ford, 1995; Bruemmer et al., 1996; Was et al., 2006):

- the corrosion potential and its change with radiation flux;
- irradiation-induced changes in grain boundary composition;
- irradiation-induced hardening; and
- for displacement loaded structures, radiation creep stress relaxation.

In contrast to the abundance of studies of IASCC in subcritical water under current WCR operating conditions, there have been few such studies under SCWR conditions. The higher temperature is expected to give rise to significant differences in microstructural response of alloys to mechanical loading. Most studies on irradiation effects on IGSCC in support of SCWR concept development have been performed using heavy ion irradiation. Proton irradiation of type 316L SS and Alloy 690 showed a significant increase in IGSCC compared to the unirradiated cases (Teysseyre et al., 2007). At both 400 and 500 °C, the maximum crack depth increased with irradiation for both alloys. At 400 °C, the fractional increase in crack depth due to irradiation was greater for Alloy 690 and at 500 °C, the fractional increase in crack depth was greater for 316L SS (Fig. 6.19).

West and Was (2013) studied the role of strain incompatibilities in IASCC. Tensile specimens were proton irradiated to 7 dpa at 400 °C then strained to 5% at 400 °C in SCW. Intergranular cracks occurred primarily at grain boundary sites that were most susceptible to slip discontinuity, suggesting that strain incompatibility promoted cracking.

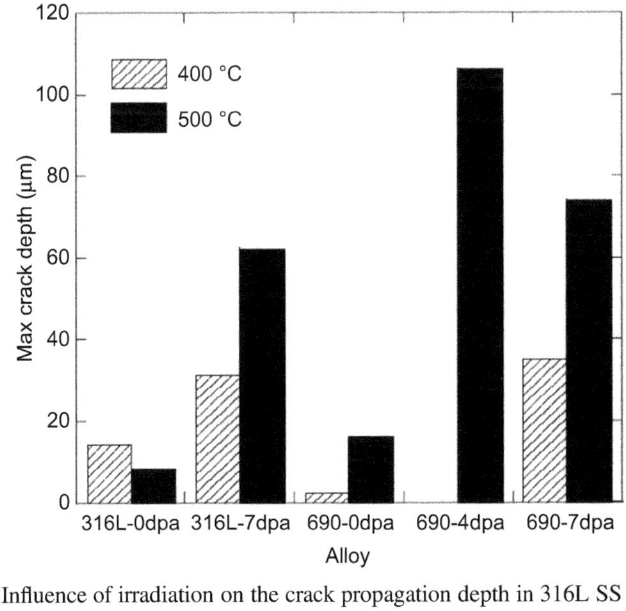

Figure 6.19 Influence of irradiation on the crack propagation depth in 316L SS and Alloy 690 in 400 and 500 °C supercritical water (Teysseyre et al., 2007).

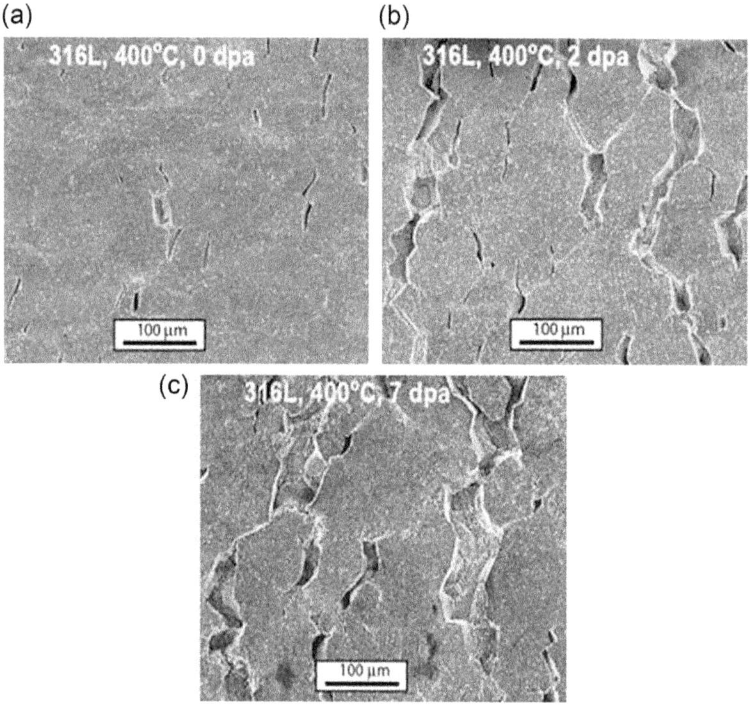

Figure 6.20 Micrographs of gauge surfaces of 316L specimens strained to failure in 400 °C supercritical water in the (a) unirradiated, (b) 2 dpa, and (c) 7 dpa conditions (Zhou et al., 2009).

Zhou et al. (2009) observed that proton irradiation of test specimens lead to SCC of austenitic alloys in SCW, the severity of cracking increasing with dose and temperature. For 316L SS, cracking increased with hardness irrespective of irradiation and test temperature. The typical appearance of secondary cracks observed on 316L SS after different levels of irradiation dose is shown in Fig. 6.20.

Teysseyre et al. (2007) reported that the density of cracks and crack depths for 316L SS and Alloy 690 preirradiated with protons up to 7 dpa at 400 and 500 °C then tested (CERT) in SCW (25 MPa, DO $< 10\ \mu g \cdot kg^{-1} < 0.1\ \mu S \cdot cm^{-1}$) at the same temperature as that of the irradiation increased over the unirradiated materials. The extent of cracking was higher at 500 °C than at 400 °C. Irradiation resulted in depletion of Cr and Fe at the grain boundaries, but enrichment of Ni for both alloys; after irradiation at 500 °C, the depleted zone was ±3.4 nm and the minimum Cr concentration was 13.8 wt% (19 wt% for the unirradiated material). Under the same irradiation and test conditions, F/M alloys were found to be resistant to cracking.

During the US nuclear steam superheat program a total of 496 fuel elements were irradiated in a superheated steam environment over about a 2-year period with only four fuel-element failures. Tests showed that fuel assemblies with 0.406 mm thick fuel cladding exposed to superheated steam at temperatures up to 738 °C for 10,292 h exhibited no SCC (Comprelli et al., 1969). One bundle failed after 6188 h on power at 493 °C, but postirradiation examination showed that the cause was likely low-cycle fatigue (Rabin et al., 1967). The stainless steels Kh18Ni10T[7] and EI-847[8] were used at the Beloyarsk NPP (Emel'yanov et al., 1972), which was operated for many years at a temperature of 550 °C and a pressure of 8 MPa.

6.3 Mechanisms and modelling

As Fig. 6.1 illustrates, SCC is the result of the combined and synergistic effects of materials, environment and mechanics. Various SCC mechanisms and controlling parameters have been proposed, such as slip-oxidation, oxidation penetration (or internal oxidation), stress-straining enhanced solid-state oxidation, hydrogen-related mechanism and dealloying (Newman, 2002; Andresen and Ford, 1988; Shoji, 2003; Scott and Combrade, 2003; Das et al., 2008). A general schematic of the oxidation/mechanics interactions for SCC in high temperature sub-critical water is shown in Fig. 6.21 (Shoji et al., 2010).

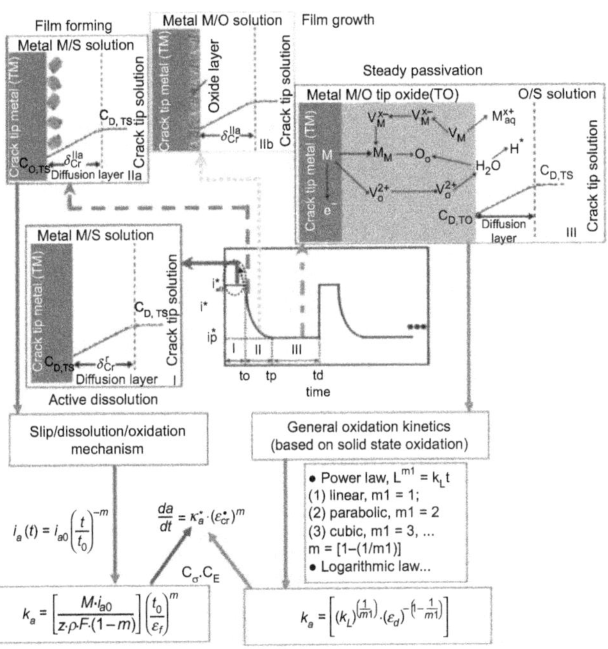

Figure 6.21 A schematic of the stress corrosion cracking subprocesses of austenitic alloys in high-temperature subcritical water (Shoji et al., 2010).

[7] Composition: C max 0.1; Si max 0.8; Mn 1–2; Ni 10–11; S max 0.02; P max 0.035; Cr 17–19; 5(C − 0. 02) < Ti < 0.6.

[8] Composition: C(0.04–0.06); Mn(0.4–0.8); Si ≤ 0.4; S ≤ 0.010; P ≤ .015; Cr(15.0–16.0); Ni(15.0–16.0); Mo(2.7–3.2); Nb≤0.9; N ≤ 0.025; B ≤ 0.001; Co ≤ 0.02; Cu ≤ 0.05; Bi ≤ 0.01; Pb ≤ 0.001; Ti ≤ 0.05.

Depending on combinations of material/environment/loading conditions, the transient oxidation can follow different kinetic rate laws, and enhancement of crack tip oxidation can be realized via either physical degradation, physical–chemical degradation or both (Shoji, 2011).

The generally accepted SCC growth mechanism in subcritical water is the slip-oxidation model, which is an electrochemical model based on separated anodic and cathodic reaction sites (Macdonald, 2009; Fig. 6.9). The crack propagates as the result of rupture of the passive oxide due to stress or strain leading to electrochemical oxidation and dissolution at the crack tip if the oxidation rate is faster than the rate at which the oxide can heal. The ionic corrosion products within the crack act as charge carriers for the anodic and cathodic processes. At the low SCW density at the higher temperatures, the extent of ion dissociation is low (cf. Fig. 4.12) and the solubility of metal oxides is low (Fig. 4.10), reducing the aggressiveness of the solution, lowering the solution conductivity and reducing metal dissolution. As discussed in Chapter 5, at $T >> T_c$ electrochemical processes that involve spatially separated anodic and cathodic sites can no longer occur and localized chemical reactions dominate. All of the processes depicted in Fig. 6.21 are affected by these changes in water properties. This suggests that the environmental factors affecting EAC in SCW change at $T > T_c$, in particular the role of oxygen and anions. At $T \gg T_c$, solid-state diffusion, creep and chemical oxidation appear to be more important factors in crack propagation.

A systematic study on SCC mechanism of austenitic stainless steels in SCW has been published by Arioka et al. (2011), who investigated mechanistic processes by which SCC and creep initiate and propagate. A great similarity between SCC and creep was observed up to 500 °C. The focus of that work was the accumulation of vacancies resulting from local stresses to produce cavities, thus causing initiation and propagation of SCC. Ru and Staehle (2013a,b,c) pointed out that since the SCW environment can be very oxidizing, it would be surprising if the crack velocity would be very high, in agreement with the results of Peng et al. (2007) and the data on DO dependence discussed in Section 6.2.1.2. Chen et al. (2017) proposed a revised slip-oxidation model that included creep; in this model creep voids nucleate along grain boundaries induced by applied stress ahead of the crack tip forming creep cracks. The main SCC crack intersects the creep cracks leading to enhanced rate of crack propagation.

The phenomenological corrosion model described in Chapter 5 can be modified to illustrate the points made above (Fig. 6.22). Below T_c, the anodic dissolution–based SCC model (denoted here slip-anodic oxidation) illustrated in Fig. 6.9 is dominant, and the SCC rate increases with temperature until the near-critical region, at which point the rapid changes in water properties lead to a dramatic decrease in crack growth (cf. Fig. 6.5). Below T_c, the occurrence of SCC depends on the DO concentration and the concentrations of anions such as chloride and sulphate. Creep also occurs but the rate is slow (Arioka et al., 2011). At $T \gg T_c$, the creep rate is higher, and while slip-oxidation still occurs, oxidation is by direct chemical interaction, and water, DO and H_2O_2 can all act as oxidants (as discussed in Chapter 5), making the DO dependence complex. Factors such as cold work and grain size that provide short-circuit diffusion pathways become much more important. In this temperature range a model such as the revised slip-oxidation model that includes creep proposed by

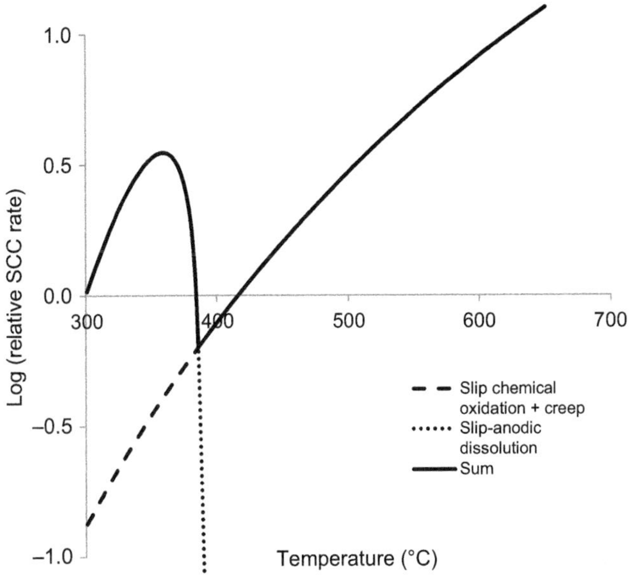

Figure 6.22 Simple phenomenological model for the relative stress corrosion cracking rate over the temperature range 300−650 °C that considers anodic slip-oxidation as the dominant mechanism below T_c and a modified slip-oxidation model that includes creep at $T \gg T_c$. In the near-critical region the two models are of equal importance and there is a crossover near T_c.

Chen et al. (2017) seems more plausible and crack growth increases with temperature. In the near-critical region the experimentally measured behaviour is expected to be highly dependent on environmental factors such as SCW density (dielectric constant) and concentrations of oxidants, and on materials factors such as surface cold work, chromium content and grain size.

Assessment of the elemental mechanistic and synergistic effects for such complex systems as EAC of austenitic stainless steels and Ni-base alloys in SCW will require systematic work and in particular mechanistic fracture mechanics-based CGR tests. More work, especially multiscale modelling based on fundamental mechanisms, is required.

References

Allen, T.R., Chen, Y., Ren, X., Sridharan, K., Tan, L., Was, G.S., West, E., Guzonas, D.A., 2012. Material performance in supercritical water. In: Konings, R.J.M. (Ed.), Comprehensive Nuclear Materials, vol. 5. Elsevier, Amsterdam, pp. 279−326.

Andresen, P.L., April 1992. Effect of Temperature on Crack Growth Rate in Sensitized Type 304 Stainless Steel and Alloy 600. Paper 89, NACE Corrosion/92, Nashville.

Andresen, P., 2002. Stress Corrosion Crack Growth Behavior of Cold Worked 304l and 316L Stainless Steels. Electric Power Research Institute Report 1007379.

Andresen, P.L., Ford, F.P., 1988. Life prediction by mechanistic modeling and system monitoring of environmental cracking of iron and nickel-alloys in aqueous systems. Mater. Sci. Eng. A 103, 167–184.

Andresen, P.L., Ford, F.P., 1995. Modeling and prediction of irradiation assisted cracking. In: Gold, R., McIlree, A. (Eds.), Proc. 7th Int. Symp. on Environmental Degradation of Materials in Nuclear Power Systems – Water Reactors, Breckenridge, August 7–10, 1995. National Association of Corrosion Engineers, pp. 893–908.

Andresen, P.L., Ford, F.P., September 2011. Prediction of stress corrosion cracking (SCC) in nuclear power systems. In: Raja, V.S., Shoji, T. (Eds.), Stress Corrosion Cracking Theory and Practice. Woodhead Publishing, ISBN 978-1-84569-673-3.

Andresen, P.L., Ford, F.P., Murphy, S.M., Perks, J.M., 1989. State of knowledge of irradiation effects on EAC in LWR core materials. In: Cubicciotti, D., Simonen, E. (Eds.), Proc. 4th Int. Symp. on Environmental Degradation of Materials in Nuclear Power Systems – Water Reactors, Jekyll Island, August 6–10, 1989. National Association of Corrosion Engineers, pp. 1.83–1.121.

Arioka, K., Yamada, T., Miyamoto, T., Terachi, T., 2011. Dependence of stress corrosion cracking of alloy 600 on temperature, cold work, and carbide precipitation, the role of diffusion of vacancies at crack tips. Corrosion 67 (3), 035006–11-035006-18.

Behnamian, Y., Dong, Z., Kohandehghan, A., Zahiri, R., Mitlin, D., Chen, W., Luo, J.L., Zheng, W., Guzonas, D., 2014. Oxidation and stress corrosion cracking of austenitic alloys in neutral pH extreme hydrothermal environments (supercritical water). In: NACE Northern Area Western Conference, Edmonton, Alberta, January 27–30, 2014.

Bevilacqua, F., Brown, G.M., 1963. Chloride Deposition from Steam onto Superheater Fuel Clad Materials. General Nuclear Engineering Corporation Report GNEC 295.

Bosch, C., Delafosse, D., 2005. Stress corrosion cracking of Ni-based and Fe-based alloys in supercritical water. In: *Corrosion 2005*. NACE International, Houston, TX. Paper 05396.

Bruemmer, S.M., Arey, B.W., Charlot, L.A., August 1–5, 1993. Grain boundary chromium concentration effects on the IGSCC and IASCC of austenitic stainless steels. In: Simonen, E., Gold, R. (Eds.), Proc. 6th Int. Symp. on Environmental Degradation of Materials in Nuclear Power Systems – Water Reactors. National Association of Corrosion Engineers, San Diego, CA, pp. 227–287.

Bruemmer, S.M., Cole, J.I., Garner, F.A., Greewood, L.R., Hamilton, M.L., Reid, B.D., Simonen, E.P., Lucas, G.E., Was, G.S., Andresen, P.L., Pettersson, K., 1996. Critical Issues Reviews for the Understanding and Evaluation of Irradiation Assisted Stress Corrosion Cracking. Electric Power Research Institute Report TR-107159.

Bull, A., Lewis, G., Owen, J., Quirk, G., Rudge, A., Woolsey, I., 2013. Mitigating the risk of stress corrosion of austenitic stainless steels in advanced gas cooled reactor boilers. In: Proceedings of the 16th International Conference on the Properties of Water and Steam, September 1–5, 2013. University of Greenwich, London, UK.

Cao, G., Firouzdor, V., Allen, T., 2011. Stress corrosion cracking of austenitic alloys in supercritical water. In: Busby, J.T., Ilevbre, G., Andresen, P.L. (Eds.), Proceedings of the 15th International Conference on Environmental Degradation of Materials in Nuclear Power Systems Water Reactors. TMS (The Minerals, Metals & Materials Society), pp. 1923–1935.

Chen, K., Du, D., Gao, W., Andresen, P.L., Zhang, L., 2017. Investigation of temperature effect on stress corrosion crack growth behavior of 310S stainless steel in supercritical water environment. In: 8th Int. Symp. on Super-Critical Water-Cooled Reactors, March 13–15, 2017, Chengdu, China.

Combrade, P., Ford, P., Nordmann, F., 2010. Key Results from Recent Conferences on Structural Materials Degradation in Water Cooled Reactors. Nuclear Technology International. LCC6 Special Report Advanced.

Comprelli, F.A., Busboom, H.J., Spalaris, C.N., 1969. Comparison of Radiation Damage Studies and Fuel Cladding Performance for Incoloy-800, Irradiation Effects in Structural Alloys for Thermal and Fast Reactors. ASTM STP 457. American Society for Testing and Materials, pp. 400–413.

Coriou, H., Grall, L., Olivier, P., Willermoz, H., 1969. Influence of carbon and nickel content on stress corrosion cracking of austenitic stainless alloys in pure or chlorinated water at 350°C. In: Staehle, R.W., Forty, A.J., van Rooyen, D. (Eds.), Proceedings of 1967 Conference: Fundamental Aspects of Stress Corrosion Cracking. NACE, Houston, TX, pp. 352–359.

Cowan, R., Ruhle, W., Hettiarachchi, S., 2011. Introduction to Boiling Water Reactor Chemistry, vol. 1. Advanced Nuclear Technology International, Sweden.

Cowan, R., Ruhle, W., Hettiarachchi, S., 2012. Introduction to Boiling Water Reactor Chemistry, vol. 2. Advanced Nuclear Technology International, Sweden.

Das, N.K., Suzuki, K., Takeda, Y., Ogawa, K., Shoji, T., 2008. Quantum chemical molecular dynamics study of stress corrosion cracking behavior for fcc Fe and Fe–Cr surfaces. Corros. Sci. 50, 1701–1706.

Devrient, B., Kilian, R., Kuster, K., Widera, M., 2011. Influence of bulk and surface cold work on crack initiation and crack growth of austenitic stainless steels under simulated BWR environmemt. In: Busby, J.T., Ilevbare, G., Andresen, P.L. (Eds.), Proc. 15th Int. Conf. on Environmental Degradation of Materials in Nuclear Power Systems Water Reactors. TMS.

Ehrlich, K., Konys, J., Heikinheimo, L., 2004. Materials for high performance light water reactors. J. Nucl. Mater. 327 (2–3), 140–147.

Emel'yanov, I., Shatskaya, O.A., Rivkin, E.Yu., Nikolenko, N.Y., 1972. Strength of construction elements in the fuel channels of the Beloyarsk power station reactors. At. Energiya 33 (3), 729–733 (in Russian) (Translated in Soviet 1972. Atomic Energy 33 (3) 842–847.

EPRI, October 2004. BWR Water Chemistry Guidelines – 2004 Revision. EPRI Technical Report TR-1008192, VIP-130.

Fournier, L., Delafosse, D., Bosch, C., Magnin, T., 2001. Stress corrosion cracking of nickel base superalloys in aerated supercritical water. In: Proceedings of the 2001 NACE Corrosion Conference, March 2001, Houston, TX, United States. NACE, Houston. Paper 01361.

Fruzzetti, K.P., Blok, J., 2005. A review of the EPRI chemistry guidelines. In: Proc. Int. Conf. on Water Chemistry of Nuclear Reactor Systems, San Francisco.

Fujisawa, R., Nishimura, K., Nishida, T., Sakaihara, M., Kurata, Y., Watanabe, Y., 2005. Cracking susceptibility of Ni base alloys and 316 stainless steel in less oxidizing or reducing SCW. In: Corrosion 2005. NACE International, Houston, TX. Paper 05395.

Garcia, S.E., Giannelli, J.F., Jarvis, M.L., 2012. Advances in BWR water chemistry. In: Proc. NPC2012 Nuclear Plant Chemistry Conference, Paris. Paper 80 001.

Garud, Y.S., Ilevbare, G.O., 2009. A review and assessment of cold-work influence on SCC of austenitic stainless steels in light water reactor environment. In: Proceedings of the 14th International Conference on Environmental Degradation of Materials in Nuclear Power Systems Water Reactors. Virginia, USA, pp. 780–790.

Gong, B., Jiang, E., Huang, Y., Zhao, Y., Liu, W., Zhou, Z., 2015. Effect of oxidation chemistry of SCW on stress corrosion cracking of austenitic steels. In: 7th International Symposium on Supercritical Water-cooled Reactors (ISSCWR-7), March 15–18, 2015, Helsinki, Finland.

Gordon, G.M., Brown, K.S., 1989. Dependence of creviced BWR component IGSCC behavior on coolant chemistry. In: Cubicciotti, D., Simonen, E. (Eds.), Proc. 4th Int. Symp. on Environmental Degradation of Materials in Nuclear Power Systems – Water Reactors, Jekyll Island, August 6–10, 1989. National Association of Corrosion Engineers, pp. 14.46–14.62.

Griess, J.C., Martin, J.M., Mravca, A.E., Impara, R.J., Weems, S., September 7, 1962. Task Force Review on Feasibility of BONUS Reactor Stainless Steel Clad Superheater Elements. memo to David F. Cope, Director. Reactor Division, ORNL.

Gruzdev, N.I., Shchapov, G.A., Tipikin, S.A., Boguslavskii, V.B., 1970. Investigating the water conditions in the second unit at Beloyarsk. Therm. Eng. 17 (12), 22–25 (Translated from Russian).

Hettiarachchi, S., Weber, C., 2010. Water chemistry improvements in an operating boiling water reactor (BWR) and associated benefits. In: Proceedings of the Nuclear Plant Chemistry Conference 2010 (NPC 2010), October 3–8, 2010, Quebec City, Canada.

Hou, J., Peng, Q.J., Shoji, T., Wang, J.Q., Han, E.-H., Ke, W., 2011. Effects of cold working path on strain concentration, grain boundary microstructure and stress corrosion cracking in alloy 600. Corros. Sci. 53, 1137–1142.

Janik, P., 2015. Acoustic Emission – A Method for Monitoring of Environmentally Assisted Cracking Processes (Doctor thesis). Institute of Chemical Technology Prague, Faculty of Environmental Technology, Department of Power Engineering.

Janik, P., Novotny, R., Nilsson, K.F., Siegl, J., Hausild, P., March 3–7, 2013. Pre-qualification of cladding materials for SCWR fuel qualification testing facility - stress corrosion cracking testing. In: 6th International Symposium on Supercritical Water-Cooled Reactors (ISSCWR-6), Shenzhen, Guangdong, China. Paper ISSCWR6–13034.

Janssen, J.S., Morra, M.M., Lewis, D.J., 2009. Sensitization and stress corrosion cracking of alloy 800H in high-temperature water. Corrosion 65, 67–78.

Jiao, Y., Kish, J., Cook, W., Zheng, W., Guzonas, D.A., 2013. Influence of thermal aging on the corrosion resistance of austenitic Fe-Cr-Ni alloys in SCW. In: 6th International Symposium on Supercritical Water-cooled Reactors (Shenzhen, Guangdong, China), March 03–07, 2013.

Kallikragas, D.T., Plugatyr, A.Yu., Guzonas, D.A., Svishchev, I.M., 2015. Effect of confinement on the hydration and diffusion of chloride at high temperatures. J. Supercrit. Fluids 97, 22–30.

Kilian, R., Hoffmann, H., Ilg, U., Küster, K., Nowak, E., Wesseling, U., Widera, M., 2005. German experience with intergranular cracking in austenitic piping in BWRs and assessment of parameters affecting the in-service IGSCC behaviour using an artificial neural network. In: Nelson, L., King, P. (Eds.), Proc. 12th Int. Conf. on Environmental Degradation of Materials in Nuclear Power Systems – Water Reactors, Skamania Lodge, August 5–9, 2005. The Metallurgical Society, pp. 803–812.

Kim, H., Mitton, D.B., Latanision, R.M., 2011. Stress Corrosion Cracking of Alloy 625 in PH 2 Aqueous Solution at High Temperature and Pressure. Corrosion 67, Paper No. 035002–2.

Lee, E.H., Yoo, M.H., Byun, T.S., Hunn, J.D., Farrell, K., Mansur, L.K., 2001. On the origin of deformation microstructures in austenitic stainless steel: Part I—Microstructures. Acta Mater. 49, 3269–3276.

Li, S.-X., He, Y.-N., Yu, S.-R., Zhang, P.-Y., 2013. Evaluation of the effect of grain size on chromium carbide precipitation and intergranular corrosion of 316L stainless steel. Corros. Sci. 66, 211–216.

Liu, J., Gong, B., 2017. The effect of dissolved oxygen on stress corrosion cracking of 310 SS in SCW. In: 8th Int. Symp. on Super-critical Water-cooled Reactors, March 13–15, 2017, Chengdu, China.

Lozano-Perez, S., Yamada, T., Terachi, T., Schröder, M., English, C.A., Smith, G.D.W., Grovenor, C.R.M., Eyre, B.L., 2009. Multi-scale characterization of stress corrosion cracking of cold-worked stainless steels and the influence of Cr content. Acta Mater. 57, 5361–5381.

Macdonald, D.D., 2009. The electrochemistry of stress corrosion cracking. In: Proceedings of the 14th International Conference on Environmental Degradation of Materials in Nuclear Power Systems Water Reactors. Virginia, USA, pp. 734–744.

Nelson, J.L., Andresen, P.L., 1991. Review of current research and understanding of irradiation assisted stress corrosion cracking. In: Cubicciotti, D., Simonen, E. (Eds.), Proc. 5th Int. Symp. on Environmental Degradation of Materials in Nuclear Power Systems – Water Reactors, Monterey, August 25–29, 1991. American Nuclear Society, pp. 10–26.

Newman, R.C., 2002. Stress Corrosion Cracking Mechanisms. Marcel Dekker, New York.

Nezakat, M., Akhiani, H., Hoseini, M., Szpunar, J., 2014. Effect of thermo-mechanical processing on texture evolution in austenitic stainless steel 316L. Mater. Charact. 98, 10–17.

Novotny, R., Hähner, P., Siegl, J., Hausild, P., Ripplinger, S., Penttilä, S., Toivonen, A., 2011. Stress corrosion cracking susceptibility of austenitic stainless steels in supercritical water conditions. J. Nucl. Mater. 409 (2), 117–123.

Oh, Y.J., Hong, J.H., 2000. Nitrogen effect on precipitation and sensitization in cold-worked type 316L(N) stainless steels. J. Nucl. Mater. 278, 242–250.

Parvathavarthini, N., Dayal, R.K., 2002. Influence of chemical composition, prior deformation and prolonged thermal aging on the sensitization characteristics of austenitic stainless steel. J. Nucl. Mater. 305, 209–219.

Parvathavarthini, N., Mulki, S., Dayal, R.K., Samajdar, I., Mani, K.V., Raj, B., 2009. Sensitization control in AISI 316L(N) austenitic stainless steel: defining the role of the nature of grain boundary. Corros. Sci. 51, 2144–2150.

Peng, Q.J., Teysseyre, S., Andersen, P.L., Was, G.S., 2007. Stress corrosion crack growth in type 316 stainless steel in supercritical water. Corrosion 63 (11), 1033–1041.

Penttilä, S., Toivonen, A., 2013. Oxidation and SCC behaviour of austenitic and ODS steels in supercritical water. In: 6th International Symposium on Supercritical Water-Cooled Reactors (ISSCWR-6), Shenzhen, Guandong, China, March 3–7, 2013. Paper ISSCWR6-13029.

Rabin, S.A., Atraz, B.G., Bader, M.B., Busboom, H.J., Hazel, V.E., 1967. Examination and Evaluation of Rupture in EVESR Superheat Fuel Rod with 0.012-inch-thick Incoloy-800 Cladding. AEC Research and Development Report GEAP-5416.

Raja, V.S., Shoji, T. (Eds.), September 2011. Corrosion Cracking Theory and Practice. Woodhead Publishing, ISBN 978-1-84569-673-3.

Ru, X., Staehle, R.W., 2013a. Historical experience providing bases for predicting corrosion and stress corrosion in emerging supercritical water nuclear technology: Part 1 – review. Corrosion 69 (3), 211–229.

Ru, X., Staehle, R.W., 2013b. Historical experience providing bases for predicting corrosion and stress corrosion in emerging supercritical water nuclear technology: Part 2 – review. Corrosion 69 (4), 319–334.

Ru, X., Staehle, R.W., 2013c. Historical experience providing bases for predicting corrosion and stress corrosion in emerging supercritical water nuclear technology: Part 3 – review. Corrosion 69 (5), 423–447.

Ru, X., Zhang, Q., Zhao, Y.-X., Tang, R., Xiao, Z., 2015. The stress corrosion cracking (SCC) behavior of Type 310S stainless steel in supercritical water (SCW). In: Proceedings 7th International Symposium on Supercritical Water-Cooled Reactors (ISSCWR-7), March 15–18, 2015, Helsinki, Finland. Paper No. ISSCWR7-2093.

Sáez-Maderuelo, A., Gómez-Briceño, D., 2016. Stress corrosion cracking behavior of annealed and cold worked 316L stainless steel in supercritical water. Nucl. Eng. Des. 307, 30–38.

Sahlaoui, H., Makhlouf, K., Philibert, J., 2004. Effects of ageing conditions on the precipitates evolution, chromium depletion and intergranular corrosion susceptibility of AISI 316L: experimental and modeling results. Mater. Sci. Eng. A 372, 98–108, 15.

Scenini, F., Newman, R.C., Cottis, R.A., Jacko, R.J., 2008. Effect of surface preparation on intergranular stress corrosion cracking of alloy 600 in hydrogenated steam. Corrosion 64, 824–835.

Scott, P., 2013. SCC of stainless steels in PWRs. In: Scott, P.M. (Ed.), INL Seminar on SCC in LWRs Idaho Falls, Idaho, USA, 2013 March 19–20. INL.

Scott, P.M., Combrade, P., 2003. On the mechanism of stress corrosion crack initiation and growth in alloy 600 exposed to PWR primary water. In: Proc. of the 11th Int. Symp. Environmental Degradation Materials Nuclear Power Systems – Water Reactors. ANS, pp. 29–38.

Shen, Z., Zhang, L., Tang, R., Zhang, Q., 2014. The effect of temperature on the SSRT behavior of austenitic stainless steels in SCW. J. Nucl. Mater. 454, 274–282.

Shoji, T., August 10–14, 2003. Progress in the mechanistic understanding of BWR SCC and its implication to the prediction of SCC growth behavior in plants. In: Proc. 11th Int. Conf. on Environmental Degradation of Materials in Nuclear Power Systems – Water Reactors, Skamania, Stevenson, WA. ANS, pp. 588–598.

Shoji, T., September 2011. Factors affecting stress corrosion cracking (SCC) and fundamental mechanistic understanding of stainless steels. In: Raja, V.S., Shoji, T. (Eds.), Stress Corrosion Cracking Theory and Practice. Woodhead Publishing, ISBN 978-1-84569-673-3.

Shoji, T., Lu, Z.P., Takeda, Y., Murakami, H., Fu, C.Y., 2008. Deterministic prediction of stress corrosion crack growth rates in high temperature water by combination of interface oxidation kinetics and crack tip asymptotic field. In: 2008 ASME Pressure Vessels and Piping Division Conference (ASME PVP 2008). ASME, pp. PVP2008–61417.

Shoji, T., Lu, Z.P., Murakami, H., 2010. Formulating stress corrosion cracking growth rates by combination of crack tip mechanics and crack tip oxidation kinetics. Corros. Sci. 52, 769–779.

Speidel, M.O., 1984. Stress corrosion and corrosion fatigue mechanics. In: Speidel, M.O., Atrens, A. (Eds.), Corrosion in Power Generating Equipment. Plenum Press, New York.

Staehle, R.W., February 2007. Predicting Failures Which Have Not yet Been Observed. Expert Panel Report on Proactive Materials Degradation Assessment, USNRC NUREG Report, CR-6923.

Staehle, R.W., Gorman, J.A., 2003. Quantitative assessment of submodes of stress corrosion cracking on the secondary side of steam generator tubing in pressurized water reactors: Part 1. Corrosion 59, 931–993.

Stellwag, B., Landner, A., Weiss, S., Huttner, F., 2011. Water chemistry control practices and data of the European BWR fleet. Powerpl. Chem. 13 (3), 167–173.

Strasser, A., Santucci, J., Lindquist, K., Yario, W., Stern, G., Goldstein, L., Joseph, L., 1982. An Evaluation of Stainless Steel Cladding for Use in Current Design LWRs. Electric Power Research Institute Report NP-2642.

Tan, L., Allen, T.R., 2009. Localized corrosion of magnetite on ferritic-martensitic steels exposed to supercritical water. Corros. Sci. 51, 2503–2507.

Terada, M., Escriba, D.M., Costa, I., Materna-Morris, E., Padilha, A.F., 2008. Investigation on the intergranular corrosion resistance of the AISI 316L(N) stainless steel after long time creep testing at 600°C. Mater. Charact. 59, 663–668.

Teysseyre, S., Was, G.S., 2006. Stress corrosion cracking of austenitic alloys in supercritical water. Corrosion 62 (12), 1100–1116.

Teysseyre, S., Jiao, Z., West, E., Was, G.S., 2007. Effect of irradiation on stress corrosion cracking in supercritical water. J. Nucl. Mater. 371 (1–3), 98–106.

Toivonen, A., Penttilä, S., Novotny, R., 2015. SCC tests in SCW at 550°C on two heats of 316L. In: 7th International Symposium on Supercritical Water-cooled Reactors (ISSCWR-7), March 15–18, 2015, Helsinki, Finland.

Tsuchiya, Y., Kano, F., Saito, N., Shioiri, A., Kasahara, S., Moriya, K., Takahashi, H., September 15–19, 2003. SCC and Irradiation Properties of Metals Under Supercritical-Water Cooled Power Reactor Conditions. GENES4/ANP2003, Kyoto, Japan. Paper 1096.

Uchida, S., Satoh, T., Tsukada, T., Miyazawa, T., Satoh, Y., Ishaa, K., 2009. Evaluation of the effects of oxide film on electrochemical corrosion potential of stainless steel in high temperature water. In: Proceedings of the 14th International Conference on Environmental Degradation of Materials in Nuclear Power Systems Water Reactors. Virginia, USA, pp. 1100–1109.

Udy, M.C., Boulger, F.W., 1953. Survey of materials for supercritical-water reactor. Battelle Memorial Inst, Columbus, Ohio. OSTI ID: 4230973, Report Number(s), BMI-890.

Was, G.S., Busby, J., Andresen, P.L., 2006. Effect of irradiation on stress corrosion cracking and corrosion in light water reactors. In: ASM Handbook on Corrosion; Environments and Industries, 13C, pp. 386–414.

Was, G.S., Ampornrat, R., Gupta, G., Teysseyre, S., West, E.A., Allen, T.R., Sridharan, K., Tan, L., Chen, Y., Ren, X., Pister, C., 2007. Corrosion and stress corrosion cracking in supercritical water. J. Nucl. Mater. 371, 176–201.

Was, G.S., Teysseyre, S., 2005. Challenges and recent progress in corrosion and stress corrosion cracking of alloys for supercritical water reactor core components. In: Allen, T.R., King, P.J., Nelson, L. (Eds.), Proceedings of the 12th International Conference on Environmental Degradation of Materials in Nuclear Power Systems Water Reactors. TMS (The Minerals, Metals & Materials Society), pp. 1343–1357.

Watanabe, Y., Abe, H., Daigo, Y., 2003. Environmentally-assisted cracking of sensitized stainless steel in supercritical water. In: Proceedings of GENES4/NP2003, Kyoto, Japan, September 15–19, 2003. Paper 1183.

Watanabe, Y., Abe, H., Daigo, Y., Nishida, T., 2004a. Corrosion Resistance and Cracking Susceptibility of 316L Stainless Steel in Sulfuric Acid-Containing Supercritical Water. Corrosion 2004, Paper 04493.

Watanabe, Y., Abe, H., Daigo, Y., Fujisawa, R., Sakaihara, M., 2004b. Effect of physical property and chemistry of water on cracking of stainless steels in sub-critical and supercritical water. Key Eng. Mater. 261–263, 1031–1036.

West, E., Was, G.S., 2013. Strain incompatibilities and their role in intergranular cracking of irradiated 316L stainless steel. J. Nucl. Mater. 441, 623–632.

Woolsey, I.S., 1989. Review of Water Chemistry and Corrosion Issues in the Steam-Water Circuits of Fossil-Fired Supercritical Plants. Central Electricity Generating Board Report RD/L/3413/R88, PubID 280118.

Zhang, Q., Tang, R., Xiong, R., Zhang, L., 2015. Effect of temperature and dissolved oxygen on the SCC behavior of 316Ti in supercritical water. In: 7th International Symposium on Supercritical Water-Cooled Reactors (ISSCWR-7), March 15–18, 2015, Helsinki, Finland.

Zheng, W., Luo, J., Li, M., Guzonas, D.A., Cook, W., 2011. Stress corrosion cracking of SCWR candidate alloys: a review of published results. In: 5th International Symposium on SCWR (ISSCWR-5), Vancouver, BC, Canada. Paper P095, March 13–16, 2011.

Zheng, W., Zeng, Y., Luo, J., Novotny, R., Li, J., Shalchi Amirkhiz, B., Guzonas, D., Matchim, M., Collier, J., Yang, L., 2014. Stress-corrosion cracking properties of candidate fuel cladding alloys for the Canadian SCWR: a summary of literature data and recent test results. In: 19th Pacific Basin Nuclear Conference (PBNC 2014) Vancouver, British Columbia, Canada, August 24–28, 2014.

Zhou, S., Turnbull, A., 2002. Steam Turbine Operating Conditions, Chemistry of Condensates, and Environment Assisted Cracking – A Critical Review. National Physical Laboratory Report MATC (A) 95.

Zhou, Z.-F., Lvov, S.N., Thakur, S., Zhou, X., Chou, P., Pathania, R., 2004. Hydrothermal deposition of zirconia coatings on BWR materials for IGSCC protection. In: Proc. Int. Conf. Water Chemistry of Nuclear Reactor Systems, October 11–14, 2004, San Francisco, CA, USA. Paper 3–4.

Zhou, R., West, E.A., Jiao, Z., Was, G.S., 2009. Irradiation-assisted stress corrosion cracking of austenitic alloys in supercritical water. J. Nucl. Mater. 395, 11–22.

Index

'*Note:* Page numbers followed by "f" indicate figures and "t" indicate tables.'